Introductory Ring Theory

Introductory Ring Theory

Ritika Chopra
Ankit Gupta

CAMBRIDGE
UNIVERSITY PRESS

Shaftesbury Road, Cambridge CB2 8EA, United Kingdom

One Liberty Plaza, 20th Floor, New York, NY 10006, USA

477 Williamstown Road, Port Melbourne, VIC 3207, Australia

314–321, 3rd Floor, Plot No. 3, Splendor Forum, Jasola District Centre, New Delhi – 110025, India

103 Penang Road, #05–06/07, Visioncrest Commercial, Singapore 238467

Cambridge University Press is part of Cambridge University Press & Assessment, a department of the University of Cambridge.

We share the University's mission to contribute to society through the pursuit of education, learning and research at the highest international levels of excellence.

www.cambridge.org
Information on this title: www.cambridge.org/9781009633536

First published 2025

Printed in India by Shree Maitrey Printech Pvt. Ltd., Noida

A catalogue record for this publication is available from the British Library

ISBN 978-1-009-63353-6 Paperback

CONTENTS

This book is designed as a textbook for a semester-long introductory course in abstract algebra, with a focus on ring theory. Yes, it's a "textbook", but it is not to be conceived as an encyclopedia, nor merely as an additional reference book for your shelf. This is an approachable tool for learners, guiding them through the thought-provoking terrain of ring theory, crafted to make the subject as clear and engaging as possible. We simply aim to provide undergraduate students with a solid understanding of the fundamentals of ring theory while igniting a lifelong curiosity in abstract algebra.

Ring theory, like most areas of modern algebra, has undergone significant changes and developments over the years. This is a subject where math and creativity intersect, and if you've believed that math is just about numbers, ring theory will soon convince you otherwise. Our experience tells us that with a clear understanding of algebraic structures, students find it easier to navigate more complex ideas in areas such as coding theory, cryptography, and advanced analysis. Rest assured, we've written this book with you, the student, in mind, so expect to find explanations that make sense without needing to pull out your hair or abandon the course.

For those who are wondering whether this book is for them, consider this: **If you're ready to understand why the algebra you learned in school was just the beginning, then welcome**. We've structured each chapter with straightforward explanations, proofs broken down step-by-step, and enough examples to keep you grounded in the concepts, even when the going gets tough. Each section is packed with problems to solve, along with hints (don't peek too early!) so you can build your skills through hands-on practice. To help you through those "I-don't-get-it-yet" moments, you'll find a few remarks and examples – think of them as little math pick-me-ups when the subject matter gets dense. Remember, the goal is not just to memorize definitions and theorems, but to see how these algebraic ideas weave together to create something extraordinary.

In creating this book, we hope it will serve as your trusted classroom companion, a resource in the late hours before exams, and a reminder that even the most abstract concepts in mathematics are easily tangible and right within your grasp. Now, buckle up as we introduce you to the rings that are far more fascinating than anything Tolkien could dream up. Wishing you every success as you explore the elegance of ring theory.

Abstract algebra, particularly ring theory, is currently being taught in all Indian universities as well as abroad as part of the mathematics curricula for various undergraduate programs, including B.Sc. (H), B.A. (Prog.), B.Sc. (Prog.), and others. A sound knowledge of binary operations and their properties, which form the core of algebra, is a prerequisite for further study of abstract mathematical ideas. With time, this has become more pertinent due to the applications of mathematical theories in several emerging fields, including computer science, biosciences, genetics, and financial modeling. Our experience shows that students with a clear understanding of the basic concepts of binary operations find it easier to grasp more abstract ideas later in these fields. An all-explanatory, student-friendly book is a basic requirement for developing interest amongst the students for complex algebraic concepts. Although there are many books written on the subject of ring theory and abstract algebra, there are still very few student-friendly books available in the current dynamic that can cater to the present syllabus of the Indian universities. With this understanding in mind, the authors conceived the idea of writing a book on ring theory, covering all the topics currently taught at the undergraduate level. The book covers a wide range of topics, including the rings, subrings, integral domains, fields, ring homomorphisms, ring isomorphism, factor rings, and various types of ideals, including prime ideals, principal ideals, maximal ideals, principal ideal domain, unique factorization domain, Noetherian domain, as well as factorization of polynomials, divisibility in integral domain. These topics have been divided over nine chapters in the book.

Each concept introduced in the book is explained in simple language, with the help of detailed examples and counterexamples so that students can grasp it without difficulty. Proofs of the results are provided in an easy, step-by-step manner. An ample number of examples are provided to facilitate an easy and better conceptual understanding for the students. Several remarks are used, if and when required, throughout the book to point out and discuss the hidden aspects of the definitions and the theorems, if there are any. Exercises are provided at the end of each chapter so that students may have sufficient practice in problem-solving. Overall, the book is student-friendly, yet full of rigour. We hope that the students, as well as the teachers, will be pleased and find this book to be a welcome addition to the classroom.

The first author would like to extend her heartfelt gratitude to her mentors and colleagues, whose insights and encouragement inspired the development of this text. She is grateful to the students, whose questions, feedback, and enthusiasm have played a crucial role in shaping the content and structure of this book.

She would like to acknowledge the support of the principal, Prof. (Dr) Payal Mago, Shaheed Rajguru College of Applied Sciences for Women, University of Delhi, and Prof. (Dr) Punita Saxena, the teacher-in-charge, for providing a stimulating academic environment and the resources necessary for completing this work. She also thanks her family and friends for their patience, understanding, and unwavering support throughout this endeavour.

The second author would like to express his heartfelt gratitude to his parents, teachers, colleagues, and especially students, each of whom has contributed in invaluable ways to his academic journey. He would like to begin by acknowledging the unwavering intellectual and moral support of his research supervisor Prof. R. D. Sarma from Rajdhani College (University of Delhi), whose mentorship has been a pillar of strength throughout this journey. No journey can be successful without the support and blessings of one's family. The second author would like to thank his parents and family members for their support, faith, and inspiration during hard times. He would also like to thank his brilliant yet delightfully unpredictable group of friends–Mr Kamal Kumar, Ms Bhavna Dada, Ms Chetna Khanna, Ms Ruchika Gupta, Ms Neha Mongia, and Ms Nisha Bohra–for their constant support, loud encouragement, and occasional reality checks. Their unsolicited advice, loud debates, and random memes somehow turned out to be the moral support he didn't know he needed. A special and heartfelt thank you goes to Dr Sunil Tiwari, whose encouragement and thoughtful insistence helped the second author stay on track and continue this academic journey even through difficult times. His quiet confidence served as a steady reminder that perseverance in scholarship is as important as passion.

He would like to acknowledge the support of, and extend special thanks to, Prof. (Dr) Saloni Gupta, Principal, Bharati College (University of Delhi), for her constant encouragement in pursuing new projects and her belief, which has been instrumental in making this project a reality.

Finally, the authors would like to express their sincere gratitude to Ms Swati Gupta for her assistance throughout the project. They also extend their heartfelt thanks to Cambridge University Press, with special appreciation to Mr Vikash, Mr Ankush, and Mr Karan, for their dedicated efforts, unwavering support, and exceptional commitment to ensuring the timely publication of this book. Their professionalism and perseverance were instrumental in bringing this project to successful completion within a remarkably timeframe. Their commitment has been instrumental in bringing this project to fruition.

Ritika Chopra
Ankit Gupta

1 PRELIMINARIES

Definition 1.1 (Binary Operation) Let G be a non-empty set. A *binary operation* on G is a function that assigns each ordered pair of elements of G an element of G.

Definition 1.2 (Group) Let G be a set together with a binary operation (usually called multiplication) that assigns to each ordered pair (a, b) of elements of G an element in G denoted by ab. We say G is a group under this operation if the following three properties are satisfied.

1. *Associativity.* The operation is associative; that is, $(ab)c = a(bc)$ for all a, b, c in G.
2. *Identity.* There is an element e (called an *identity*) in G such that $ae = ea = a$ for all a in G.
3. *Inverse.* For each element a in G, there is an element b in G (called an *inverse* of a) such that $ab = ba = e$.

Theorem 1.3 (Uniqueness of the Identity) *In a group G, there is only one identity element. Therefore, now onwards, we refer it the identity.*

Theorem 1.4 (Cancellation) *In a group G, the right and left cancellation laws hold; that is, $ba = ca$ implies $b = c$, and $ab = ac$ implies $b = c$.*

Theorem 1.5 (Uniqueness of Inverses) *For each element a in a group G, there is a unique element b in G such that $ab = ba = e$.*

Theorem 1.6 (Socks–Shoes Property) *For group elements a and b, we have $(ab)^{-1} = b^{-1}a^{-1}$.*

Definition 1.7 (Order of a Group) The number of elements of a group (finite or infinite) is called its *order*. We will use $|G|$ or $O(G)$ to denote the order of G.

Definition 1.8 (Order of an Element) The order of an element g in a group G is the smallest positive integer n such that $g^n = e$. (In additive notation, this would be $ng = 0$.) If no such integer exists, we say that g has *infinite order*. The order of an element g is denoted by $|g|$ or $O(g)$.

Definition 1.9 (Subgroup) If a non-empty subset H of a group G is itself a group under the operation of G, we say that H is a subgroup of G. A subgroup H of group G is denoted by $H \leq G$.

Theorem 1.10 (One-Step Subgroup Test) *Let G be a group and H a non-empty subset of G. H is a subgroup of G if and only if ab^{-1} is in H whenever a and b are in H. (In additive notation, H is a subgroup of G if and only if $a - b$ is in H whenever a and b are in H.)*

Theorem 1.11 (Two-Step Subgroup Test) *Let G be a group and H be a non-empty subset of G. H is a subgroup of G if and only if ab and a^{-1} are in H whenever a and b are in H. (In additive notation, H is a subgroup of G if and only if $a + b$ and $-a$ are in H whenever a and b are in H.)*

Theorem 1.12 (Finite Subgroup Test) *Let H be a non-empty finite subset of a group G. H is a subgroup of G if and only if H is closed under the operation.*

Theorem 1.13 ($\langle a \rangle$ Is a Subgroup) *Let $\langle a \rangle$ denote the set $\{a^n \mid n \in \mathbb{Z}\}$. Let G be a group, and a be any element of G. Then, $\langle a \rangle$ is a subgroup of G called the cyclic subgroup of G generated by a.*

Definition 1.14 (Center of a Group) The *center*, $Z(G)$, of a group G is the subset of elements in G that commute with every element of G. In symbols,

$$Z(G) = \{a \in G \mid ax = xa \text{ for all } x \text{ in } G\}.$$

Theorem 1.15 (Center Is a Subgroup) *The center of a group G is a subgroup of G.*

Definition 1.16 (Centralizer of a in G) Let a be a fixed element of a group G. The *centralizer* of a in G, $C(a)$, is the set of all elements in G that commute with a. In symbols, $C(a) = \{g \in G \mid ga = ag\}$.

Theorem 1.17 ($C(a)$ **Is a Subgroup**) *For each a in a group G, the centralizer of a is a subgroup of G.*

Definition 1.18 (Cyclic Group) A group G is called *cyclic* if there is an element a in G such that $G = \{a^n \mid n \in \mathbb{Z}\}$. Such an element a is called a *generator* of G. A cyclic group G generated by a is denoted by $G = \langle a \rangle$.

Theorem 1.19 (Criterion for $a^i = a^j$) *Let G be a group, and let a belong to G. If a has infinite order, then $a^i = a^j$ if and only if $i = j$. If a has finite order, say, n, then $\langle a \rangle = \{e, a, a^2, \cdots, a^{n-1}\}$ and $a^i = a^j$ if and only if n divides $i - j$.*

Corollary 1.20 ($|a| = |\langle a \rangle|$) For any group element a, $|a| = |\langle a \rangle|$.

Corollary 1.21 ($a^k = e$ **Implies that $|a|$ Divides k**) Let G be a group and let a be an element of order n in G. If $a^k = e$, then n divides k.

Theorem 1.22 ($\langle a^k \rangle = \langle a^{\gcd(n,k)} \rangle$ **and** $|a^k| = n/\gcd(n,k)$) *Let a be an element of order n in a group and let k be a positive integer. Then $\langle a^k \rangle = \langle a^{\gcd(n,k)} \rangle$ and $|a^k| = n/\gcd(n,k)$.*

Corollary 1.23 (Orders of Elements in Finite Cyclic Groups) In a finite cyclic group, the order of an element divides the order of the group.

Corollary 1.24 (Criterion for $\langle a^i \rangle = \langle a^j \rangle$ and $|a^i| = |a^j|$) Let $|a| = n$. Then $\langle a^i \rangle = \langle a^j \rangle$ if and only if $\gcd(n,i) = \gcd(n,j)$, and $|a^i| = |a^j|$ if and only if $\gcd(n,i) = \gcd(n,j)$.

Corollary 1.25 (Generators of Finite Cyclic Groups) Let $|a| = n$. Then $\langle a \rangle = \langle a^j \rangle$ if and only if $\gcd(n,j) = 1$, and $|a| = |a^j|$ if and only if $\gcd(n,j) = 1$.

Corollary 1.26 (Generators of $\mathbb{Z}_n$) An integer k in $\mathbb{Z}_n$ is a generator of $\mathbb{Z}_n$ if and only if $\gcd(n,k) = 1$.

Theorem 1.27 (Fundamental Theorem of Cyclic Groups) *Every subgroup of a cyclic group is cyclic. Moreover, if $|\langle a \rangle| = n$, then the order of any subgroup of $\langle a \rangle$ is a divisor of n; and, for each positive divisor k of n, the group $\langle a \rangle$ has exactly one subgroup of order k—namely, $\langle a^{n/k} \rangle$.*

Corollary 1.28 (Subgroups of $\mathbb{Z}_n$) For each positive divisor k of n, the set $\langle n/k \rangle$ is the unique subgroup of $\mathbb{Z}_n$ of order k; moreover, these are the only subgroups of $\mathbb{Z}_n$.

Theorem 1.29 (Number of Elements of Each Order in a Cyclic Group) *If d is a positive divisor of n, the number of elements of order d in a cyclic group of order n is $\phi(d)$.*

Corollary 1.30 (Number of Elements of Order d in a Finite Group) In a finite group, the number of elements of order d is a multiple of $\phi(d)$.

Definition 1.31 (Permutation of A, Permutation Group of A) A *permutation* of a set A is a function from A to A that is both one-one and onto. A *permutation group* of a set A is a set of permutations of A that forms a group under function composition.

Definition 1.32 (Symmetric Group S_n) Let $A = \{1, 2, \cdots, n\}$. The set of all permutations of A is called the *symmetric group* of degree n and is denoted by S_n. The order of S_n is $n!$.

Theorem 1.33 (Products of Disjoint Cycles) *Every permutation of a finite set can be written as a cycle or as a product of disjoint cycles.*

Theorem 1.34 (Disjoint Cycles Commute) *If the pair of cycles $\alpha = (a_1, a_2, \cdots, a_m)$ and $\beta = (b_1, b_2, \cdots, b_n)$ have no entries in common, then $\alpha\beta = \beta\alpha$.*

Theorem 1.35 (Order of a Permutation) *The order of a permutation of a finite set written in disjoint cycle form is the least common multiple of the lengths of the cycles.*

Theorem 1.36 (Product of 2-Cycles) *Every permutation in $S_n, n > 1$, is a product of 2-cycles.*

Definition 1.37 (Even and Odd Permutations) A permutation that can be expressed as a product of an even number of 2-cycles is called an *even* permutation. A permutation that can be expressed as a product of an odd number of 2-cycles is called an *odd* permutation.

Lemma 1.38 If $\epsilon = \beta_1\beta_2 \cdots \beta_r$, where the β's are 2-cycles, then r is even.

Theorem 1.39 (Always Even or Always Odd) *If a permutation α can be expressed as a product of an even (odd) number of 2-cycles, then every decomposition of α into a product of 2-cycles must have an even (odd) number of 2-cycles. In symbols, if $\alpha = \beta_1\beta_2 \cdots \beta_r$ and $\alpha = \gamma_1\gamma_2 \cdots \gamma_s$, where the β's and the γ's are 2-cycles, then r and s are both even or both odd.*

Theorem 1.40 (Even Permutations Form a Group) *The set of even permutations in S_n forms a subgroup of S_n.*

Definition 1.41 (Alternating Group of Degree n) The group of even permutations of n symbols is denoted by A_n and is called the *alternating group of degree n.*

Theorem 1.42 *For $n > 1, A_n$ has order $n!/2$.*

Definition 1.43 (Group Isomorphism) An *isomorphism* ϕ from a group G to a group $\overline{G}$ is a one-to-one mapping (or function) from G onto $\overline{G}$ that preserves the group operation. That is, $\phi(ab) = \phi(a)\phi(b)$ for all a, b in G. If there is an isomorphism from G onto $\overline{G}$, we say that G and $\overline{G}$ are *isomorphic* and denoted by $G \approx \overline{G}$.

Theorem 1.44 (Cayley's Theorem) *Every group is isomorphic to a group of permutations.*

Theorem 1.45 (Properties of Isomorphisms Acting on Elements) *Suppose that f is an isomorphism from a group G onto a group $\overline{G}$. Then*

1. *ϕ carries the identity of G to the identity of $\overline{G}$.*
2. *For every integer n and for every group element a in G, $\phi(a^n) = [\phi(a)]^n$.*
3. *For any elements a and b in G, a and b commute if and only if $\phi(a)$ and $\phi(b)$ commute.*
4. *$G = \langle a \rangle$ if and only if $\overline{G} = \langle \phi(a) \rangle$.*
5. *$|a| = |\phi(a)|$ for all a in G (isomorphisms preserve orders).*
6. *For a fixed integer k and a fixed group element b in G, the equation $x^k = b$ has the same number of solutions in G as does the equation $x^k = \phi(b)$ in $\overline{G}$.*
7. *If G is finite, then G and $\overline{G}$ have exactly the same number of elements of every order.*

Theorem 1.46 (Properties of Isomorphisms Acting on Groups) *Suppose that ϕ is an isomorphism from a group G onto a group $\overline{G}$. Then*

1. *ϕ^{-1} is an isomorphism from $\overline{G}$ onto G.*
2. *G is Abelian if and only if $\overline{G}$ is Abelian.*
3. *G is cyclic if and only if $\overline{G}$ is cyclic.*
4. *If K is a subgroup of G, then $\phi(K) = \{\phi(k) \mid k \in K\}$ is a subgroup of $\overline{G}$.*
5. *If $\overline{K}$ is a subgroup of $\overline{G}$, then $\phi^{-1}(\overline{K}) = \{g \in G \mid \phi(g) \in \overline{K}\}$ is a subgroup of G.*
6. *$\phi(Z(G)) = Z(\overline{G})$.*

Definition 1.47 (Automorphism) An isomorphism from a group G onto itself is called an *automorphism* of G.

Definition 1.48 (Inner Automorphism Induced by a) Let G be a group, and let $a \in G$. The function $\phi(a)$ defined by $\phi_a(x) = axa^{-1}$ for all x in G is called the *inner automorphism of G induced by a.*

Theorem 1.49 ($\mathrm{Aut}(G)$ and $\mathrm{Inn}(G)$ Are Groups) *The set of automorphisms of a group and the set of inner automorphisms of a group are both groups under the operation of function composition.*

Theorem 1.50 ($\mathrm{Aut}(Z_n) \approx U(n)$) *For every positive integer n, $\mathrm{Aut}(Z_n)$ is isomorphic to $U(n)$.*

Definition 1.51 (Coset of H in G) Let G be a group and H be a non-empty subset of G. For any $a \in G$, the set $\{ah \mid h \in H\}$ is denoted by aH. Analogously, $Ha = \{ha \mid h \in H\}$ and $aHa^{-1} = \{aha^{-1} \mid h \in H\}$.

When H is a subgroup of G, the set aH is called the *left coset of H in G containing a*, whereas Ha is called the *right coset of H in G containing a*. In this case, the element a is called the *coset representative of aH (or Ha)*.

We use $|aH|$ to denote the number of elements in the set aH, and $|Ha|$ to denote the number of elements in Ha.

Lemma 1.52 (Properties of Cosets) Let H be a subgroup of G, and let a and b belong to G. Then

1. $a \in aH$.
2. $aH = H$ if and only if $a \in H$.
3. $(ab)H = a(bH)$ and $H(ab) = (Ha)b$.
4. $aH = bH$ if and only if $a \in bH$.
5. $aH = bH$ or $aH \cap bH = \emptyset$.
6. $aH = bH$ if and only if $a^{-1}b \in H$.
7. $|aH| = |bH|$.
8. $aH = Ha$ if and only if $H = aHa^{-1}$.
9. aH is a subgroup of G if and only if $a \in H$.

Theorem 1.53 (Lagrange's Theorem: $|H|$ Divides $|G|$) *If G is a finite group and H is a subgroup of G, then $|H|$ divides $|G|$. Moreover, the number of distinct left (right) cosets of H in G is $|G|/|H|$.*

Corollary 1.54 ($|G{:}H| = |G|/|H|$) If G is a finite group and H is a subgroup of G, then $|G{:}H| = |G|/|H|$.

Corollary 1.55 ($|a|$ Divides $|G|$) In a finite group, the order of each element of the group divides the order of the group.

Corollary 1.56 (Groups of Prime Order Are Cyclic) A group of prime order is cyclic.

Corollary 1.57 ($a^{|G|} = e$) Let G be a finite group, and let $a \in G$. Then, $a^{|G|} = e$.

Corollary 1.58 (Fermat's Little Theorem) For every integer a and every prime p, a^p mod $p = a$ mod p.

Theorem 1.59 ($|HK| = |H||K|/|H \cap K|$) *For two finite subgroups H and K of a group, define the set $HK = \{hk \mid h \in H, k \in K\}$. Then $|HK| = |H||K|/|H \cap K|$.*

Theorem 1.60 (Classification of Groups of Order $2p$) *Let G be a group of order $2p$, where p is a prime greater than 2. Then G is isomorphic to Z_{2p} or D_p.*

Definition 1.61 (Stabilizer of a Point) Let G be a group of permutations of a set S. For each i in S, let

$$\mathrm{stab}_G(i) = \{\sigma \in G \mid \sigma(i) = i\}.$$

We call $\mathrm{stab}_G(i)$ the *stabilizer of i in G.*

Definition 1.62 (Orbit of a Point) Let G be a group of permutations of a set S. For each s in S, let

$$\mathrm{orb}_G(s) = \{\sigma(s) \mid \sigma \in G\}.$$

The set $\mathrm{orb}_G(s)$ is a subset of S called the *orbit of s under G.* We use $|\mathrm{orb}_G(s)|$ to denote the number of elements in $\mathrm{orb}_G(s)$.

Theorem 1.63 (Orbit-Stabilizer Theorem) *Let G be a finite group of permutations of a set S. Then, for any i from S,*

$$|G| = |orb_G(i)|\,|stab_G(i)|.$$

Theorem 1.64 (The Rotation Group of a Cube) *The group of rotations of a cube is isomorphic to S_4.*

Definition 1.65 (External Direct Product) Let $G_1, G_2, \cdots, G_n$ be a finite collection of groups. The *external direct product* of $G_1, G_2, \cdots, G_n$, written as $G_1 \oplus G_2 \oplus \cdots \oplus G_n$, is the set of all n-tuples for which the ith component is an element of G_i and the operation is componentwise.

Theorem 1.66 (Order of an Element in a Direct Product) *The order of an element in a direct product of a finite number of finite groups is the least common multiple of the orders of the components of the element. In symbols, $|(g_1, g_2, \cdots, g_n)| = lcm\ (|g_1|, |g_2|, \cdots, |g_n|)$.*

Theorem 1.67 (Criterion for $G \oplus H$ to Be Cyclic) *Let G and H be finite cyclic groups. Then $G \oplus H$ is cyclic if and only if $|G|$ and $|H|$ are relatively prime.*

Corollary 1.68 (Criterion for $G_1 \oplus G_2 \oplus \cdots \oplus G_n$ to Be Cyclic) An external direct product $G_1 \oplus G_2 \oplus \cdots \oplus G_n$ of a finite number of finite cyclic groups is cyclic if and only if $|G_i|$ and $|G_j|$ are relatively prime when $i \neq j$.

Corollary 1.69 (Criterion for $Z_{n_1 n_2 \cdots n_k} \approx Z_{n_1} \oplus Z_{n_2} \oplus \cdots \oplus Z_{n_k}$) Let $m = n_1 n_2 \cdots n_k$. Then Z_m is isomorphic to $Z_{n_1} \oplus Z_{n_2} \oplus \cdots \oplus Z_{n_k}$ if and only if n_i and n_j are relatively prime when $i \neq j$.

Theorem 1.70 ($U(n)$ as an External Direct Product) *Suppose s and t are relatively prime. Then $U(st)$ is isomorphic to the external direct product of $U(s)$ and $U(t)$. In short, $U(st) \approx U(s) \oplus U(t)$. Moreover, $U_s(st)$ is isomorphic to $U(t)$ and $U_t(st)$ is isomorphic to $U(s)$.*

Corollary 1.71 Let $m = n_1 n_2 \cdots n_k$, where $\gcd(n_i, n_j) = 1$ for $i \neq j$. Then, $U(m) \approx U(n_1) \oplus U(n_2) \oplus \cdots \oplus U(n_k)$.

Definition 1.72 (Normal Subgroup) A subgroup H of a group G is called a *normal subgroup* of G if $aH = Ha$ for all a in G. We denote this by $H \lhd G$.

Theorem 1.73 (Normal Subgroup Test) *A subgroup H of G is normal in G if and only if $xHx^{-1} \subseteq H$ for all x in G.*

Theorem 1.74 (Factor Groups) *Let G be a group and let H be a normal subgroup of G. The set $G/H = \{aH \mid a \in G\}$ is a group under the operation $(aH)(bH) = abH$.*

Theorem 1.75 (G/Z Theorem) *Let G be a group and let $Z(G)$ be the center of G. If $G/Z(G)$ is cyclic, then G is Abelian.*

Theorem 1.76 ($G/Z(G) \approx Inn(G)$) *For any group G, $G/Z(G)$ is isomorphic to $Inn(G)$.*

Theorem 1.77 (Cauchy's Theorem for Abelian Groups) *Let G be a finite Abelian group and let p be a prime that divides the order of G. Then G has an element of order p.*

Definition 1.78 (Internal Direct Product of H and K) We say that G is the *internal direct product* of its normal subgroup H and K if $G = HK$ and $H \cap K = \{e\}$ and it is written as $G = H \times K$.

Definition 1.79 (Internal Direct Product $H_1 \times H_2 \times \cdots \times H_n$) Let $H_1, H_2, \cdots, H_n$ be a finite collection of normal subgroups of G. We say that G is the *internal direct product* of $H_1, H_2, \cdots, H_n$ and write $G = H_1 \times H_2 \times \cdots \times H_n$ if

1. $G = H_1 \times H_2 \times \ldots \times H_n = \{h_1 \cdot h_2 \cdot \cdots \cdot h_n \mid h_i \in H_i\}$,
2. $(H_1 \times H_2 \times \ldots \times H_i) \cap H_{i+1} = \{e\}$ for $i = 1, 2, \cdots, n-1$.

Theorem 1.80 ($H_1 \times H_2 \times \cdots \times H_n \approx H_1 \oplus H_2 \oplus \cdots \oplus H_n$) *If a group G is the internal direct product of a finite number of subgroups $H_1, H_2, \cdots, H_n$, then G is isomorphic to the external direct product of $H_1, H_2, \cdots, H_n$.*

Theorem 1.81 (Classification of Groups of Order p^2) *Every group of order p^2, where p is a prime, is isomorphic to Z_{p^2} or $Z_p \oplus Z_p$.*

Corollary 1.82 If G is a group of order p^2, where p is a prime, then G is Abelian.

Definition 1.83 (Group Homomorphism) A *homomorphism* ϕ from a group G to a group $\overline{G}$ is a mapping from G into $\overline{G}$ that preserves the group operation; that is, $\phi(ab) = \phi(a)\phi(b)$ for all a, b in G.

Definition 1.84 (Kernel of a Homomorphism) The *kernel* of a homomorphism ϕ from a group G to a group $\overline{G}$ with identity e is the set $\{x \in G \mid \phi(x) = e\}$. The kernel of ϕ is denoted by Ker ϕ.

Theorem 1.85 (Properties of Elements under Homomorphisms) *Let ϕ be a homomorphism from a group G to a group $\overline{G}$ and let g be an element of G. Then*

1. *ϕ carries the identity of G to the identity of $\overline{G}$.*
2. *$\phi(g^n) = (\phi(g))^n$ for all n in $\mathbb{Z}$.*
3. *If $|g|$ is finite, then $|\phi(g)|$ divides $|g|$.*
4. *Ker ϕ is a subgroup of G.*
5. *$\phi(a) = \phi(b)$ if and only if aKer $\phi = b$Ker ϕ.*
6. *If $\phi(g) = g'$, then $\phi^{-1}(g') = \{x \in G \mid \phi(x) = g'\} = g$Ker ϕ.*

Theorem 1.86 (Properties of Subgroups under Homomorphisms) *Let ϕ be a homomorphism from a group G to a group $\overline{G}$ and let H be a subgroup of G. Then*

1. $\phi(H) = \{\phi(h) \mid h \in H\}$ *is a subgroup of $\overline{G}$.*
2. *If H is cyclic, then $\phi(H)$ is cyclic.*
3. *If H is Abelian, then $\phi(H)$ is Abelian.*
4. *If H is normal in G, then $\phi(H)$ is normal in $\phi(G)$.*
5. *If $|Ker\ \phi| = n$, then ϕ is an n-to-1 mapping from G onto $\phi(G)$.*
6. *If $|H| = n$, then $|\phi(H)|$ divides n.*
7. *If $\overline{K}$ is a subgroup of $\overline{G}$, then $\phi^{-1}(\overline{K}) = \{k \in G \mid \phi(k) \in \overline{K}\}$ is a subgroup of G.*
8. *If $\overline{K}$ is a normal subgroup of $\overline{G}$, then $\phi^{-1}(\overline{K}) = \{k \in G \mid \phi(k) \in \overline{K}\}$ is a normal subgroup of G.*
9. *If ϕ is onto and $Ker\ \phi = \{e\}$, then ϕ is an isomorphism from G to $\overline{G}$.*

Corollary 1.87 (Kernels Are Normal) Let ϕ be a group homomorphism from G to $\overline{G}$. Then Ker ϕ is a normal subgroup of G.

Theorem 1.88 (First Isomorphism Theorem) *Let ϕ be a group homomorphism from G to $\overline{G}$. Then the mapping from $G/Ker\ \phi$ to $\phi(G)$, given by $gKer\ \phi \mapsto \phi(g)$, is an isomorphism. In symbols, $G/Ker\ \phi \approx \phi(G)$.*

Corollary 1.89 If ϕ is a homomorphism from a finite group G to $\overline{G}$, then $|\phi(G)|$ divides $|G|$ and $|\overline{G}|$.

Theorem 1.90 (Normal Subgroups Are Kernels) *Every normal subgroup of a group G is the kernel of a homomorphism of G. In particular, a normal subgroup N is the kernel of the mapping $g \to gN$ from G to G/N.*

2 INTRODUCTION TO RINGS

Rings are algebraic structures that originated from the theory of algebraic integers. The concept was introduced by *Richard Dedekind*, taking inspiration from the algebraic structure of integers over complex numbers. Rings were first formalized as a generalization of *Dedekind domains* that occur in number theory, and of polynomial rings and rings of invariants that occur in algebraic geometry and invariant theory. The concept of ring is an extension of groups and has a wide range of applications in mathematics and computer science. Rings are applied in the study of geometric objects, topology, cryptography and various other branches of algebra.

In the vast arena of mathematics, the emergence of ring theory as a pivotal branch of algebra owes much to the visionary insights of two more extraordinary minds: David Hilbert and Emmy Noether. At the beginning of the 20th century, Hilbert provided a unifying framework for understanding sets of numbers endowed with specific algebraic properties. Concurrently, Noether's seminal contributions, notably expounded in her 1921 paper "Ideal Theory in Rings", brought forth profound advancements in the understanding of commutative rings, laying a robust theoretical foundation for subsequent explorations. Despite encountering formidable obstacles, including entrenched gender biases and political upheavals, Hilbert and Noether remained steadfast in their pursuit of mathematical truth.

In this chapter, the concept of rings is formally defined and illustrated through various examples. The study is contemplated through various questions to enable students to develop the foundation of abstract knowledge of rings and different solution techniques. The chapter is concluded with an exercise for learners to test their solving capabilities.

Definition 2.1 (Ring) A non-empty set R equipped with two binary operations '$+$' (addition) and '$\cdot$' (multiplication) is said to be a *ring* if it satisfies the following axioms:

$R1$ Closure w.r.t. '$+$'

$$x + y \in R \qquad \text{for all } x,\, y \in R$$

R2 Commutativity w.r.t. '+'

$$x + y = y + x \qquad \text{for all } x, y \in R$$

R3 Associativity w.r.t. '+'

$$x + (y + z) = (x + y) + z \qquad \text{for all } x, y, z \in R$$

R4 Existence of additive identity

there exists $0 \in R$ such that $x + 0 = x = 0 + x$ for all $x \in R$. The element 0 is called the *additive identity* of R.

R5 Existence of additive inverse

for every $x \in R$, there exists $-x \in R$ such that $x + (-x) = 0 = (-x) + x$. The element $-x$ is called the *additive inverse* of x.

R6 Closure w.r.t. '·'

$$x \cdot y \in R \qquad \text{for all } x, y \in R$$

R7 Associativity w.r.t. '·'

$$x \cdot (y \cdot z) = (x \cdot y) \cdot z \qquad \text{for all } x, y, z \in R$$

R8 Distributive Laws

$$x \cdot (y + z) = x \cdot y + x \cdot z \qquad \text{for all } x, y, z \in R$$
$$(x + y) \cdot z = x \cdot z + y \cdot z \qquad \text{for all } x, y, z \in R$$

For the sake of simplicity, now onwards, we will write $x \cdot y$ as xy.

A ring is usually denoted by the triplet $(R, +, \cdot)$ or simple R if the operations are understood.

A ring can be further characterized by additional axioms that may hold true which leads to the delineation of various subclasses within the framework. In the subsequent definitions, we enumerate two significant classes of rings.

Definition 2.2 (Commutative Ring) A ring R is said to be a *commutative ring* if the following axiom holds:

Commutativity w.r.t. '·'

$$x \cdot y = y \cdot x \qquad \forall \, x, y \in R.$$

In other words, we can say that a ring R is commutative if multiplication '·' is commutative.

Example 2.3 The set of integers, $\mathbb{Z}$ is a commutative ring under usual addition and multiplication.

Definition 2.4 (Ring with Unity) A ring R is said to be a *ring with unity* if there exists $1 \in R$ such that $1 \cdot x = x \cdot 1 = x$ for every $x \in R$.

In other words, we can say that R is a ring with unity if multiplicative identity exists in R.

The element 1(mulitplicative identity) is called the *unity of ring R* and it can be easily shown that the unity (if it exists) is unique.

Example 2.5 The set of integers, $\mathbb{Z}$ under usual addition and usual multiplication is a commutative ring with unity 1.

Remark In general, a ring may not necessarily possess unity.

One can easily observe that the ring of even integers, $2\mathbb{Z}$, is a commutative ring without unity under usual addition and multiplication.

The definition of ring does not guarantee the multiplicative inverse of an element. There may be element x in R which does not possess any multiplicative inverse. If an element x in R has a multiplicative inverse then x is known as a *unit* in R.

Definition 2.6 (Units) Let $(R, +, \cdot)$ be a ring with unity 1 and x be an element in R. Then, x is said to be a *unit* in R if x has a multiplicative inverse. That is, there exists an element $y \in R$ such that

$$x \cdot y = y \cdot x = 1.$$

Here, x and y are the inverses of each other and thus y is a unit as well and the set of units of R is denoted by $U(R)$.

One can easily verify that the multiplicative inverse of an element (if it exists) is always unique.

Example 2.7 Let $R = \mathbb{Z}$, ring of integers with usual addition and multiplication. Then the only units in R are 1 and -1.

Now, we introduce the concept of a division ring—a fundamental idea in algebra.

Definition 2.8 (Division Ring) Let $(R, +, \cdot)$ be a ring with unity. Then, R is said to be a *division ring* if every nonzero element of R is a unit in R.

There are numerous real-world instances, where abstract algebra, particularly ring theory, finds relevance. The following examples illustrate the concept of ring theory in various contexts and show that the concept of ring theory is ambient.

Example 2.9 Let us consider the set of integers, $\mathbb{Z}$ with usual addition and multiplication. It is a commutative ring under usual addition and multiplication with unity 1. Also, the set of units of $\mathbb{Z}$, are $U(\mathbb{Z}) = \{1, -1\}$.

Example 2.10 Let us consider $\mathbb{R}$, the set of real numbers, $\mathbb{Q}$ the set of rationals and $\mathbb{C}$ the set of complex numbers. These all are commutative rings under usual addition and multiplication. Also they have the unity as well which is 1.

It is easy to conclude that the all nonzero elements are units in these rings. That is, we have $U(\mathbb{R}) = \mathbb{R}^*$, $U(\mathbb{Q}) = \mathbb{Q}^*$, $U(\mathbb{C}) = \mathbb{C}^*$.

Example 2.11 Let us consider the set of even integers, $2\mathbb{Z}$ under usual addition and multiplication which also forms a ring structure. Here, one can note that, $2\mathbb{Z}$ is a commutative ring without unity as $1 \notin 2\mathbb{Z}$. Also, it has no units. Thus, we have

$$U(2\mathbb{Z}) = \emptyset$$

In general, the set $n\mathbb{Z}$, for $n > 1$, is a ring without unity.

Example 2.12 Let us consider $\mathbb{Z}[x]$, the set of polynomials with integral coefficients, which is a commutative ring under usual addition and multiplication with unity 1. The units in $\mathbb{Z}[x]$ are 1 and -1, which can be realized as follows:

$$\mathbb{Z}[x] = \{a_0 + a_1 x + a_2 x^2 + \ldots \mid a_i \in \mathbb{Z}\}$$

Let, $f(x) = a_0 + a_1 x + a_2 x^2 + \ldots \in \mathbb{Z}[x]$ such that $f(x)$ has inverse in $\mathbb{Z}[x]$. That is, $[f(x)]^{-1} \in \mathbb{Z}[x]$. Thus, we can say that $\dfrac{1}{f(x)} \in \mathbb{Z}[x]$.

We have $\dfrac{1}{a_0 + a_1 x + a_2 x^2 + \ldots} \in \mathbb{Z}[x]$ which is only possible if $a_0 = \pm 1$, $a_1 = a_2 = \ldots = 0$. Therefore, the possible values of $f(x)$ are $1, -1$ only. Hence, the set of units of $\mathbb{Z}[x]$, that is, $U(\mathbb{Z}[x]) = \{1, -1\}$.

In the following example, we will discuss an important algebraic structure known as $\mathbb{Z}_n$.

Example 2.13 Let $R = \mathbb{Z}_n$, the set of positive integers under addition modulo n, that is $\oplus_n$ and multiplication modulo n, that is $\odot_n$. It is a commutative ring with unity 1. The set of units of $\mathbb{Z}_n$, is $U(\mathbb{Z}_n) = U(n)$, where $U(n)$ is the set of all positive integers relatively prime to n. (Refer Corollary 4.16)

In particular, for any prime number p, the set of units of $\mathbb{Z}_p$, is $U(\mathbb{Z}_p) = Z_p - \{0\}$.

Example 2.14 The set $C[0, 1]$, the set of continuous real-valued functions defined on the interval $[0, 1]$ is a commutative ring with unity 1 under point-wise addition and multiplication defined as follows:

$$\begin{aligned} (f + g)(x) &= f(x) + g(x) \\ (f \cdot g)(x) &= f(x) \cdot g(x) \end{aligned}$$

The set of units of $C[0, 1]$ contains all continuous real-valued functions $f(x)$ such that $f(x) \neq 0$ for all $x \in [0, 1]$.

Example 2.15 Consider $R = M_n(\mathbb{R})$, the set of $n \times n$ matrices with real entries under matrix addition and multiplication with unity $\begin{pmatrix} 1 & 0 \\ 0 & 1 \end{pmatrix}$ is a noncommutative ring. Here, the units in $M_n(\mathbb{R})$ are all invertible matrices. That is, all non-singular matrices of $M_n(\mathbb{R})$ are units in $M_n(\mathbb{R})$.

Example 2.16 Consider the set of 2×2 matrices with entries from $M_n(\mathbb{Z}_2)$. Here, $M_n(\mathbb{Z}_2)$ is a finite noncommutative ring under matrix addition and multiplication with unity $\begin{pmatrix} 1 & 0 \\ 0 & 1 \end{pmatrix}$.

Example 2.17 The set of Gaussian integers $\mathbb{Z}[i]$ defined by

$$\mathbb{Z}[i] = \{a + bi \mid a, b \in \mathbb{Z}\}$$

is a commutative ring with unity 1. The set of units of $\mathbb{Z}[i]$ is $U(\mathbb{Z}[i]) = \{1, -1, i, -i\}$.

Example 2.18 The set $\mathbb{Z}[\sqrt{2}] = \{a + b\sqrt{2} \mid a, b \in \mathbb{Z}\}$ is a commutative ring with unity.

A ring possesses some special properties which we explore in the following theorems.

Theorem 2.19 (Properties of a Ring) *Let R be a ring with additive and multiplicative identities 0 and 1 respectively. Then, for $a, b \in R$, the following properties hold.*

  (i) $0a = a0 = 0$;
  (ii) $(-a)b = a(-b) = -(ab)$;
  (iii) $(-a)(-b) = ab$
  (iv) $(na)b = a(nb) = n(ab)$ *for all $n \in \mathbb{Z}$.*

Proof   **(i)** Clearly $a0 = a(0 + 0) = a0 + a0$ then on adding $-(a0)$ to both sides, we get
$0 = a0 - (a0) = a0 + a0 - a0 = a0$.

  (ii) From (i) we have $0 = 0b$ for all $b \in R$. Thus, we can write $0 = 0b = (-a + a)b = (-a)b + ab$. On adding $-(ab)$ to both sides, one can have $-(ab) = (-a)b$.
  Similarly, we can prove that $a(-b) = -(ab)$.

  (iii) From (ii), we have $(-a)(-b) = -(a(-b)) = -(-(ab)) = ab$.

  (iv) $(na)b = (a + ... + a)b = (ab + ... + ab) = n(ab) = a(b + ... + b) = a(nb)$.

$\square$

Like the cyclic groups, we also have cyclic rings in the literature. Next, we study cyclic rings, as there are several familiar structures that are cyclic in nature. Among these are the rings $\mathbb{Z}_n$ and $n\mathbb{Z}$ for any positive integer n. Cyclic rings are relatively easy to analyze because their additive structure forces them to have certain properties. For instance, if R is an infinite cyclic ring under addition, and if r is a generator of R, and $a, b \in \mathbb{Z}$, then we have $ar = br$ if and only if $a = b$. Also, if R is a finite cyclic ring of order n, and r is a generator of R, $a, b \in \mathbb{Z}$, then $ar = br$ if and only if $a \cong b \bmod n$. This observation motivates the proceeding result.

Proposition 2.20 *A ring which is cyclic under addition is commutative.*

Proof Let R be a cyclic ring under addition, which is generated by an element $a \in R$. Let, $x, y \in R$ then we have $x = ma$ and $y = na$ for some $m, n \in \mathbb{Z}$.

We need to show that $x \cdot y = y \cdot x$ that is, $(m \cdot a)(n \cdot a) = (n \cdot a)(m \cdot a)$.
If m or n is 0, then the statement is vacuously true.

Case 1. Suppose m and n both are positive, then we have

$$(ma)(na) \;=\; \underbrace{(a + a + ... + a)}_{m \text{ times}} \cdot \underbrace{(a + a + ... + a)}_{n \text{ times}}$$

$$=\; \underbrace{(aa + aa + ... + aa)}_{mn \text{ times}}$$

$$=\; (mn)(aa) = (mn)(a^2)$$

Case 2. Suppose m is positive and n is negative. Therefore, $-n$ is positive, then we have

$$(ma)(na) = \underbrace{(a + a + \ldots + a)}_{m \text{ times}} \cdot \underbrace{((-a) + (-a) + \ldots + (-a))}_{-n \text{ times}}$$

$$= \underbrace{a(-a) + a(-a) + \ldots + a(-a)}_{m(-n) \text{ times}} = (m(-n))(a(-a))$$

$$= -(mn)(-(aa)) = (mn)(aa)$$

Thus, we have $(ma)(na) = (mn)(a^2)$.

Case 3. Suppose m is negative and n is positive. Then, $-m$ and n both are positive. Consider

$$(ma)(na) = ((-m)(-a))(na)$$

$$= \underbrace{((-a) + (-a) + \ldots + (-a))}_{-m \text{ times}} \cdot \underbrace{(a + a + \ldots + a)}_{n \text{ times}}$$

$$= \underbrace{(-a)a + (-a)a + \ldots + (-a)a}_{(-m)n \text{ times}}$$

$$= ((-m)n)((-a)a) = (-mn)(-aa) = (mn)(a^2)$$

Therefore, we have $(ma)(na) = (mn)(a^2)$.

Case 4. Suppose m and n both are negative. Then, $-m$ and $-n$ both are positive. Therefore

$$(ma)(na) = ((-m)(-a))((-n)(-a))$$

$$= \underbrace{((-a) + (-a) + \ldots + (-a))}_{-m \text{ times}} \cdot \underbrace{((-a) + (-a) + \ldots + (-a))}_{-n \text{ times}}$$

$$= \underbrace{(-a)(-a) + (-a)(-a) + \ldots + (-a)(-a)}_{(-m)(-n) \text{ times}}$$

$$= ((-m)(-n))((-a)(-a)) = (mn)(aa)$$

Hence, we have $(ma)(na) = (mn)(a^2)$. Therefore, from the above cases, we can say that

$$(ma)(na) = (mn)(a^2) = (nm)(a^2)$$

$$= \underbrace{(a^2 + a^2 + \ldots + a^2)}_{nm \text{ times}}$$

$$= \underbrace{(a \cdot a + a \cdot a + \ldots + a \cdot a)}_{nm \text{ times}}$$

$$= \underbrace{(a + a + \ldots + a)}_{n \text{ times}} \cdot \underbrace{(a + a + \ldots + a)}_{m \text{ times}} = (na) \cdot (ma)$$

Hence, we conclude that $x \cdot y = y \cdot x \ \forall \ x, y \in R$. Therefore, every cyclic ring under addition is commutative. $\square$

Remark The converse of the above result need not be true. That is, every commutative ring need not be cyclic under addition.

To the support the above remark, we have the following example.

Example 2.21 Let us consider $R = C[0, 1]$. The set of all continuous real-valued functions defined on the interval $[0, 1]$. Here, R is a commutative ring but it is not cyclic.

In the following result, we discuss an important property of a commutative ring.

Proposition 2.22 *Let R be a ring. Then R is commutative if and only if $a^2 - b^2 = (a + b)(a - b)$ for all a, b in R.*

Proof Let R be a ring. Now, we need to prove that R is commutative if and only if $a^2 - b^2 = (a + b)(a - b)$ for all a, b in R.

Consider,

$$(a + b)(a - b) = a^2 + ba - ab - b^2 = a^2 - b^2 \quad \text{for all} \quad a, b \in R$$

if and only if we have $ba - ab = 0$, that is, $ba = ab$ for all $a, b \in R$,

Thus, we have R is commutative if and only if $a^2 - b^2 = (a + b)(a - b)$ for all $a, b \in R$. $\square$

Let us consider $R = \mathbb{Z}$, which is a commutative ring with unity under usual addition and multiplication. Also, the set of units of $\mathbb{Z}$ is $U(\mathbb{Z}) = \{1, -1\}$, which forms a group under multiplication. In the next result, we generalize this property to any commutative ring with unity. Consequently, we have the following theorem.

Theorem 2.23 *Let R be a commutative ring with unity. Then, the set of units of R, $U(R)$ forms a group under the multiplication of R.*

(This group is called the *group of units of R*.)

Proof Let R be a ring and we define

$$U(R) = \{a \in R \,|\, a^{-1} \in R\}$$

Now, we will prove that $U(R)$ forms a group under multiplication.

Closure: Let $a, b \in U(R)$ then $a^{-1}, b^{-1} \in R$ such that

$$aa^{-1} = 1 = a^{-1}a \quad \text{and} \quad bb^{-1} = 1 = b^{-1}b.$$

We need to show that $ab \in U(R)$.
Consider

$$
\begin{aligned}
(ab)(b^{-1}a^{-1}) &= abb^{-1}a^{-1} \\
&= a1a^{-1} \\
&= aa^{-1} \\
&= 1.
\end{aligned}
$$

Similarly, we have $(b^{-1}a^{-1})ab = 1$. Hence, ab has an inverse in R, which is given by $b^{-1}a^{-1} \in R$. Therefore, ab is a unit, and thus $ab \in U(R)$.

Hence, $U(R)$ is closed under multiplication.

Associativity: Since, $U(R) \subseteq R$ and R is associative. Therefore, associativity holds in $U(R)$ as well.

Multiplicative identity: As R is a ring with unity 1
Then,

$$a.1 = 1.a = a \ \forall \ a \in R$$

In particular, take $a = 1$, we have $1.1 = 1$. Thus, $1 \in U(R)$. Hence, $U(R)$ has the multiplicative identity.

Multiplicative inverse: Let $a \in U(R)$, we need to show that $a^{-1} \in U(R)$.

Suppose $a \in U(R)$, therefore $a^{-1} \in R$ such that $aa^{-1} = 1 = a^{-1}a$. Also, $(a^{-1})^{-1} = a \in R$. Thus, we can say that a^{-1} also has an inverse in R. Thus, a^{-1} is a unit and hence $a^{-1} \in U(R)$.

For all $a \in U(R)$, we can say that $a^{-1} \in U(R)$ such that $aa^{-1} = 1 = a^{-1}a$. Thus, $U(R)$ has multiplicative inverses.

Hence, $U(R)$ forms a group under the multiplication of R. $\qquad\qquad\qquad\square$

Observe that $\mathbb{Z}_3$ and $\mathbb{Z}_4$, both are commutative rings with unity 1. Here, $U(\mathbb{Z}_3) = \{1, 2\}$ and $U(\mathbb{Z}_4) = \{1, 3\}$. Then,

$$U(\mathbb{Z}_3 \oplus \mathbb{Z}_4) = \{(1, 1), (1, 3), (2, 1), (2, 3)\}.$$

Thus, if the group of units of commutative rings $R_1, R_2, \ldots, R_n$ with unity is known, one can easily find the group of units of the direct product of these rings by the proceeding theorem.

Theorem 2.24 *Let $R_1, R_2, \ldots, R_n$ be commutative rings with unity. Then*

$$U(R_1 \oplus R_2 \oplus \ldots \oplus R_n) = U(R_1) \oplus U(R_2) \oplus \ldots \oplus U(R_n).$$

(Here, we recall that $R_1 \oplus R_2 \oplus \ldots \oplus R_n = \{(r_1, r_2, \ldots, r_n) \mid r_i \in R_i\}$ with the operations: $(r_1, \ldots, r_n) + (s_1, \ldots, s_n) = (r_1 + s_1, \ldots, r_n + s_n)$ and $(r_1, \ldots, r_n)(s_1, \ldots, s_n) = (r_1 s_1, \ldots, r_n s_n)$ is called the direct product of $R_1, R_2, \ldots, R_n$.)

Proof Let $R_1, R_2, \ldots R_n$ be rings and $x \in U(R_1 \oplus R_2 \oplus \ldots \oplus R_n)$. That is, x is a unit of $R_1 \oplus R_2 \oplus \ldots \oplus R_n$.

Let, $x = (r_1, r_2, \ldots, r_n)$; therefore x^{-1} exists in $R_1 \oplus R_2 \oplus \ldots \oplus R_n$.
Suppose $x^{-1} = (s_1, s_2, \ldots, s_n)$.
Then,

$$
\begin{aligned}
(1, 1, \ldots, 1) &= xx^{-1} \\
&= (r_1, r_2, \ldots, r_n)(s_1, s_2, \ldots, s_n) \\
&= (r_1 s_1, r_2 s_2, \ldots, r_n s_n)
\end{aligned}
$$

Similarly, we have $(1, 1, \dots, 1) = (s_1 r_1, s_2 r_2, \dots, s_n r_n)$

Therefore, we have $r_i s_i = 1 = s_i r_i$ where r_i, $s_i \in R_i$ for all i. Thus, we can say that r_i has an inverse given by s_i in R_i for all i. Hence r_i is a unit of R_i.

We have $(r_1, r_2, \dots, r_n) \in U(R_1) \oplus U(R_2) \oplus \dots \oplus U(R_n)$, that is, $x = (r_1, r_2, \dots, r_n) = \in U(R_1) \oplus U(R_2) \oplus \dots \oplus U(R_n)$. Therefore, we have

$$U(R_1 \oplus R_2 \oplus \dots \oplus R_n) \subseteq U(R_1) \oplus U(R_2) \oplus \dots \oplus U(R_n). \tag{2.1}$$

Let, $y \in U(R_1) \oplus U(R_2) \oplus \dots \oplus U(R_n)$ and let, $y = (r_1', r_2', \dots, r_n')$ where $r_i' \in U(R_i)$. As $r_i' \in U(R_i)$ for all i, therefore r_i' has an inverse in R_i say s_i'.

Then, we have $r_i' s_i' = 1 = s_i' r_i'$.

Let us consider $z = (s_1', s_2', \dots, s_n')$. Then, $yz = (r_1', r_2', \dots, r_n')(s_1', s_2', \dots, s_n') = (1, 1, \dots, 1)$ Similarly, $zy = 1$. Thus, y has an inverse in $R_1 \oplus R_2 \oplus \dots \oplus R_n$ and therefore, y is a unit of $R_1 \oplus R_2 \oplus \dots \oplus R_n$.

Since y was an arbitrary element of $U(R_1) \oplus U(R_2) \oplus \dots \oplus U(R_n)$. Thus,

$$U(R_1) \oplus U(R_2) \oplus \dots \oplus U(R_n) \subseteq U(R_1 \oplus R_2 \oplus \dots \oplus R_n) \tag{2.2}$$

From Relations (2.1) and (2.2), we have

$$U(R_1 \oplus R_2 \oplus \dots \oplus R_n) = U(R_1) \oplus U(R_2) \oplus \dots \oplus U(R_n)$$

$\square$

On similar lines, we can characterize the unity of external direct product of rings, $R_1 \oplus R_2 \oplus \dots \oplus R_n$ in terms of unity of R_i, which follows from the next result.

Proposition 2.25 *Suppose $R_1, R_2, \dots, R_n$ are rings that contain more than one element. Then, the ring $R_1 \oplus R_2 \oplus \dots \oplus R_n$ has a unity if and only if R_i has unity for all $i = 1, 2, \dots n$.*

Proof Suppose $R_1 \oplus R_2 \oplus \dots \oplus R_n$ has unity $(1, 1, 1, \dots, 1)$.
Then, consider

$$
\begin{aligned}
(r_1, r_2, \dots, r_n)(1, 1, \dots, 1) &= (r_1, r_2, \dots, r_n) &= (1, 1, \dots, 1)(r_1, r_2, \dots, r_n) \\
(r_1 \cdot 1, r_2 \cdot 1, \dots, r_n \cdot 1) &= (r_1, r_2, \dots, r_n) &= (1 \cdot r_1, 1 \cdot r_2, \dots, 1 \cdot r_n) \\
r_i \cdot 1 &= 1 &= 1 \cdot r_i \text{ for all } i = 1, 2, \dots n.
\end{aligned}
$$

Thus, R_i has unity for all $i = 1, 2, \dots, n$. Hence, R_i has unity for all $i = 1, 2, \dots, n$ if $R_1 \oplus R_2 \oplus \dots \oplus R_n$ has a unity.

Converse is left for readers. $\square$

It can be noticed that, in $\mathbb{Z}_4$, 1 and 3 are units and 3 divides every element as $3 \odot_4 3 = 1$ which implies 3 divides 1, $3 \odot_4 2 = 2$, that is, 3 divides 2 and $3 \odot_4 1 = 3$ implies 3 divides 3. However, $2 \in \mathbb{Z}_4$ does not divide 1 and 3 as $2 \odot_4 x \neq 1$ and $2 \odot_4 x \neq 3$ for any $x \in \mathbb{Z}_4$. This observation leads the succeeding result.

Theorem 2.26 *A unit of a ring divides every element of the ring.*

Proof Let R be a ring and $a \in R$ be a unit of the ring. Let, $b \in R$ be any element in R. Since, a is a unit, then, we have $a^{-1} \in R$ such that $a \cdot a^{-1} = 1$.

Now

$$
\begin{aligned}
b &= 1 \cdot b \\
&= (a \cdot a^{-1})\, b \\
&= a\,(a^{-1} b)
\end{aligned}
$$

Thus, we can say that b is divisible by a. Since b was an arbitrary element of ring R. Therefore, unit a of the ring divides every element of the ring. $\qquad\square$

It is interesting to observe that in the ring $\mathbb{Z}_4$, $2 \in \mathbb{Z}_4$ such that $2^2 = 0$. Here, the element 2 is called a *nilpotent* element in $\mathbb{Z}_4$. Also, in the ring $\mathbb{Z}_6$, $3 \in \mathbb{Z}_6$ such that $3^2 = 3$. Here, the element 3 is called an *idempotent* element in $\mathbb{Z}_6$. We now define some special elements in a ring, which are given as follows:

Definition 2.27 (Nilpotent Element) Let R be a ring, then an element $x \in R$ is said to be a *nilpotent* element if there exists a positive integer k such that $x^k = 0$.

Here, the exponent of x denotes $\underbrace{x \cdot x \dots x}_{k \text{ times}}$ (where '$\cdot$' is the multiplication operation defined on R).

Definition 2.28 (Idempotent Element) Let R be a ring, then an element $x \in R$ is said to be an *idempotent* element if $x^2 = x$.

In the following examples, we will discuss about the nilpotent and idempotent elements in rings.

Example 2.29 Let R be any ring with unity. Then, 0 and 1 are (trivially) idempotent elements of ring R. Also, 0 is (trivially) nilpotent element.

Example 2.30 Let $R = \mathbb{Z}$ be ring of integers. Then 0 is the only nilpotent element in R. The only idempotent elements of R are 0 and 1

Example 2.31 Let $R = \mathbb{Q}[x]$, the set of all polynomials with rational coefficient. There are no non-trivial idempotent or nilpotent elements in R.

One can observe that in the ring $\mathbb{Z}_2 \oplus \mathbb{Z}_2$, every element is an idempotent element. This motivates to expound the next definition.

Definition 2.32 (Boolean Ring) A ring R is said to be a *Boolean ring* if $a^2 = a$ for all a in R.

In other word, we can say that, in a ring R, if every element of R is idempotent then it is known as Boolean Ring.

Example 2.33 The rings $\mathbb{Z}_2$ and $\mathbb{Z}_2 \oplus \mathbb{Z}_2$ are finite Boolean rings. While, the ring $\mathbb{Z}_2 \oplus \mathbb{Z}_2 \oplus \dots$ is an infinite Boolean ring.

In the next theorem, we discuss an important property of the Boolean ring.

Theorem 2.34 *Every boolean ring is commutative, that is, if $a^2 = a$ for all a in R. Then, R is commutative.*

Proof Let R be a Boolean ring. That is, for all $a \in R$, we have $a^2 = a$.

We need to show that R is commutative, i.e., $ab = ba$ for all $a, b \in R$.

For this, we need to show the following:

(i) $a = -a$ for all $a \in R$

(ii) $ab = ba$ for all $a, b \in R$

(i) To show that $a = -a$ for all $a \in R$.

Consider, $a \in R$

$\Rightarrow a + a \in R$

$\Rightarrow (a + a)^2 = a + a$

$\Rightarrow (a + a)(a + a) = a + a$

$\Rightarrow a^2 + a^2 + a^2 + a^2 = a + a$

$\Rightarrow a + a = 0$

$\Rightarrow a = -a$ for all $a \in R$.
$\hfill (2.3)$

(ii) To show that $ab = ba \ \forall \ a, b \in R$.

Consider, $a, b \in R$

$\Rightarrow a + b \in R$

$\Rightarrow (a + b)^2 = a + b$

$\Rightarrow (a + b)(a + b) = a + b$

$\Rightarrow a^2 + ab + ba + b^2 = a + b$

$\Rightarrow ab + ba = 0$

$\Rightarrow ab = -ba$

$\Rightarrow ab = ba \ \forall \ a, b \in R.$
$\hfill \text{(Using (2.3))}$

Hence, R is commutative.

$\hfill \square$

Illustrative Problems

Problem 1 Give an example of a finite noncommutative ring. Also give an example of an infinite noncommutative ring that does not have a unity.

<u>**Solution**</u> Let us consider

$$M_2(\mathbb{Z}_p) = \left\{ \begin{bmatrix} a & b \\ c & d \end{bmatrix} \middle| \ a, b, c, d \in \mathbb{Z}_p \right\}$$

for any prime p.

Indeed, $M_2(\mathbb{Z}_p)$ is a commutative group under addition. Also, distributivity can be easily verified.

Since, matrix multiplication is not commutative, therefore $M_2(\mathbb{Z}_p)$ is a finite ring, which is noncommutative.

Consider

$$M_2(2\mathbb{Z}) = \left\{ \begin{bmatrix} a & b \\ c & d \end{bmatrix} \,\middle|\, a, b, c, d \in 2\mathbb{Z} \right\}.$$

Then, $M_2(2\mathbb{Z})$ is an infinite noncommutative ring without unity.

Problem 2 Let R be the set of 2×2 matrices over $\mathbb{R}$, with the usual matrix addition and matrix multiplication. Check whether R is a commutative ring.

<u>Solution</u> One can easily observe that R is not a commutative ring, because matrix multiplication is not commutative in general.

For example, let us consider $a = \begin{pmatrix} 1 & 0 \\ 0 & 0 \end{pmatrix}$ and $b = \begin{pmatrix} 0 & 1 \\ 0 & 0 \end{pmatrix}$ then $ab = b$ (verify) and $ba = 0$ (verify). Thus, we have $ab \neq ba$. All the other properties are satisfied, however.

Problem 3 Find the unity of the ring $\{0, 2, 4, 6, 8\}$ under addition and multiplication modulo 10.

<u>Solution</u> Recall that $a(\neq 0) \in R$ is said to be unity of a ring R such that $a.r = r.a = r$ for all $r \in R$.

Let us consider the Cayley table for $\{0, 2, 4, 6, 8\}$ under $\otimes_{10}$:

$\otimes_{10}$	2	4	6	8
2	4	8	2	6
4	8	6	4	2
6	2	4	6	8
8	6	2	8	4

It can be verified from the above Cayley table that the unity of the given ring is 6.

Problem 4 Let R be a ring with unity 1. Suppose the product of any pair of nonzero elements of R is nonzero. Then show that $ab = 1$ implies $ba = 1$.

<u>Solution</u> Let, R be a ring with unity 1. Let, $ab = 1$, which implies $a \neq 0$ and $b \neq 0$. Let us consider, $ab = 1$. Then post multiplying by a, we get $aba = a$, which implies $aba - a = 0$. Thus,

$$a(ba - 1) = 0 \tag{2.4}$$

Since $a \neq 0$ and it is given that product of any pair of nonzero elements is nonzero. Therefore, by Equation (2.4), we have $ba - 1 = 0$. Hence, $ba = 1$.

Problem 5 Let R be a ring and $a \in R$ such that $a^4 = a^2$. Then show that $a^{2n} = a^2$ for all $n \geq 1$.

<u>Solution</u> Let, R be a ring and $a \in R$ such that $a^4 = a^2$. We need to prove that $a^{2n} = a^2$ for all $n \geq 1$.

We shall prove it by using *Principle of Mathematical Induction.*
For $n = 1$, we have $a^{2 \cdot 1} = a^2$. Thus, we have $a^2 = a^2$. Therefore the result is true for $n = 1$.
Suppose the result holds for $n = k$ that is,

$$a^{2k} = a^2 \tag{2.5}$$

Now, we have to show that the result holds for $n = k + 1$.
   For this, consider

$$
\begin{aligned}
a^{2(k+1)} &= a^{2k} \cdot a^2 \\
&= a^2 \cdot a^2 \\
&= a^4 = a^2.
\end{aligned}
$$

Hence, by the *Principle of Mathematical Induction*, we obtain $a^{2n} = a^2$ for all $n \geq 1$.

Problem 6 Let R be a ring such that $x^3 = x$ for all x in R. Prove that $6x = 0$ for all x in R.

<u>Solution</u> Let R be a ring such that $x^3 = x$ for all $x \in R$. Here, we need to prove that $6x = 0$ for all $x \in R$. Let $x \in R$, then $2x \in R$.
   We have $2x = (2x)^3 = 8x^3 = 8x$ $\hfill (x^3 = x)$
Hence $0 = 8x - 2x = 6x$ for all $x \in R$.

Problem 7 Determine an integer $n > 1$ such that $a^n = a$ for every a in $\mathbb{Z}_6$. Also, find such an n for $\mathbb{Z}_{10}$. Show that there does not exist any $n \in \mathbb{Z}_m$, such that $a^n = a$ if m is divisible by the square of some prime.

<u>Solution</u> Let us consider $\mathbb{Z}_6 = \{0, 1, 2, 3, 4, 5\}$ and consider $n = 7$. We have $0^7 = 0$; $1^7 = 1$; $2^7 = 2$; $3^7 = 3$; $4^7 = 4$ and $5^7 = 5$.
Hence, required integer is 7.
   Now, for $\mathbb{Z}_{10}$, where $\mathbb{Z}_{10} = \{0, 1, 2, 3, 4, 5, 6, 7, 8, 9\}$. We have $n = 11$, such that $0^{11} = 0$; $1^{11} = 1$; $2^{11} = 2$; $3^{11} = 3$; $4^{11} = 4$; $5^{11} = 5$; $6^{11} = 6$; $7^{11} = 7$; $8^{11} = 8$ and $9^{11} = 9$.
Therefore, required integer is $n = 11$.
   In general, if $m = p^2 r$ for some $r \in \mathbb{Z}$, then $p \in \mathbb{Z}_m$. We show that there does not exist any n such that $a^n = a$ for all $a \in \mathbb{Z}_m$.
Here, one can observe that, $n \neq 2$, because, if it is so, then we have $a^2 = a$ for all $a \in \mathbb{Z}_m$.
In particular, we have $p^2 = p$ which is not possible as if $p^2 = p$ in $\mathbb{Z}_m$ then, we have $p = 1$.
Also, $n \neq 3$, as if $n = 3$, then, we have $p^3 = p$. Therefore, $p^2 = 1$ in $\mathbb{Z}_m$.
Hence, we have $p^2 = 1 \bmod p^2 r$, that is, $1 = p^2(1 - rq)$ for some $q \in \mathbb{Z}$, which is not possible.
Let $n > 3$, then, we have $p^n = p$ in $\mathbb{Z}_m$. Thus, we have $p^{n-1} = 1 \bmod m$. Hence $p^{n-1} - 1 = p^2 rq$ for some $q \in \mathbb{Z}$. We have, $1 = p^2(p^{n-3} - rq)$, which is not possible. Hence, there does not any n for $\mathbb{Z}_m$, when m is divisible by the square of some prime.

Problem 8 Let R be a ring, possibly noncommutative, such that for $x, y \in R$, $xy = 0$ implies that either $x = 0$ or $y = 0$. If $a, b \in R$ and $a^n = b^n$ and $a^m = b^m$ for two relatively prime positive integers m and n. Then, prove that $a = b$.

<u>Solution</u> Let R be a ring and $m, n \in \mathbb{Z}$ such that m and n are relatively prime, that is, $gcd(m, n) = 1$. This implies, there exist integers $s, t \in \mathbb{Z}$ such that $ms + nt = 1$. Since $m, n > 0$, then either $s > 0$ or $t > 0$.

Without loss of generality, assume that $s > 0$. Then, we have $ms - nt = 1$ where $t > 0$. Hence, we get $ms = nt + 1$.
Now, let us consider
$$a^{ms} = (a^m)^s = (b^m)^s = b^{ms}.$$
Also, we have $b^{nt+1} = a^{ms} = a^{nt+1} = aa^{nt} = a(a^n)^t = a(b^n)^t = ab^{nt}$. Hence, we have $bb^{nt} = ab^{nt}$ implies that $(b - a)b^{nt} = 0$.

Then, we get either $b^{nt} = 0$ or $b = a$. The first possibility $b^{nt} = 0$ is impossible if $b \neq 0$ because, if $k \in \mathbb{Z}$ is the smallest positive integer such that $b^{k-1} \neq 0$, but $b^k = 0$. Then, we have $0 = b^k = b^{k-1}.b \neq 0$. This implies that the assumption $b^k = 0$ is not possible, whenever $b \neq 0$. Hence b must be equal to a.

Problem 9 Let R be a commutative ring with unity. Prove that the ring $R[x]$ is a commutative ring with unity.

<u>Solution</u> Let R be a commutative ring with unity and $R[x]$ be the polynomial ring, defined as the set of all formal sums $a_n x^n + a_{n-1} x^{n-1} + ... + a_1 x + a_0$ where each $a_i \in R$, (here $a_1, a_2, ...$ are called the coefficients of the polynomial; a_i is the coefficient of x^i).
Let $f(x)$ and $g(x)$ be two elements in $R[x]$ say $f(x) = a_n x^n + ... + a_1 x + a_0$ and $g(x) = b_m x^m + ... + b_1 x + b_0$.
Then, we define

$$f(x)g(x) = a_n b_m x^{n+m} + (a_n b_{m-1} + a_{n-1} b_m)x^{n+m-1} + ... + (a_1 b_0 + a_0 b_1)x + a_0 b_0.$$

Since R is commutative. Thus, we have,
$$f(x)g(x) = b_m a_n x^{n+m} + (b_{m-1} a_n + b_m a_{n-1})x^{n+m-1} + ... + (b_0 a_1 + b_1 a_0)x + b_0 a_0 = g(x)f(x).$$
Let $1 \in R$ be the unity of R. Then for any $f(x) \in R[x]$, we have $f(x)1 = 1f(x) = f(x)$. Hence, 1 is also the unity in $R[x]$.

Problem 10 Let R and S be two rings. Equip the Cartesian product $R \times S$ with the following operations

$$
\begin{aligned}
(r, s) + (r', s') &= (r + r', s + s') \\
(r, s)(r', s') &= (rr', ss') \\
0_{R \times S} &= (0_R, 0_S).
\end{aligned}
$$

Show that $R \times S$ is commutative if R and S are both commutative. Also, Show that $R \times S$ has unity if R and S each have unity.

<u>Solution</u> Let R and S be two rings, then $R \times S$ will be a ring with respect to the given operation, if it satisfies the 8 axioms, R1—R8, of a ring.

(1) Closure under addition in $R \times S$ is guaranteed by definition.

(2) Associativity under addition:

Let $(r_1, s_1), (r_2, s_2).(r_3, s_3) \in R \times S$. Then, consider

$$
\begin{aligned}
(r_1, s_1) + ((r_2, s_2) + (r_3, s_3)) &= (r_1, s_1) + (r_2 + r_3, s_2 + s_3) \\
&= (r_1 + (r_2 + r_3), s_1 + (s_2 + s_3)) \\
&= (((r_1 + r_2) + r_3), (s_1 + s_2) + s_3) \\
&= (r_1 + r_2, s_1 + s_2) + (r_3, s_3) \\
&= ((r_1, s_1) + (r_2, s_2)) + (r_3, s_3)
\end{aligned}
$$

(3) Commutativity under addition:

Let $(r_1, s_1), (r_2, s_2) \in R \times S$. Then, consider

$$
(r_1, s_1) + (r_2, s_2) = (r_1 + r_2, s_1 + s_2) = (r_2 + r_1, s_2 + s_1) = (r_2, s_2) + (r_1, s_1).
$$

(4) Existence of $0_{R \times S}$: Let $0_{R \times S} = (0_R, 0_S)$. Then, we have

$$
(r, s) + 0_{R \times S} = (r, s) + (0_R, 0_S) = (r + 0_R, s + 0_S) = (r, s) \text{ for all } (r, s) \in R \times S.
$$

(5) Existence of inverse of $(r, s) \in R \times S$.

Consider,

$$
(r, s) + (-r, -s) = (r - r, s - s) = (0_R, 0_S) = 0_{R \times S} = (-r, -s) + (r, s).
$$

for all $(r, s) \in R \times S$. Hence, $(-r, -s) \in R \times S$ is the additive inverse of (r, s).

(6) Closure of multiplication in $R \times S$ is guaranteed by its definition.

(7) Associativity under multiplication:

$$
\begin{aligned}
(r_1, s_1)((r_2, s_2)(r_3, s_3)) &= (r_1, s_1)(r_2 r_3, s_2 s_3) \\
&= (r_1(r_2 r_3), s_1(s_2 s_3)) \\
&= ((r_1 r_2) r_3, (s_1 s_2) s_3) \\
&= (r_1 r_2, s_1 s_2)(r_3, s_3) \\
&= ((r_1, s_1)(r_2, s_2))(r_3, s_3).
\end{aligned}
$$

(8) Distributive laws:

$$
\begin{aligned}
(r, s)((r', s') + (r'', s'')) &= (r, s)(r' + r'', s' + s'') \\
&= (r(r' + r''), s(s' + s'')) \\
&= (rr' + rr'', ss' + ss'') \\
&= (rr', ss') + (rr'', ss'') \\
&= (r, s)(r', s') + (r, s)(r'', s'')
\end{aligned}
$$

$$
\begin{aligned}
((r, s) + (r', s'))(r'', s'') &= (r + r', s + s')(r'', s'') \\
&= ((r + r')r'', (s + s')s'')
\end{aligned}
$$

$$
\begin{aligned}
&= (rr'' + r'r'', ss'' + s's'') \\
&= (rr'', ss'') + (r'r'', s's'') \\
&= (r,s)(r'',s'') + (r',s')(r'',s'').
\end{aligned}
$$

Thus, we can say that $R \times S$ is a ring.

Let R and S be commutative. Then, we have

$$
(r_1, s_1)(r_2, s_2) = (r_1 r_2, s_1 s_2) = (r_2 r_1, s_2 s_1) = (r_2, s_2)(r_1, s_1).
$$

Therefore, $R \times S$ is commutative.

Now, let R and S have unity say 1_R and 1_S respectively. Then, we have $1_{R \times S} = (1_R, 1_S)$ is unity for $R \times S$; since

$$
(1_R, 1_S)(r, s) = (1_R r, 1_S s) = (r, s) = (r 1_R, s 1_S) = (r, s)(1_R, 1_S).
$$

Problem 11 Let R be a commutative ring and $a, b \in R$ then find

(a) $(a+b)(a-b)$
(b) $(a+b)^3$

<u>Solution</u>   (a) Consider

$$
\begin{aligned}
(a+b)(a-b) &= (a+b)a - (a+b)b \\
&= aa + ba - ab - bb \\
&= a^2 - b^2 \qquad \text{(because } R \text{ is commutative.)}
\end{aligned}
$$

(b) Consider

$$
\begin{aligned}
(a+b)^3 &= (a+b)(a+b)(a+b) \\
&= (a+b)(a+b)a + (a+b)(a+b)b \\
&= (a+b)(aa+ba) + (a+b)(ab+bb) \\
&= (a+b)aa + (a+b)ba + (a+b)ab + (a+b)bb \\
&= aaa + baa + aba + baa + aab + bab + abb + bbb \\
&= a^3 + 3a^2 b + 3ab^2 + b^3 \qquad \text{(because } R \text{ is commutative.)}
\end{aligned}
$$

Problem 12 Let R be a commutative ring with unity. Let $x, y \in R$ such that x is a unit of R and $y^2 = 0$. Show that $x + y$ is a unit.

<u>Solution</u> It is given that R is a commutative ring with unity 1 (say). If x is a unit then there exists x^{-1} such that

$$
xx^{-1} = x^{-1}x = 1. \text{ Also, we have } y^2 = 0
$$

We will show that $x + y$ is a unit, that is, $(x+y)^{-1}$ exists.
Let us take $z = x^{-1} - yx^{-2} \in R.$

Consider

$$\begin{aligned} z(x+y) &= (x^{-1} - yx^{-2})(x+y) \\ &= x^{-1}x + x^{-1}y - xyx^{-2} - y^2x^{-2} \\ &= x^{-1}x + x^{-1}y - x^{-1}y - y^2x^{-2} \\ &= x^{-1}x - y^2x^{-2} \\ &= x^{-1}x \\ &= 1. \end{aligned}$$

Since R is commutative, therefore $(x+y)z = 1$. Thus, $z(x+y) = (x+y)z = 1$. Hence, z is the inverse of $x+y$. Hence, $x+y$ is a unit.

Problem 13 Let m and n be two positive integers. Let k be the least common multiple of m and n. Then, show that $m\mathbb{Z} \cap n\mathbb{Z} = k\mathbb{Z}$.

Solution Let $x \in k\mathbb{Z}$, that is, $k|x$, where $k = lcm(m,n)$.

Then, we have $m|k$ and $n|k$. Hence, $m|x$ and $n|x$. Therefore, $x \in m\mathbb{Z}$ and $x \in n\mathbb{Z}$, thus $x \in m\mathbb{Z} \cap n\mathbb{Z}$. Since x is an arbitrary element of $k\mathbb{Z}$, therefore

$$k\mathbb{Z} \subseteq m\mathbb{Z} \cap n\mathbb{Z} \tag{2.6}$$

Now, we will show that

$$m\mathbb{Z} \cap n\mathbb{Z} \subseteq k\mathbb{Z}$$

Consider $y \in m\mathbb{Z} \cap n\mathbb{Z}$, then we have $y \in m\mathbb{Z}$ and $y \in n\mathbb{Z}$. Hence, $m|y$ and $n|y$. Therefore, y is a common multiple of m and n both. We have k, the least common multiple of m and n. Therefore, we have $k|y$. Hence $y \in k\mathbb{Z}$. As y was an arbitrary element of $m\mathbb{Z} \cap n\mathbb{Z}$. Therefore

$$m\mathbb{Z} \cap n\mathbb{Z} \subseteq k\mathbb{Z} \tag{2.7}$$

From Equations (2.6) and (2.7), we have $k\mathbb{Z} = m\mathbb{Z} \cap n\mathbb{Z}$.

Problem 14 Determine all the units of $R[x]$, the ring of polynomials.

Solution The ring of polynomials is given by

$$R[x] = \{a_0 + a_1x + a_2x^2 + ... \mid a_i \in R\}$$

Let $f(x) = a_0 + a_1x + a_2x^2 + ... \in R[x]$ be a unit of $R[x]$. Then, $f(x)$ has inverse in $R[x]$, that is, $[f(x)]^{-1} \in R[x]$. We have, $\dfrac{1}{f(x)} \in R[x]$, that is,

$$\frac{1}{a_0 + a_1x + a_2x^2 + ...} \in R[x]$$

which is only possible if $a_0 \in R\backslash\{0\}$, and $a_1 = a_2 = ... = 0$. Therefore the possible values of $f(x)$ are c such that $c \in R, c \neq 0$.

Hence, the set of unit of $R[x]$ are,

$$U(R[x]) = \{c \mid c \in R, c \neq 0\}.$$

Problem 15 Let a, b and c be elements of a commutative ring R. Let a be a unit then prove that b divides c if and only if ab divides c.

Solution Let R be a commutative ring and a, b and $c \in R$.
Suppose $b|c$ then, we have $c = bq$ for some $q \in R$. Therefore, we have $c = 1 \cdot bq = (a^{-1} \cdot a)bq$.
Thus, $c = ab(a^{-1}q)$. $\hspace{4cm}$ (R is commutative.)
Hence, one can say that $ab|c$.
$\hspace{0.5cm}$ Converse is obvious, hence left for the readers.

Problem 16 Discuss all the units of $M_2(\mathbb{Z})$.

Solution Let, x_1, x_2, x_3 and $x_4 \in \mathbb{Z}$, then units of $M_2(\mathbb{Z})$ are of the form

$$\left\{ \begin{bmatrix} a & b \\ c & d \end{bmatrix} \,\middle|\, ad - bc = \pm 1 \; \forall \; a, b, c \text{ and } d \in \mathbb{Z} \right\}$$

Problem 17 Show that

(a) In the ring $\mathbb{Z}_6$, $4|2$.
(b) In $\mathbb{Z}_8$, $3|7$.
(c) In $\mathbb{Z}_{15}$, $9|12$.

Solution $\hspace{0.5cm}$ (a) We have $\mathbb{Z}_6 = \{0, 1, 2, 3, 4, 5\}$.
$\hspace{1cm}$ Then $4 \cdot 2 = 8(\text{mod } 6) = 2$ Therefore, we can say that $2 = 4 \cdot 2$ in $\mathbb{Z}_6$, thus $4|2$ in $\mathbb{Z}_6$.
(b) Consider $\mathbb{Z}_8 = \{0, 1, 2, 3, 4, 5, 6, 7\}$.
$\hspace{1cm}$ Then, $3 \cdot 5 = 15(\text{mod } 8) = 7$, thus, we have $7 = 3 \cdot 5$ in $\mathbb{Z}_8$. Therefore, we can say that $3|7$ in $\mathbb{Z}_8$.
(c) In $\mathbb{Z}_{15} = \{0, 1, 2, \dots, 14\}$.
$\hspace{1cm}$ We have $9 \cdot 3 = 27(\text{mod } 15) = 12$, therefore $9|12$.

Problem 18 Let R be a ring and $a, b \in R$. Let k be any integer, then show that $k \cdot (ab) = (k \cdot a)b = a(k \cdot b)$.

Solution Let R be a ring and $a, b \in R$. The result is clearly true for $k = 0$.

Case 1. Let k be a positive integer. We need to show that

$$k \cdot (ab) = (k \cdot a)b = a(k \cdot b)$$

$\hspace{1cm}$ Consider

$$
\begin{aligned}
k \cdot (ab) &= \underbrace{ab + ab + \dots + ab}_{k \text{ times}} \\
&= \underbrace{a(b + b + \dots + b)}_{k \text{ times}} \\
&= a(k \cdot b) \hspace{4cm} (2.8)
\end{aligned}
$$

Again, we have

$$k \cdot (ab) = \underbrace{ab + ab + \ldots + ab}_{k \text{ times}}$$

$$= \underbrace{(a + a + \ldots + a)}_{k \text{ times}} \cdot b$$

$$= (k \cdot a)b \tag{2.9}$$

From Equations (2.8) and (2.9), we get $k \cdot (ab) = (k \cdot a)b = a(k \cdot b)$.

Case 2. Let k be a negative integer, then we have some $m > 0$ such that $k = -m$. Then, rest of the proof is similar to Case 1.

Problem 19 Let R be a ring and $a, b \in R$. Then, show that $(m \cdot a)(n \cdot b) = (mn) \cdot (ab)$ for $m, n \in \mathbb{Z}$.

<u>**Solution**</u> Let, R be a ring and $a, b \in R$. Let, m and n are integers.
We need to show that $(m \cdot a)(n \cdot b) = (mn) \cdot (ab)$.

Case 1. Either m or n is 0, then result holds true.
Case 2. Suppose m and n both are positive, then we have

$$(ma)(nb) = \underbrace{(a + a + \ldots + a)}_{m \text{ times}} \cdot \underbrace{(b + b + \ldots + b)}_{n \text{ times}}$$

$$= \underbrace{(ab + ab + \ldots + ab)}_{mn \text{ times}}$$

$$= (mn)(ab)$$

Thus, we have

$$(ma)(mb) = (mn)(ab) \tag{2.10}$$

Case 3. Suppose m is positive and n is negative. Then, we have $-n$ is positive. Thus, consider

$$(ma)(nb) = \underbrace{(a + a + \ldots + a)}_{m \text{ times}} \cdot \underbrace{((-b) + (-b) + \ldots + (-b))}_{-n \text{ times}}$$

$$(ma)((-n)(-b)) = \underbrace{a(-b) + a(-b) + \ldots + a(-b)}_{m(-n) \text{ times}}$$

$$= (m(-n))(a(-b)) = -(mn)(-(ab)) = (mn)(ab)$$

Hence, we have

$$(ma)(nb) = (mn)(ab) \tag{2.11}$$

Case 4. Suppose m is negative, that is, $-m$ is positive and n is positive, then on the lines of Case 3, we have

$$(ma)(nb) = (mn)(ab) \tag{2.12}$$

Case 5. Suppose m and n both are negative, that is $-m$ and $-n$ both are positive. Then, there exist k_1 and k_2, both are positive such that $k_1 = -m$ and $k_2 = -n$. Likewise, as Case 2, we have

$$
\begin{aligned}
(ma)(nb) &= (-k_1 a)(-k_2 b) \\
&= (-(k_1 a))\,(-(k_2 b)) \\
&= -(-(k_1 k_2)(ab)) \\
&= (k_1 k_2)(ab)
\end{aligned}
$$

Therefore, from all the above cases, we can conclude that $(ma)(nb) = (mn)(ab)$ for all $a, b \in R$ and for all $m, n \in \mathbb{Z}$.

Problem 20 Show that for a ring R, with $a \in R$ and for $n \in \mathbb{Z}$, we have $n \cdot (-a) = -(n \cdot a)$.

<u>Solution</u> Similar to Problem 19.

Problem 21 Let R be a ring such that $x^n = x$ for all $x \in R$ where n is an integer greater than 1. Show that whenever $ab = 0$, we have $ba = 0$.

<u>Solution</u> Let, $n > 1$ be an integer such that $x^n = x$ for all $x \in R$. Here, we need to show that $ba = 0$, provided $ab = 0$.

Let, $a, b \in R$ such that $ab = 0$.

Now,

$$
\begin{aligned}
ba &= (ba)^n \\
&= (ba)^{n-1}(ba) \\
&= (ba)^{n-2}(ba)(ba) \\
&= (ba)^{n-2}b(ab)a \\
&= 0
\end{aligned}
$$

Thus, we have $ba = 0$.

Problem 22 Suppose that there is a positive even integer n such that $a^n = a$ for all $a \in R$. Show that $-a = a$ for all a in the ring.

<u>Solution</u> Let R be a ring and there exists a positive even integer n such that $a^n = a$ for all $a \in R$. We need to show that $-a = a$ for all $a \in R$. Since, n is a positive even integer, then, we have $n = 2m$ for some integer m. As $a \in R$, therefore, there exists $b = (-a) \in R$. Then, consider

$$
\begin{aligned}
-a = b &= (b)^n \\
&= (-a)^n \\
&= (-a)^{2m} \\
&= ((-a)^2)^m = (a^2)^m \\
&= a^{2m} = a^n \\
&= a
\end{aligned}
$$

Therefore, we have $-a = a$ for all a in the ring.

Problem 23 Determine the value of n for which the ring $\mathbb{Z}_n$ does not possess the following properties that the ring of integers possess:

   (a) $a^2 = a$ implies either $a = 0$ or $a = 1$.
   (b) $ab = 0$ implies either $a = 0$ or $b = 0$.
   (c) $ab = ac$ and $a \neq 0$ imply $b = c$.

<u>Solution</u> The appropriate n for all the above parts is $n = 10$ and the ring is $R = \mathbb{Z}_{10} = \{0, 1, 2, 3, \ldots, 9\}$

   (a) For $R = \mathbb{Z}_{10}$ and $a = 6 \in \mathbb{Z}_{10}$, we have $6^2 = 36 = 6 \pmod{10}$ and one can observe that, here neither $6 \neq 0$ nor $6 \neq 1$.
   (b) Let us consider $a = 2$ and $b = 5$. Then, we have $2 \cdot 5 = 10 = 0 \pmod{10}$. Here, obviously, $2 \neq 0$ and $5 \neq 0$.
   (c) For $a = 2$, $b = 1$, $c = 6$, we have $2 \cdot 1 = 2 \cdot 6$ and obviously, $b \neq c$ and $a \neq 0$.

Problem 24 Let R be a ring and suppose that there is an integer $n > 1$ such that $x^n = x$ for all $x \in R$. Let m be a positive integer such that $a^m = 0$ for some $a \in R$. Show that $a = 0$.

<u>Solution</u> Let, R be a ring and let there exists an integer $n > 1$ such that $x^n = x$ for all $x \in R$. Let m be a positive integer with $a^m = 0$ for some $a \in R$. Then we need to show that $a = 0$. We will divide the solution in the following cases:

Case 1. Suppose $m < n$, then, we have

$$a = a^n = a^{m-m+n} = a^m \cdot a^{n-m} = 0 \cdot a^{n-m} = 0$$

      Thus, we have $a = 0$.

Case 2. Suppose $m = n$. Then, we have

$$a = a^n = a^m = 0$$

      Hence, we have $a = 0$.

Case 3. Suppose $m > n$. Then, consider

$$
\begin{aligned}
a &= a^n \\
&= (a \cdot a \ldots a) \\
&= (a^n \cdot a^n \ldots a^n) \\
&= (a^n)^n \\
&= \ldots \\
&= ((a^n)^n)^n \\
&= \ldots \\
&= (a^k) \text{ for some } k \in \mathbb{Z} \text{ and } k > 1
\end{aligned}
$$

Therefore, we can see that there are arbitrary large k such that $a = a^k$ for some k. Thus, we get $m < k$. Hence, $a = a^k = a^{m-m+k} = (a^m)a^{k-m} = 0 \cdot a^{k-m} = 0$. Thus, we get $a = 0$.

Hence, from all the above cases we conclude that $a = 0$.

Problem 25 Let R be a ring and let $x, y \in R$. Show that $1 + xy$ is a unit if and only if $1 + yx$ is a unit.

Solution One can observe that

$$y(1 + xy) = y + yxy = (1 + yx)y.$$

Similarly, we have

$$(1 + xy)x = x + xyx = x(1 + yx)$$

Let $1 + xy$ be a unit. Then, we need to show that $1 + yx$ is also a unit. Let us take $z = 1 - y(1 + xy)^{-1}x$.

We claim that z is the inverse of $1 + yx$ in R. For this, consider

$$
\begin{aligned}
(1 + yx)z &= (1 + yx)\left(1 - y(1 + xy)^{-1}x\right) \\
&= (1 + yx) - (1 + yx)y(1 + xy)^{-1}x \\
&= 1 + yx - y(1 + xy)(1 + xy)^{-1}x \\
&= 1 + yx - yx \\
&= 1
\end{aligned}
$$

Similarly, we have $z(1 + yx) = (1 + yx)z = 1$. Thus $(1 + yx)z = z(1 + yx) = 1$. Hence $1 + yx$ is a unit. This shows that if $1 + xy$ is a unit then $1 + yx$ is a unit, as required.

The converse can be proved on similar lines.

Problem 26 Let R be a ring and $x, y \in R$. Let x, y, and $x + y$ be units in R. Then, show that $x^{-1} + y^{-1}$ is also a unit. Also, give a formula for $(x^{-1} + y^{-1})^{-1}$ in terms of x, y and $(x + y)^{-1}$.

Solution Let us consider

$$x(x^{-1} + y^{-1})y = (1 + xy^{-1})y = y + x = x + y$$

Therefore, we have

$$x^{-1} + y^{-1} = x^{-1}(x + y)y^{-1}$$

is a product of three units. Hence, $x^{-1} + y^{-1}$ is also a unit with

$$(x^{-1} + y^{-1})^{-1} = (y^{-1})^{-1}(x + y)^{-1}(x^{-1})^{-1} = y(x + y)^{-1}x.$$

Problem 27 Determine the set of units of $\mathbb{Z}[i]$, where $\mathbb{Z}[i]$ is the set of Gaussian integers defined as

$$\mathbb{Z}[i] = \{a + bi \mid a, b \in \mathbb{Z}\}$$

Solution Suppose $a + bi \in \mathbb{Z}[i]$ and suppose $a + bi$ has an inverse in $\mathbb{Z}[i]$. That is,

$$(a + bi)^{-1} = \frac{1}{a + bi} \in \mathbb{Z}[i] \tag{2.13}$$

Now, consider

$$\frac{1}{a + bi} = \frac{1}{a + bi} \times \frac{a - bi}{a - bi} = \frac{a - bi}{a^2 + b^2} = \frac{a}{a^2 + b^2} - \frac{bi}{a^2 + b^2} \tag{2.14}$$

From Equations (2.13) and (2.14), we have $\dfrac{a}{a^2 + b^2}, \dfrac{-b}{a^2 + b^2} \in \mathbb{Z}$, which is only possible provided $a = 0$ and $b = \pm 1$ or $a = \pm 1$ and $b = 0$. Thus, the possible values of $a + bi$ are $1, -1, i, -i$. Hence, $U(\mathbb{Z}[i]) = \{1, -1, i, -i\}$.

Problem 28 Let H be the set of real quaternions, where $\mathbf{1}, \mathbf{i}, \mathbf{j}$ and $\mathbf{k}$ are the matrices defined as

$$\mathbf{1} = \begin{pmatrix} 1 & 0 \\ 0 & 1 \end{pmatrix}, \quad \mathbf{i} = \begin{pmatrix} i & 0 \\ 0 & -i \end{pmatrix}, \quad \mathbf{j} = \begin{pmatrix} 0 & 1 \\ -1 & 0 \end{pmatrix}, \quad \mathbf{k} = \begin{pmatrix} 0 & i \\ i & 0 \end{pmatrix}.$$ Then, prove the following:

(a)

$$\begin{aligned}
\mathbf{i}^2 &= \mathbf{j}^2 = \mathbf{k}^2 = -1 \\
\mathbf{jk} &= -\mathbf{kj} = \mathbf{i} \\
\mathbf{ij} &= -\mathbf{ji} = \mathbf{k} \\
\mathbf{ki} &= -\mathbf{ik} = \mathbf{j}
\end{aligned}$$

(b) H is a noncommutative ring with identity.
(c) H is a division ring.

Solution   (a) By direct calculations, we get: $\mathbf{i}^2 = \begin{pmatrix} i & 0 \\ 0 & -i \end{pmatrix} \begin{pmatrix} i & 0 \\ 0 & -i \end{pmatrix} =$

$\begin{pmatrix} -1 & 0 \\ 0 & -1 \end{pmatrix} = -1.$

Similarly, $\mathbf{j}^2 = \mathbf{k}^2 = -1.$

Now, consider

$$\mathbf{jk} = \begin{pmatrix} 0 & 1 \\ -1 & 0 \end{pmatrix} \begin{pmatrix} 0 & i \\ i & 0 \end{pmatrix} = \begin{pmatrix} i & 0 \\ 0 & -i \end{pmatrix} = \mathbf{i}$$

and

$$-\mathbf{kj} = -\begin{pmatrix} 0 & i \\ i & 0 \end{pmatrix}\begin{pmatrix} 0 & 1 \\ -1 & 0 \end{pmatrix}$$

$$= \begin{pmatrix} 0 & -i \\ -i & 0 \end{pmatrix}\begin{pmatrix} 0 & 1 \\ -1 & 0 \end{pmatrix}$$

$$= \begin{pmatrix} i & 0 \\ 0 & -i \end{pmatrix} = \mathbf{i}$$

Similarly, $\mathbf{ij} = -\mathbf{ji} = \mathbf{k}$ and $\mathbf{ki} = -\mathbf{ik} = \mathbf{j}$.

(b) Here, H is a subset of $M_2(\mathbb{C})$, the set of all 2×2 matrices with complex entries. Since $M_2(\mathbb{C})$ is a ring under usual matrix addition and matrix multiplication (verify !). Thus, in order to show H is a ring, it is sufficient to show that H is closed under addition and multiplication, contains $0 = \begin{pmatrix} 0 & 0 \\ 0 & 0 \end{pmatrix}$ and additive inverses. All these properties are easily confirmed by direct calculation.

(c) Also, every nonzero element of H has a multiplicative inverse, therefore H is a division ring.

Problem 29 Determine all elements of a ring that are both units and idempotents.

<u>**Solution**</u> Let R be a ring and $a(\neq 0) \in R$ be an element which is both a unit and an idempotent. Then, we have $a^2 = a$ and a^{-1} exists. Since, we have $a^2 = a$, thus we have $a^2 - a = 0$. Hence, $a(a-1) = 0$. As, a is a unit, therefore, $a - 1 = 0$ and hence $a = 1$.

Problem 30 If a is a nilpotent in R, show that $1 - a$ and $1 + a$ are units.

<u>**Solution**</u> If $a^n = 0$ where $n \geq 1$, then, we have $a^k = 0$ for all $k \geq n$ by the definition of nilpotent elements.

Hence, we have $u = 1 + a + a^2 + a^3 + \dots$ is a finite sum, and so is an element of R. Then $(1-a)u = 1 = u(1-a)$. Therefore $1 - a$ is a unit. As $(-a)^n = 0$ thus $1 + a$ is a unit too.

Exercise

(1) Let $R_1, R_2, \dots, R_n$ be rings. Show that $R = R_1 \oplus R_2 \oplus \dots \oplus R_n$ is a ring with pointwise addition and multiplication.

(2) Let R be a ring such that $a^3 = a$ for all $a \in R$. Show that R is commutative.

(3) Show that a ring is commutative if satisfies the property that $ab = ca$ implies $b = c$ when $a \neq 0$.

(4) Let R be a ring.
   (a) If $ab + ba = 1$ and $a^3 = a$ for $a, b \in R$. Then, show that $a^2 = 1$.
   (b) If $ab = a$ and $ba = b$ for $a, b \in R$. Then, show that $a^2 = a$ and $b^2 = b$.

(5) Give an example to show that for fixed nonzero elements a and b in a ring, the equation $ax = b$ can have more than one solution. Compare this property of rings to groups.

(6) Show that $\mathbb{Z}_{10}^{\times}$, is a cyclic group. Also, find an element $a \in \mathbb{Z}_{10}^{\times}$ which generates it.

(7) Let R be a ring. Then, prove, or disprove the assertion: for all $x, y \in R$, we have $(x + y)^2 = x^2 + 2xy + y^2$.

(8) Let us denote the set of invertible elements of the ring $\mathbb{Z}_n$ by U_n.

    (a) List all the elements of U_{18} and U_{24}.

    (b) Are U_{18} and U_{24} both cyclic groups under multiplication? Justify your answer.

(9) Let R be a ring and a, b and c be elements of a ring R. Does the equation $ax + b = c$ always have a solution for $a, b, c \in R$? Must the solution be unique, if it exists? What happens when a is a unit?

(10) Let R be a ring with unity 1 and $a, b \in R$ such that $ab = 1$. Also, let $X = \{x \in R \mid ax = 1\}$. Then show that

    (a) $b + (1 - xa) \in X$ for all $x \in X$;

    (b) the mapping $\phi \colon X \to X$ defined as

$$\phi(x) = b + (1 - xa)$$

    is one-to-one.

    (c) the set X is an infinite set provided that X has more than one element.

(11) Let $R = C[0, 1]$ be the set of all continuous real-valued functions defined on $[0, 1]$. For $f, g \in R$ and $x \in [0, 1]$, we define addition and multiplication on R as follows:

$$(f + g)(x) = f(x) + g(x) \text{ and } (fg)(x) = f(x)g(x)$$

Then,

    (a) show that R is a commutative ring with unity.

    (b) find the units of R.

    (c) for all $f \in R$ and $f^2 = f$. Show that either $f = 0_R$ or $f = 1_R$.

    (d) for $f \in R$ such that $f^n = 0_R$, for $n \in \mathbb{N}$, show that $f = 0_R$.

(12) Let R be a ring and $a, b \in R$ ($a \neq b$) such that $a^3 = b^3$ and $a^2 b = b^2 a$. Show that $a^2 + b^2$ is not a unit.

(13) Let R be a commutative ring with unity 1. Show that if a is a unit in R and b is nilpotent, then $a + b$ is a unit.

(14) Let R be a ring. Show that 0 is the only nilpotent in R if and only if $a^2 = 0$ implies $a = 0$.

3 SUBRINGS

In the arena of abstract algebra, there's a neat idea that helps us understand rings better: subrings. They're like smaller pieces within bigger algebraic structures. A ring may be characterized by a variety of its subsets, known as subrings. This characterization also gives a geometrical interpretation to a usual abstract ring. The concept of a subring is analogous to the concept of a subgroup in group theory. These concepts came into existence by the efforts of mathematicians like David Hilbert and Emmy Noether, who laid the foundation for modern algebra. Just like how you might break down a big puzzle into smaller parts to understand it better, subrings help us to see the patterns and properties within rings.

In this chapter, the concept of subring is explored through various examples. Then, subring test is stated to check if a given subset of a ring is a subring. The study is examined through various problems to enable readers to apprehend the notion of subring. In this chapter, we'll take a closer look at these ideas, exploring where they came from, why they're important, and how they're used in real-life problems. Let's dive in and uncover the secrets of subrings together!

Definition 3.1 (Subring) Let $(R, +, \cdot)$ be a ring. A non-empty subset S of a ring R is said to be *subring* of R if S is a ring in itself restricted to the same binary operations as of R.

Example 3.2 Let R be a ring, then the set $\{0\}$ and the ring R are subrings of R and are called the *trivial subrings*. These are also known as *improper subrings* of ring R. Apart from these subrings, if any, of R are known as *proper subrings* of R.

Example 3.3 The set of all integers, $\mathbb{Z}$ is a subring of $\mathbb{Q}$; $\mathbb{Q}$ is a subring of $\mathbb{R}$ and $\mathbb{R}$ is a subring of $\mathbb{C}$ under usual addition and usual multiplication.

Remark Suppose R is a ring then, by definition R is an abelian group with respect to addition. Thus, if S is any subring of R then S is a subgroup of additive group R. Therefore, if S is a subring of R then it is subgroup of the additive group R and must be closed under multiplication.

We will prove this fact in the following result.

Theorem 3.4 (Subring Test) *A non-empty subset S of a ring R is a subring if and only if S is closed under subtraction and multiplication. That is, for $a, b \in S$, we have*

(i) $a - b \in S$;
(ii) $ab \in S$.

Proof The conditions are necessary:
Let S be a subring of a ring $(R, +, .)$. Then, by definition $(S, +, .)$ is a ring itself. Therefore, for all $a, b \in S$, we have $a \cdot b \in S$. Also, as $b \in S$, thus $-b \in S$. Therefore, $a + (-b) = a - b \in S$.

The conditions are sufficient:
Let S be a non-empty subset of R such that for all $a, b \in S$, we have $a \cdot b \in S$ and $a - b \in S$. Now, we need to prove that S is a subring of R, that is, S is itself a ring with respect to '+' and '·'.

(i) As $S \neq \emptyset$ (given), therefore, we have $a \in S$. Thus, $a, a \in S$, which implies $a - a = 0 \in S$. Hence, additive identity exists in S.
(ii) Now, $0 \in S$ and $a \in S$; therefore, we have $0 - a = 0 + (-a) = -a \in S$. Since, a was any arbitrary element of S. Thus, the result holds for all $a \in S$. Therefore, every element of S has an additive inverse in S.
(iii) Now, $a, b \in S$, then, we have $a, -b \in S$. Hence, $a - (-b) = a + b \in S$. Therefore, S is closed with respect to '+'.
(iv) Let $a, b, c \in S$ as $S \subset R$. Therefore, for all $a, b, c \in R$, we have $a + (b + c) = (a + b) + c$. Hence, S is associative with respect to '+'.
(v) Now, for all $a, b \in S$, we have $a \cdot b \in S$. Therefore, S is closed with respect to '·'.
(vi) Since $S \subseteq R$ and for all $a, b, c \in S \subseteq R$, we have $(a \cdot b) \cdot c = a \cdot (b \cdot c)$. This implies, S is associative with respect to '·'.
(vii) Also, for all $a, b, c \in S$. We have $a, b, c \in S \subseteq R$. Since R is a ring which implies $a \cdot (b + c) = a \cdot b + a \cdot c$. Therefore, distributive law holds in S.

Therefore, $(S, +, \cdot)$ is a ring. Hence, S is a subring of R. $\qquad \square$

In the next example, we use the above theorem to check if a given subset is a subring.

Example 3.5 The set $S = \{0, 2, 4\}$ is a subring of the ring of integers modulo 6, $\mathbb{Z}_6$.
For this, let us consider the Cayley table for $\{0, 2, 4\}$ under $\oplus_6$ and $\otimes_6$:

$\oplus_6$	0	2	4		$\otimes_6$	0	2	4
0	0	2	4		0	0	0	0
2	2	4	0		2	0	4	2
4	4	0	2		4	0	2	4

From the Cayley table, it can be easily verified that for $a, b \in S$, we have $a \oplus_6 (-b) = a - b \in S$ and $a \otimes_6 b \in S$ for all $a, b \in S$. Hence, S is a subring of $\mathbb{Z}_6$.

From the above example, one can observe that a unity of a subring may differ from the unity of the ring. Here, one can see that 1 is the unity of $\mathbb{Z}_6$ but $1 \notin S$. Also, from the Cayley table, it is clear that 4 is the unity of S.

Example 3.6 Let us consider

$$\mathbb{Z}[i] = \{a + bi \mid a, b \in \mathbb{Z}\}$$

which is also known as *the set of Gaussian integers*. It is a subring of the complex numbers $\mathbb{C}$. Since, $0 + i0 \in \mathbb{Z}[i]$, therefore $\mathbb{Z}[i] \neq \emptyset$. For any $a + bi, c + di \in \mathbb{Z}[i]$, where $a, b, c, d \in \mathbb{Z}$, we have $(a + bi) - (c + di) = (a - c) + (b - d)i \in \mathbb{Z}[i]$, since $a - c, b - d \in \mathbb{Z}$. Also, $(a + bi)(c + di) = (ac - bd) + (ad + bc)i \in \mathbb{Z}[i]$, since $ac - bd, ad + bc \in \mathbb{Z}$. Hence $\mathbb{Z}[i]$ is a subring of $\mathbb{C}$.

Example 3.7 The set $n\mathbb{Z} = \{0, \pm n, \pm 2n, \pm 3n, ...\}$, for each positive integer n, is a subring of the integers, $\mathbb{Z}$.

Since, $0 = n \cdot 0 \in n\mathbb{Z}$, therefore $n\mathbb{Z} \neq \emptyset$.

Suppose $x, y \in n\mathbb{Z}$, where, we have $x = nk_1$ and $y = nk_2$ for some $k_1, k_2 \in \mathbb{Z}$. Subsequently, we have

$x - y = nk_1 - nk_2 = n(k_1 - k_2) \in n\mathbb{Z}$. Thus, $x - y \in n\mathbb{Z}$. Also, we have $xy = nk_1 \cdot nk_2 = n(nk_1 k_2) \in n\mathbb{Z}$. Therefore, $xy \in n\mathbb{Z}$.

Hence, $n\mathbb{Z}$ is a subring of $\mathbb{Z}$.

The above example clearly states that $n\mathbb{Z}$ is a subring of $\mathbb{Z}$ for $n \in \mathbb{Z}$. Similarly, we can show that $4\mathbb{Z}$ is a subring of of $2\mathbb{Z}$ ($\because 2\mathbb{Z}$ and $4\mathbb{Z}$ are subrings of $\mathbb{Z}$ and $4\mathbb{Z} \subseteq 2\mathbb{Z}$).

In general, $m\mathbb{Z}$ is a subring of $n\mathbb{Z}$ if and only if n divides m. This result is proved in the following theorem.

Theorem 3.8 *The ring $m\mathbb{Z}$ is a subring of the ring $n\mathbb{Z}$ if and only if n divides m.*

Proof Since, $m \cdot 0 = 0 \in m\mathbb{Z}$, thus $m\mathbb{Z} \neq \emptyset$.

Let us assume that $m\mathbb{Z}$ is a subring of $n\mathbb{Z}$. Then, we have $m\mathbb{Z} \subseteq n\mathbb{Z}$. Now, $m = m.1 \in m\mathbb{Z}$, therefore, $m \in n\mathbb{Z}$. Thus, there exists some $x \in \mathbb{Z}$ such that $m = nx$ and hence, n divides m.

Conversely, let us assume that n divides m. Then there must exist some $d \in \mathbb{Z}$ such that $m = dn$.

Let $x, y \in m\mathbb{Z}$ be arbitrary. Then, we have to show the following:

(a) $x \in n\mathbb{Z}$ (such that $m\mathbb{Z} \subset n\mathbb{Z}$)
(b) $x - y \in m\mathbb{Z}$
(c) $xy \in m\mathbb{Z}$.

(a) Since $x \in m\mathbb{Z}$, there exists $b \in \mathbb{Z}$ such that $x = bm$. Also, we have $m = dn$; then $x = b(dn) = (bd)n$. Therefore, $x \in n\mathbb{Z}$. Since x was arbitrary, $m\mathbb{Z} \subseteq n\mathbb{Z}$.
(b) Since $y \in m\mathbb{Z}$, there exists $c \in \mathbb{Z}$ such that $y = cm$. Then $x - y = bm - cm = (b - c)m \in m\mathbb{Z}$. So, $m\mathbb{Z}$ is closed under subtraction.
(c) Moreover, $xy = (bm)(cm) = (bcm)m \in m\mathbb{Z}$ because $bcm \in \mathbb{Z}$. Thus, $m\mathbb{Z}$ is closed under multiplication.

Hence, by subring test, $m\mathbb{Z}$ is a subring of $n\mathbb{Z}$. $\qquad\square$

It would be interesting to observe different set-theoretic operations like intersection, union, sum and product of subrings and further analyze the resultant sets and their properties.

Suppose that S_1 and S_2 are both subrings of a ring R. Let $S = S_1 \cap S_2$. Since S_1 and S_2 are subrings of R then they must contain zero (identity with respect to addition), so $S_1 \cap S_2$ must have at least zero. Therefore, $S = S_1 \cap S_2 \neq \emptyset$. Further, let $x, y \in S$ be arbitrary. Then $x, y \in S_1$ and $x, y \in S_2$. Since S_1 and S_2 are both subrings, $x - y, xy \in S_1$ and $x - y$, $xy \in S_2$. By definition of intersection, $x - y, xy \in S_1 \cap S_2 = S$. Since x and y are arbitrary in S, S is closed under subtraction and multiplication. Hence, by subring test, S is a subring. Thus, the intersection of two subrings is a subring. This observation can be generalised to arbitrary collection of subrings that is shown in the following theorem.

Theorem 3.9 *Let $\{S_i \mid i \in I\}$ be an arbitrary class of subrings of R. Then, intersecion of collection of these subrings is a subring of R.*

Proof Let $\{S_i \mid i \in I\}$ be an arbitrary class of subrings of R. Let $S = \bigcap_{i \in I} S_i$.

We need to show that S is a subring of R.

Since, $0 \in S_i \ \forall \ i$ $\hfill$ (S_i is a subring $\forall \ i$.)

$\Rightarrow 0 \in \bigcap_i S_i = S$. Hence, S is non-empty.

Consider $x, y \in S = \bigcap_i S_i$.

$\Rightarrow x, y \in S_i \ \forall \ i$

$\Rightarrow x - y, \ xy \in S_i \ \forall \ i$ $\hfill$ (S_i is a subring.)

$\Rightarrow x - y, xy \in \bigcap_i S_i$

$\Rightarrow x - y, \ xy \in S$.

Hence, S is a subring of R. $\hfill \square$

Remark Union of subrings of a ring R need not be a subring. For example, in the ring of integers, $\mathbb{Z}$, $2\mathbb{Z}$ and $3\mathbb{Z}$ are subrings of $\mathbb{Z}$. However, $2\mathbb{Z} \cup 3\mathbb{Z}$ is not a subring of $\mathbb{Z}$. Here, $2, 3 \in 2\mathbb{Z} \cup 3\mathbb{Z}$ but $3 - 2 = 1 \notin 2\mathbb{Z} \cup 3\mathbb{Z}$. Therefore, $2\mathbb{Z} \cup 3\mathbb{Z}$ is not a subring of $\mathbb{Z}$.

However, union of two subrings can be a subring if one of the subrings is contained in the other.

Accordingly we have the following theorem that states the necessary and sufficient condition under which the union of subrings will be a subring.

Theorem 3.10 *Let S_1, S_2 be two subrings of a ring R. Then, union of these subrings is a subring of R if and only if either $S_1 \subseteq S_2$ or $S_2 \subseteq S_1$.*

Proof Let $(R, +, \cdot)$ be a ring. Let, S_1 and S_2 be its two subrings. If $S_1 \subseteq S_2$ or $S_2 \subseteq S_1$ then $S_1 \cup S_2 = S_2$ or $S_1 \cup S_2 = S_1$. This implies, $S_1 \cup S_2$ is a subring.

Conversely, suppose that S_1 and S_2 are two subrings such that $S_1 \cup S_2$ is also a subring. Now, we need to show that either $S_1 \subseteq S_2$ or $S_2 \subseteq S_1$. Let, if possible, this is not true; then, there exists an element $a \in S_1$ such that a is not in S_2, and there exists an element $b \in S_2$ such that $b \notin S_1$.

Now, $a \in S_1, b \in S_2$ implies that $a, b \in S_1 \cup S_2$ which is a subring.

Thus, $ab, a - b \in S_1 \cup S_2$.

Now, $a - b \in S_1 \cup S_2 \Rightarrow a - b \in S_1$ or $a - b \in S_2$.

Case 1. $a - b \in S_1$

Also, $a \in S_1$. Then, $a - (a - b) \in S_1$. Thus, $b \in S_1$, which is a contradiction.

Case 2. $a - b \in S_2$

Also, $b \in S_2$. Then, $b + (a - b) \in S_2$. Thus, $a \in S_2$, which again leads to a contradiction.

Hence, we get either $S_1 \subseteq S_2$ or $S_2 \subseteq S_1$.

Hence, union of two subrings is a subring if and only if one of them is contained in another. $\qquad\square$

We know that $2\mathbb{Z}$ and $3\mathbb{Z}$ are subrings of $\mathbb{Z}$. Also, $2\mathbb{Z} \oplus 3\mathbb{Z} = \{(2x, 3y) : x, y \in \mathbb{Z}\}$ is a subring of $\mathbb{Z}$ (for $(2x_1, 3y_1), (2x_2, 3y_2) \in 2\mathbb{Z} \oplus 3\mathbb{Z}$, we have $(2x_1, 3y_1) - (2x_2, 3y_2) = (2(x_1 - x_2), 3(y_1 - y_2)) \in 2\mathbb{Z} \oplus 3\mathbb{Z}$ and $(2x_1, 3y_1) \cdot (2x_2, 3y_2) = (2x_1 \cdot 2x_2, 3y_1 \cdot 3y_2) \in 2\mathbb{Z} \oplus 3\mathbb{Z}$). In general, if S_1 and S_2 are subrings of a ring R then the external direct product $S_1 \oplus S_2$ will also be a subring, which is proved in the next theorem.

Theorem 3.11 *Let, S_1 and S_2 be subrings of a ring R. Then, external direct product of these subrings, $S_1 \oplus S_2$ is a subring of $R \oplus R$.*

Proof Let $(R, +, \cdot)$ be a ring. Let, S_1 and S_2 be its subrings. We define

$$S_1 \oplus S_2 = \{(a, b) \mid a \in S_1, b \in S_2\}.$$

Clearly, $(0, 0) \in S_1 \oplus S_2$. So, $S_1 \oplus S_2$ is non-empty.
Let, $(a, b), (c, d) \in S_1 \oplus S_2$ for some $a, c \in S_1$ and $b, d \in S_2$. Then,
$(a, b) - (c, d) = (a - c, b - d) \in S_1 \oplus S_2$. ($a - c \in S_1$ and $b - d \in S_2$ as S_1 and S_2 are subrings.)

Also, $(a, b)(c, d) = (ac, bd) \in S_1 \oplus S_2$. ($ac \in S_1$ and $bd \in S_2$ as S_1 and S_2 are subrings.)
Hence, $S_1 \oplus S_2$ is a subring of $R \oplus R$. $\qquad\square$

Although the direct product of two subrings is a subring, the same need not be true for the usual sum.

Remark Sum of two subrings need not be a subring of a ring R.

Let $M = \left\{ \begin{pmatrix} a & b \\ c & d \end{pmatrix} \,\middle|\, a, b, c, d \in \mathbb{Z} \right\}$.

Then, M is a ring with unity under matrix addition and matrix multiplication with unity $\begin{pmatrix} 1 & 0 \\ 0 & 1 \end{pmatrix}$ and additive identity $\begin{pmatrix} 0 & 0 \\ 0 & 0 \end{pmatrix}$.

Let, $S_1 = \left\{ \begin{pmatrix} a & 0 \\ c & 0 \end{pmatrix} \,\middle|\, a, c \in \mathbb{Z} \right\}$, which is a subset of M.

Since, $\begin{pmatrix} 0 & 0 \\ 0 & 0 \end{pmatrix} \in S_1$, S_1 is a non-empty subset of M.

Let, $\begin{pmatrix} a_1 & 0 \\ b_1 & 0 \end{pmatrix}$ and $\begin{pmatrix} a_2 & 0 \\ b_2 & 0 \end{pmatrix} \in S_1$ for some $a_1, a_2, b_1, b_2 \in \mathbb{Z}$.

Then, $\begin{pmatrix} a_1 & 0 \\ b_1 & 0 \end{pmatrix} - \begin{pmatrix} a_2 & 0 \\ b_2 & 0 \end{pmatrix} = \begin{pmatrix} a_1 - a_2 & 0 \\ b_1 - b_2 & 0 \end{pmatrix} \in S_1$ as $a_1 - a_2, b_1 - b_2 \in \mathbb{Z}$.

Also, $\begin{pmatrix} a_1 & 0 \\ b_1 & 0 \end{pmatrix} \cdot \begin{pmatrix} a_2 & 0 \\ b_2 & 0 \end{pmatrix} = \begin{pmatrix} a_1 a_2 & 0 \\ b_1 a_2 & 0 \end{pmatrix} \in S_1$ as $a_1 a_2, b_1 b_2 \in \mathbb{Z}$.

Since, $\begin{pmatrix} a_1 & 0 \\ b_1 & 0 \end{pmatrix}$ and $\begin{pmatrix} a_2 & 0 \\ b_2 & 0 \end{pmatrix}$ were arbitrary elements of S_1.

Therefore, S_1 is a subring of M.

Now consider, $S_2 = \left\{ \begin{pmatrix} 0 & b \\ 0 & 0 \end{pmatrix} \middle| b \in \mathbb{Z} \right\}$, which is a subset of M.

Since, $\begin{pmatrix} 0 & 0 \\ 0 & 0 \end{pmatrix} \in S_2$, S_2 is a non-empty subset of M.

Similarly, we can show that S_2 is a subring of M.

Thus, S_1 and S_2 are subrings of a ring M. However, $S_1 + S_2$ is not a subring of M.
Here,

$$S_1 + S_2 = \left\{ \begin{pmatrix} a & b \\ c & 0 \end{pmatrix} \middle| a, b, c \in \mathbb{Z} \right\}.$$

For $\begin{pmatrix} 2 & 3 \\ 4 & 0 \end{pmatrix}, \begin{pmatrix} 3 & 5 \\ 1 & 0 \end{pmatrix} \in S_1 + S_2$

we have

$$\begin{pmatrix} 2 & 3 \\ 4 & 0 \end{pmatrix} \cdot \begin{pmatrix} 3 & 5 \\ 1 & 0 \end{pmatrix} = \begin{pmatrix} 9 & 10 \\ 12 & 20 \end{pmatrix} \notin S_1 + S_2$$

Therefore, $S_1 + S_2$ is not a subring of M.
Hence, sum of two subrings need not be a subring of a ring R.

The next theorem illustrates the Cartesian product of subrings.

Theorem 3.12 *Let S_1 and S_2 be subrings of the rings R_1 and R_2, respectively. Then $S_1 \times S_2$ is a subring of $R_1 \times R_2$. (The Cartesian product $R_1 \times R_2$ of two rings R_1 and R_2 is a ring under componentwise addition and multiplication.)*

Proof Let S_1 and S_2 be subrings of the rings R_1 and R_2, respectively. Since S_1 and S_2 are subrings of R_1 and R_2, $S_1 \neq \emptyset$ and $S_2 \neq \emptyset$. Therefore, $S_1 \times S_2 \neq \emptyset$.
Now, let (a, b) and $(a', b') \in S_1 \times S_2$. Then, $a, a' \in S_1$ and $b, b' \in S_2$. As S_1 and S_2 are subrings, $a - a', a \cdot a' \in S_1$ and $b - b', b \cdot b' \in S_2$. Thus,
$(a, b) - (a', b') = (a - a', b - b') \in S_1 \times S_2$, and
$(a, b) \cdot (a', b') = (aa', bb') \in S_1 \times S_2$.
Hence, by subring test, $S_1 \times S_2$ is a subring of $R_1 \times R_2$. $\qquad\square$

We know that commutativity with respect to multiplication does not hold in general in

a ring. For example, the ring of 2×2 matrices, $M = \begin{pmatrix} a & b \\ c & d \end{pmatrix}$ is noncommutative as for elements $\begin{pmatrix} 1 & 0 \\ 1 & 0 \end{pmatrix}$ and $\begin{pmatrix} 1 & 1 \\ 0 & 0 \end{pmatrix}$, we have

$$\begin{pmatrix} 1 & 0 \\ 1 & 0 \end{pmatrix}\begin{pmatrix} 1 & 1 \\ 0 & 0 \end{pmatrix} = \begin{pmatrix} 1 & 1 \\ 1 & 1 \end{pmatrix} \text{ while } \begin{pmatrix} 1 & 1 \\ 0 & 0 \end{pmatrix}\begin{pmatrix} 1 & 0 \\ 1 & 0 \end{pmatrix} = \begin{pmatrix} 2 & 0 \\ 0 & 0 \end{pmatrix}.$$

However, we have subring S of noncommutative ring R which is commutative. Example, the subset $S = \begin{pmatrix} a & 0 \\ 0 & 0 \end{pmatrix}$ of 2×2 matrices, $M = \begin{pmatrix} a & b \\ c & d \end{pmatrix}$ is a commutative subring of M.

Since, $\begin{pmatrix} 0 & 0 \\ 0 & 0 \end{pmatrix} \in S$, S is non-empty.

Let, $\begin{pmatrix} a_1 & 0 \\ 0 & 0 \end{pmatrix}$ and $\begin{pmatrix} a_2 & 0 \\ 0 & 0 \end{pmatrix} \in S$ for some $a_1, a_2 \in \mathbb{Z}$.

Then, $\begin{pmatrix} a_1 & 0 \\ 0 & 0 \end{pmatrix} - \begin{pmatrix} a_2 & 0 \\ 0 & 0 \end{pmatrix} = \begin{pmatrix} a_1 - a_2 & 0 \\ 0 & 0 \end{pmatrix} \in S$ as $a_1 - a_2 \in \mathbb{Z}$.

Also, $\begin{pmatrix} a_1 & 0 \\ 0 & 0 \end{pmatrix} \cdot \begin{pmatrix} a_2 & 0 \\ 0 & 0 \end{pmatrix} = \begin{pmatrix} a_1 a_2 & 0 \\ 0 & 0 \end{pmatrix} \in S$ as $a_1 a_2 \in \mathbb{Z}$.

Since, $\begin{pmatrix} a_1 & 0 \\ 0 & 0 \end{pmatrix}$ and $\begin{pmatrix} a_2 & 0 \\ 0 & 0 \end{pmatrix}$ were arbitrary elements of S.

Therefore, S is a subring of M.

Now, since
$$\begin{pmatrix} a_2 & 0 \\ 0 & 0 \end{pmatrix} \cdot \begin{pmatrix} a_1 & 0 \\ 0 & 0 \end{pmatrix} = \begin{pmatrix} a_2 a_1 & 0 \\ 0 & 0 \end{pmatrix}$$

$$= \begin{pmatrix} a_1 a_2 & 0 \\ 0 & 0 \end{pmatrix}$$

$$= \begin{pmatrix} a_1 & 0 \\ 0 & 0 \end{pmatrix}\begin{pmatrix} a_2 & 0 \\ 0 & 0 \end{pmatrix} \ \forall \ a_1, a_2 \in \mathbb{Z}.$$

Hence, S is commutative.

Remark Every subring of a commutative ring is commutative.

Example 3.13 Consider, a ring $R = \mathbb{Z}$ and its subring $S = 2\mathbb{Z}$. Here, R has unity 1 while S has no unity. Thus, there exist rings with unity having subrings without unity.

Example 3.14 Consider, a ring $R = \mathbb{Z} \oplus 2\mathbb{Z}$ and its subring $S = \mathbb{Z} \oplus \{0\}$. Here, R is a ring without unity while S has unity $(1, 0)$. Thus, there exist rings without unity but having subrings with unity.

Observe that the ring may be noncommutative in general, but there are some elements that commute with every element of the ring. For example, in the ring of $M_2(\mathbb{R})$, elements

$\begin{pmatrix} k & 0 \\ 0 & k \end{pmatrix}$ commute with every element of $M_2(\mathbb{R})$. These special elements constitute integral subset of a ring, which is known as center of a ring.

Definition 3.15 (Center of a Ring) Let R be a ring. The center of R is $Z(R) = \{r \in R \mid rx = xr$ for all $x \in R\}$.

Thus, for the ring $M_2(\mathbb{R})$,

$$Z(M_2(\mathbb{R})) = \left\{ \begin{pmatrix} k & 0 \\ 0 & k \end{pmatrix} : k \in \mathbb{R} \right\}.$$

Also, $Z(M_2(\mathbb{R}))$ is a subring of $M_2(\mathbb{R})$. In fact $Z(M_2(\mathbb{R}))$ is a commutative subring of a noncommutative ring $M_2(\mathbb{R})$.

In the next theorem, it is proved that $Z(R)$ is a commutative subring of a ring R in general.

Theorem 3.16 *Let, R be a ring. Then, the center of a ring R, $Z(R)$ is a commutative subring of R.*

Proof Let, R be a ring. Then, $Z(R) = \{a \in R \mid ax = xa$ for all $x \in R\}$ is the center of R.
Since, $0 \cdot x = x \cdot 0 = 0$ for all $x \in R$
Thus, $0 \in Z(R)$ therefore, $Z(R)$ is non-empty.
Suppose, $a, b \in Z(R)$. which implies that $ax = xa$ and $bx = xb$ for all $x \in R$.
We need to show that $a - b \in Z(R)$ and $ab \in Z(R)$.

  (i) $(a - b)x = ax - bx = xa - xb = x(a - b) \in Z(R) \ \forall \ a, b \in R.$
  (ii) $(ab)x = a(bx) = (ax)b = (xa)b = x(ab) \in Z(R) \ \forall \ a, b \in R.$

Hence, the center of a ring is a subring.
It follows directly from the definition of center of a ring that $Z(R)$ is commutative. $\qquad\square$

Note that, the center contains those elements that commute with every element of the ring. But, we may have elements that commute with some elements of a ring R. For example, in the ring $M_2(\mathbb{R})$, $\begin{pmatrix} k & 0 \\ 0 & 0 \end{pmatrix}$ does not commute with every element of $M_2(\mathbb{R})$ but commutes with elements like $\begin{pmatrix} k' & 0 \\ 0 & 0 \end{pmatrix}$ and $\begin{pmatrix} k' & 0 \\ 0 & k' \end{pmatrix}$. All such elements that commute with $\begin{pmatrix} k & 0 \\ 0 & 0 \end{pmatrix}$ also form a subring of $M_2(\mathbb{R})$, which is called centralizer of

$$S = \left\{ \begin{pmatrix} k & 0 \\ 0 & 0 \end{pmatrix} : k \in \mathbb{R} \right\}.$$

Definition 3.17 (Centralizer of S in R) Let S be a non-empty subset of a ring R. Then,

$$C(S) = \{c \in R \mid cx = xc \text{ for all } x \in S\}$$

is called the *centralizer* of S in R.

Let us now prove that $C(S)$ is a subring.

Theorem 3.18 *Centralizer of S, $C(S) = \{c \in R \mid cx = xc$ for all $x \in S\}$ in R is a subring of R.*

Proof Let R be a ring. Let S be a non-empty subset of a ring R. Then, centralizer of S in R is $C(S) = \{c \in R \mid cx = xc$ for all $x \in S\}$.
We need to show that $C(S)$ is a subring of R.

Clearly $0 \in C(S)$ as $0x = 0 = x0$ for all $x \in S$.
Let, $c, d \in C(S)$ then $cx = xc$ and $dx = xd$ for all $x \in S$.
Now, we will show that $c - d \in C(S)$ and $cd \in C(S)$.
Consider, $\quad (c - d)x = cx + (-d)x = cx - (dx) = xc - (xd) = xc + x(-d) = x(c - d)$,
thus we have $c - d \in C(S)$
and $(cd)x = c(dx) = c(xd) = (cx)d = (xc)d = x(cd)$. Hence, $cd \in C(S)$.
Therefore, $C(S)$ is a subring of R. $\qquad\qquad\square$

Another class of subsets which play an important role in characterization of rings is normalizer of a subset.

Definition 3.19 (Normalizer of S in R) Let S be a non-empty subset of a ring R. Then,

$$N(S) = \{c \in R \mid Sc = cS\}$$

is called the *normalizer* of S in R.

Theorem 3.20 *Normalizer of S, $N(S) = \{c \in R \mid Sc = cS\}$ in R is a subring of R.*

Proof Let R be a ring. Let S be a non-empty subset of a ring R. Then, normalizer of S in R is $N(S) = \{c \in R \mid Sc = cS\}$.
We need to show that $N(S)$ is a subring of R.
Clearly, $0 \in N(S)$ as $S \cdot 0 = \{0\} = 0 \cdot S \Rightarrow N(S) \neq \emptyset$.
Let, $c, d \in N(S)$ then $Sc = cS$ and $Sd = dS$.
Now, we will show that $c - d \in N(S)$ and $cd \in N(S)$.
Consider, $S(c - d) = Sc - Sd = cS - dS = (c - d)S \Rightarrow c - d \in N(S)$
and $S(cd) = (Sc)d = (cS)d = c(Sd) = c(dS) = (cd)S \Rightarrow cd \in N(S)$.
Hence, $N(S)$ is a subring of R. $\qquad\qquad\square$

The following theorem shows that the ring with unity always has a non-trivial subring.

Theorem 3.21 *Let R be a ring with unity 1. Let $S = \{n \cdot 1 \mid n \in \mathbb{Z}\}$. Then, S is a subring of R.*

Proof Suppose R be a ring with unity 1.
We need to show that, $S = \{n \cdot 1 \mid n \in \mathbb{Z}\}$ is a subring of R.
Let, $n, m \in \mathbb{Z}$.
First we need to show that $n \cdot 1 - m \cdot 1 \in S$.
Consider, $n \cdot 1 - m \cdot 1 = (n - m) \cdot 1 \in S \ \forall \ n, m \in \mathbb{Z}$.
Next, we need to show that $(n \cdot 1) \cdot (m \cdot 1) \in S$.
$(n \cdot 1) \cdot (m \cdot 1) = (n \cdot m) \cdot 1^2 = (nm) \cdot 1 \in S \ \forall \ n, m \in \mathbb{Z}$. $\qquad$ (1 is the unity.)
Thus, S is a subring of R. $\qquad\qquad\square$

A subring of R can also be obtained if the ring under consideration has an element a such that $a^2 = 1$. This can be put in the form of the following theorem.

Theorem 3.22 *Suppose that R is a ring with unity, 1. Let $a \in R$ such that $a^2 = 1$. Define $S = \{ara \mid r \in R\}$. Then, S is a subring of R.*

Proof Note that $a \cdot 0 \cdot a \in S \Rightarrow S \neq \emptyset$.
Let $x, \, y \in S$ such that $x = ar_1 a$ and $y = ar_2 a$ for some $r_1, \, r_2 \in R$.
First we need to show that $x - y \in S$.
Consider $\quad x - y = ar_1 a - ar_2 a = a(r_1 - r_2)a = ar_3 a \in S$ (where $r_3 = r_1 - r_2 \in R$).
Thus $x - y \in S$.
Now, we need to show that $xy \in S$.
Consider $xy = ar_1 a \cdot ar_2 a = ar_1 a^2 r_2 a = ar_1 r_2 a$ that is, $xy = ar_4 a$. Hence, $xy \in S$ where $r_4 = r_1 r_2 \in R$.
Therefore, S is a subring of R. $\qquad\qquad\square$

Remark The set S defined in Theorem 3.22 contains unity because $1 \in R$, which implies $a \cdot 1 \cdot a \in S$, that is $a^2 \in S \Rightarrow 1 \in S$. $\hfill (a^2 = 1).$

In the next set of theorems, we categorize some special forms of subrings.

Theorem 3.23 *Let R be a commutative ring of characteristic 2. Then, the set of all idempotents in R is a subring of R.*

Let R be a ring. If there exists a positive integer n such that $na = 0_R$ for all $a \in R$, then the smallest such positive integer is called the *characteristic* of R and denoted by $\mathrm{Char}(R)$. If no such positive integer exists, then R is said to be characteristic zero.

Proof Let R be a commutative ring with $\mathrm{char}(R) = 2$. Then, $2x = 0$ for all $x \in R$ (note that this is equivalent to saying that every element is the self additive inverse, that is, $x = -x$ for all $x \in R$).
Recall that $x \in R$ is an idempotent if $x^2 = x$.
Let $S = \{x \in R \mid x^2 = x\}$ be the set of all idempotents in R.
Clearly, $S \neq \emptyset$ since $0 \in S$ (for $0^2 = 0$).
Let $x, y \in S$ (thus $x^2 = x$ and $y^2 = y$). Then
$(x - y)^2 = (x - y)(x - y) = x^2 - xy - yx + y^2 = x^2 - 2xy + y^2$ (since R is commutative).
But this is equivalent to $\hfill$ (because char $R{=}2$)

$$(x - y)^2 = x^2 + y^2$$
$$= x + y = x - y,$$

since $y = -y$. Hence, $x - y \in S$.
Moreover, $(xy)^2 = x^2 y^2$ (since R is commutative), which is equivalent to $(xy)^2 = xy$.
Therefore, $xy \in S$.
Hence, S is a subring of R, by the subring test. $\qquad\qquad\square$

Note that the nilpotent elements of a ring do not form a subring in general as the class of nilpotent elements of R need not be closed. Let us analyze this through the following example.

Example 3.24 Let R be the noncommutative ring of 2×2 matrices with real coefficients. Consider the matrices $A = \begin{pmatrix} 0 & 1 \\ 0 & 0 \end{pmatrix}$, $B = \begin{pmatrix} 0 & 0 \\ 1 & 0 \end{pmatrix}$ in R. Then, $A^2 = O$ and $B^2 = O$, where O is the zero matrix. Hence, A and B are nilpotent elements.

However, the product $AB = \begin{pmatrix} 0 & 1 \\ 0 & 0 \end{pmatrix} \cdot \begin{pmatrix} 0 & 0 \\ 1 & 0 \end{pmatrix} = \begin{pmatrix} 1 & 0 \\ 0 & 0 \end{pmatrix}$ is not nilpotent as $\begin{pmatrix} 1 & 0 \\ 0 & 0 \end{pmatrix}^n \neq \begin{pmatrix} 0 & 0 \\ 0 & 0 \end{pmatrix}$ for any $n \in \mathbb{N}$.

The next theorem confers the condition under which the set of nilpotent elements, called the *nilradical of ring R*, forms subring of a ring R.

Theorem 3.25 *The nilpotent elements of a commutative ring form a subring.*

Proof Let R be a commutative ring. Recall that $x \in R$ is nilpotent if $x^n = 0$ for some $n \in \mathbb{N}$.

Let $N = \{x \in R \mid x^n = 0 \text{ for some } n \in \mathbb{N}\}$ be the set of all nilpotents in R.

Now, we need to show that N is a subring of R.

Clearly, $0^1 = 0 \Rightarrow N \neq \emptyset$. Let, $a, b \in N$. Then there exists $n, m \in \mathbb{Z}$ such that $a^n = 0 = b^m$.

Now, we will show that $a - b \in N$ and $ab \in N$.

Consider,

$$
\begin{aligned}
(a - b)^{n+m} &= \sum_{k=0}^{n+m} \binom{n+m}{k} a^k b^{n+m-k} && (R \text{ is commutative}) \\
&= \sum_{k=0}^{n-1} \binom{n+m}{k} a^k b^{n+m-k} + \sum_{k=n}^{n+m} \binom{n+m}{k} a^k b^{n+m-k} \\
&= \sum_{k=0}^{n-1} \binom{n+m}{k} a^k \cdot 0 + \sum_{k=n}^{n+m} \binom{n+m}{k} 0 \cdot b^{n+m-k} \\
&= 0.
\end{aligned}
$$

(for $n \leq k \leq n + m, a^k = a^n a^{k-n} = 0$. For $0 \leq a < n, b^{n+m-k} = b^m b^{n-k} = 0$.) Therefore, $a - b \in N$.

Also, $(ab)^{nm} = a^{nm} b^{nm} = (a^n)^m (b^m)^n = (0)(0) = 0 \Rightarrow ab \in N$.

Since N is closed under subtraction and multiplication, then, by subring test it is a subring of R. $\qquad\square$

Illustrative Problems

Problem 1 Let $M_2(\mathbb{Z})$ be the ring of all 2×2 matrices over the integers. Show that the set $S = \left\{ \begin{bmatrix} a & 0 \\ 0 & b \end{bmatrix} \mid a, b \in \mathbb{Z} \right\}$ of diagonal matrices is a subring of the ring $M_2(\mathbb{Z})$.

__Solution__ Since $0 = \begin{bmatrix} 0 & 0 \\ 0 & 0 \end{bmatrix} \in S \Rightarrow S \neq \emptyset$.

Let, $X, Y \in S$ where $X = \begin{bmatrix} a_1 & 0 \\ 0 & b_1 \end{bmatrix}$, $Y = \begin{bmatrix} a_2 & 0 \\ 0 & b_2 \end{bmatrix}$, where a_1, a_2, b_1 and $b_2 \in \mathbb{Z}$.

Firstly, we have to show that $X - Y \in S$.

$$X - Y = \begin{bmatrix} a_1 & 0 \\ 0 & b_1 \end{bmatrix} - \begin{bmatrix} a_2 & 0 \\ 0 & b_2 \end{bmatrix} = \begin{bmatrix} a_1 - a_2 & 0 \\ 0 & b_1 - b_2 \end{bmatrix} \in S \ \forall \ a_1, a_2, b_1, b_2 \in \mathbb{Z}.$$

Secondly, we have to show that $X \cdot Y \in S$

$$X \cdot Y = \begin{bmatrix} a_1 & 0 \\ 0 & b_1 \end{bmatrix} \cdot \begin{bmatrix} a_2 & 0 \\ 0 & b_2 \end{bmatrix} = \begin{bmatrix} a_1 a_2 & 0 \\ 0 & b_1 b_2 \end{bmatrix} \in S \ \forall \ a_1, a_2, b_1, b_2 \in \mathbb{Z}.$$

Hence, S is a subring.

Problem 2 Let $\mathbb{Z}[\sqrt{2}] = \{a + b\sqrt{2} : a, b \in \mathbb{Z}\}$. Show that $\mathbb{Z}[\sqrt{2}]$ is a commutative ring by showing it is a subring of $\mathbb{R}$.

__Solution__ It is clear that $\mathbb{Z}[\sqrt{2}] \subset \mathbb{R}$. We apply subring test to show that it is a subring. Let $x, y \in \mathbb{Z}[\sqrt{2}]$. Then there exist, $a, b, c, d \in \mathbb{Z}$ such that $x = a + b\sqrt{2}$ and $y = c + d\sqrt{2}$ Observe that $x - y = (a + b\sqrt{2}) - (c + d\sqrt{2}) = (a - c) + (b - d)\sqrt{2}$.
Since $a - c, \ b - d \in \mathbb{Z}, x - y \in \mathbb{Z}[\sqrt{2}]$.
Because x, y were arbitrary in $\mathbb{Z}[\sqrt{2}]$, $\mathbb{Z}[\sqrt{2}]$ is closed under subtraction.
Also, $xy = (a + b\sqrt{2})(c + d\sqrt{2}) = (ac + 2bd) + (ad + bc)\sqrt{2}$.
As $ac + 2bd, \ ad + bc \in \mathbb{Z}, xy \in \mathbb{Z}[\sqrt{2}]$.
Because x, y were arbitrary in $\mathbb{Z}[\sqrt{2}]$, $\mathbb{Z}[\sqrt{2}]$ is also closed under multiplication as well.
Hence, $\mathbb{Z}[\sqrt{2}]$ is a subring of $\mathbb{R}$.
Since $\mathbb{R}$ is a commutative ring, $\mathbb{Z}[\sqrt{2}]$ is also a commutative ring.

Problem 3 Let F denote the ring of all real-valued functions $f : \mathbb{R} \to \mathbb{R}$, where $(f + g)(x) = f(x) + g(x)$ and $(fg)(x) = f(x)g(x)$ for all $f, g \in F$ for each $x \in \mathbb{R}$. Show that $S = \{f \in F : f(1) = 0\}$ is a subring of F.

__Solution__ For the zero map, we have $O_F(1) = 0$. Thus, $O_F \in S \Rightarrow S \neq \emptyset$.
Let $f, g \in S$. Then, $f(1) = 0 = g(1)$. Then, $(f - g)(1) = f(1) - g(1) = 0 - 0 = 0 \Rightarrow f - g \in S$.
Also, $(fg)(1) = f(1)g(1) = 0 \cdot 0 = 0 \Rightarrow fg \in S$.
Therefore, S is a subring of F.

Problem 4 Is $\mathbb{Z}_6$ a subring of $\mathbb{Z}_{12}$?

__Solution__ The ring $\mathbb{Z}_6$ is not a subring of $\mathbb{Z}_{12}$ under addition and multiplication modulo 12 because $4, 5 \in \mathbb{Z}_6$ but $4 \oplus_{12} 5 = 9 \notin \mathbb{Z}_6$.

Remark Although $\mathbb{Z}_6 \subseteq \mathbb{Z}_{12}$ forms a ring itself but under $\oplus_6$ and $\otimes_6$, not under integer modulo 12. Therefore, $\mathbb{Z}_6$ is a ring but not a subring of $\mathbb{Z}_{12}$.

Problem 5 Show that the set of matrices, $S = \left\{ \begin{pmatrix} a & b \\ -b & a \end{pmatrix} \middle| a, b \in \mathbb{R} \right\}$. Prove that S is a subring of $M_2(\mathbb{R})$.

<u>**Solution**</u> Since the additive identity of $M_2(\mathbb{R})$, $\begin{pmatrix} 0 & 0 \\ 0 & 0 \end{pmatrix} \in S$

Therefore, $S \neq \emptyset$.

Let, $A, B \in S$ where $A = \begin{pmatrix} a_1 & b_1 \\ -b_1 & a_1 \end{pmatrix}$, $B = \begin{pmatrix} a_2 & b_2 \\ -b_2 & a_2 \end{pmatrix}$ for some $a_1, a_2, b_1, b_2 \in \mathbb{R}$.

Firstly, we need to show that $A - B \in S$.

$$A - B = \begin{pmatrix} a_1 & b_1 \\ -b_1 & a_1 \end{pmatrix} - \begin{pmatrix} a_2 & b_2 \\ -b_2 & a_2 \end{pmatrix}$$

$$= \begin{pmatrix} a_1 - a_2 & b_1 - b_2 \\ -b_1 + b_2 & a_1 - a_2 \end{pmatrix} \in S \text{ for } a_1, a_2, b_1, b_2 \in \mathbb{R}.$$

Secondly, we need to show that $A \cdot B \in S$.

$$A \cdot B = \begin{pmatrix} a_1 & b_1 \\ -b_1 & a_1 \end{pmatrix} \cdot \begin{pmatrix} a_2 & b_2 \\ -b_2 & a_2 \end{pmatrix}$$

$$= \begin{pmatrix} a_1 a_2 - b_1 b_2 & a_1 b_2 + b_1 a_2 \\ -b_1 a_2 - b_2 a_1 & -b_1 b_2 + a_1 a_2 \end{pmatrix} \in S \text{ for } a_1, a_2, b_1, b_2 \in \mathbb{R}.$$

Hence, S is a subring of $M_2(\mathbb{R})$.

Problem 6 Prove that $R = \left\{ \begin{bmatrix} a & a \\ b & b \end{bmatrix} \middle| a, b \in \mathbb{Z} \right\}$ is a subring of $M_2(\mathbb{Z})$.

<u>**Solution**</u> Since $0 = \begin{bmatrix} 0 & 0 \\ 0 & 0 \end{bmatrix} \in R \Rightarrow R \neq \emptyset$.

Let, $X, Y \in R$ where $X = \begin{bmatrix} a_1 & a_1 \\ b_1 & b_1 \end{bmatrix}$, $Y = \begin{bmatrix} a_2 & a_2 \\ b_2 & b_2 \end{bmatrix}$ for some $a_1, a_2, b_1, b_2 \in \mathbb{Z}$.

Firstly, we need to show that $X - Y \in R$.

$$X - Y = \begin{bmatrix} a_1 & a_1 \\ b_1 & b_1 \end{bmatrix} - \begin{bmatrix} a_2 & a_2 \\ b_2 & b_2 \end{bmatrix}$$

$$= \begin{bmatrix} a_1 - a_2 & a_1 - a_2 \\ b_1 - b_2 & b_1 - b_2 \end{bmatrix} \in S \text{ for } a_1, a_2, b_1, b_2 \in \mathbb{Z}.$$

Secondly, we need to show that $X \cdot Y \in R$.

$$X \cdot Y = \begin{bmatrix} a_1 & a_1 \\ b_1 & b_1 \end{bmatrix} \cdot \begin{bmatrix} a_2 & a_2 \\ b_2 & b_2 \end{bmatrix}$$

$$= \begin{bmatrix} a_1 a_2 + a_1 b_2 & a_1 a_2 + a_1 b_2 \\ a_2 b_1 + b_1 b_2 & a_2 b_1 + b_1 b_2 \end{bmatrix} \in R \text{ for } a_1, a_2, b_1, b_2 \in \mathbb{Z}.$$

Hence, R is a subring of $M_2(\mathbb{Z})$.

Problem 7 Let, $R = \mathbb{Z} \oplus \mathbb{Z} \oplus \mathbb{Z}$. Let, $S = \{(a, b, c) \mid c = a + b\}$. Is S a subring of R?

Solution No, S is not a subring of R because $(1, 0, 1), (0, 1, 1) \in S$. But $(1, 0, 1) \cdot (0, 1, 1) = (0, 0, 1) \notin S$.

Problem 8 Describe all the subrings of $\mathbb{Z}$.

Solution Here, $\mathbb{Z} = \{\ldots, -2, -1, 0, 1, 2, \ldots\}$. Subrings of $\mathbb{Z}$ are given by $n\mathbb{Z}$, $n \in \mathbb{Z}$.

Problem 9 Explain why every subgroup of $\mathbb{Z}_n$ under addition is also a subring of $\mathbb{Z}_n$.

Solution Every subgroup of $\mathbb{Z}_n$ is closed under multiplication and hence every subgroup of $\mathbb{Z}_n$ under addition is also a subring of $\mathbb{Z}_n$.

Problem 10 Let $M_2(\mathbb{Z})$ be the ring of all 2×2 matrices over the integers and let

$$R = \left\{ \begin{bmatrix} a & a - b \\ a - b & b \end{bmatrix} \,\middle|\, a, b \in \mathbb{Z} \right\}.$$

Prove that R is a subring of $M_2(\mathbb{Z})$.

Solution Since, $0 = \begin{bmatrix} 0 & 0 - 0 \\ 0 - 0 & 0 \end{bmatrix} = \begin{bmatrix} 0 & 0 \\ 0 & 0 \end{bmatrix} \in R \Rightarrow R \neq \emptyset.$

Let $X, Y \in R$ where, $X = \begin{bmatrix} a_1 & a_1 - b_1 \\ a_1 - b_1 & b_1 \end{bmatrix}$ and $Y = \begin{bmatrix} a_2 & a_2 - b_2 \\ a_2 - b_2 & b_2 \end{bmatrix}$

$\forall\, a_1, a_2, b_1, b_2 \in \mathbb{Z}.$

(i) We need to check whether $X - Y \in R$.

$$\text{Consider, } X - Y = \begin{bmatrix} a_1 & a_1 - b_1 \\ a_1 - b_1 & b_1 \end{bmatrix} - \begin{bmatrix} a_2 & a_2 - b_2 \\ a_2 - b_2 & b_2 \end{bmatrix}$$

$$= \begin{bmatrix} a_1 - a_2 & (a_1 - a_2) - (b_1 - b_2) \\ (a_1 - a_2) - (b_1 - b_2) & b_1 - b_2 \end{bmatrix}$$

$\in R$ for $a_1, a_2, b_1, b_2 \in \mathbb{Z}$.

(ii) We need to check whether $X \cdot Y \in R$.

$$\text{Consider, } X \cdot Y = \begin{bmatrix} a_1 & a_1 - b_1 \\ a_1 - b_1 & b_1 \end{bmatrix} \cdot \begin{bmatrix} a_2 & a_2 - b_2 \\ a_2 - b_2 & b_2 \end{bmatrix}$$

$$= \begin{bmatrix} a_1 a_2 + a_1 a_2 - a_1 b_2 - a_2 b_1 + b_1 b_2 & a_1 a_2 - a_1 b_2 + a_1 b_2 - b_1 b_2 \\ a_1 a_2 - a_2 b_1 + a_2 b_1 - b_1 b_2 & a_1 a_2 - a_1 b_2 - a_2 b_1 + b_1 b_2 + b_1 b_2 \end{bmatrix}$$

$$= \begin{bmatrix} a_1 a_2 + a_1 a_2 - a_1 b_2 - a_2 b_1 + b_1 b_2 & a_1 a_2 - b_1 b_2 \\ a_1 a_2 - b_1 b_2 & a_1 a_2 - a_1 b_2 - a_2 b_1 + b_1 b_2 + b_1 b_2 \end{bmatrix}$$

$\in R$ for $a_1, a_2, b_1, b_2 \in \mathbb{Z}$.

Hence, R is a subring.

Problem 11 Let $M_2(\mathbb{Z})$ be the ring of all 2×2 matrices over the integers and let

$$R = \left\{ \begin{bmatrix} a & a + b \\ a + b & b \end{bmatrix} \,\middle|\, a, b \in \mathbb{Z} \right\}.$$

Prove that R is a subring of $M_2(\mathbb{Z})$.

Solution R is not a subring of $M_2(\mathbb{Z})$.

Consider, two 2×2 matrices X and Y such that

$$X = \begin{bmatrix} 1 & 1 \\ 1 & 0 \end{bmatrix} \text{ and } Y = \begin{bmatrix} 0 & 1 \\ 1 & 1 \end{bmatrix}.$$

$$\text{Then, } X - Y = \begin{bmatrix} 1 & 1 \\ 1 & 0 \end{bmatrix} - \begin{bmatrix} 0 & 1 \\ 1 & 1 \end{bmatrix} = \begin{bmatrix} 1 & 0 \\ 0 & -1 \end{bmatrix} \in R$$

$$\text{and } X \cdot Y = \begin{bmatrix} 1 & 1 \\ 1 & 0 \end{bmatrix} \cdot \begin{bmatrix} 0 & 1 \\ 1 & 1 \end{bmatrix} = \begin{bmatrix} 1 & 2 \\ 0 & 1 \end{bmatrix} \notin R.$$

Thus, R is not a subring of $M_2(\mathbb{Z})$ as the set is not closed under multiplication.

Problem 12 Determine the smallest subring of Q that contains $\dfrac{1}{2}$. (That is, find the subring S with the property that S contains $\dfrac{1}{2}$ and if T is any subring containing $\dfrac{1}{2}$, then T contains S.)

Solution Let $S = \left\{ \dfrac{m}{2^n} \;\middle|\; m \in \mathbb{Z}, n \in \mathbb{Z}^+ \right\}$. Then, we show that S is the smallest subring

containing $\dfrac{1}{2}$. Clearly, $\dfrac{1}{2} \in S$. We now show that S is a subring.

Let, $\dfrac{m_1}{2^{n_1}}, \dfrac{m_2}{2^{n_2}} \in S$ for $m_1, m_2 \in \mathbb{Z}$ and $n_1, n_2 \in \mathbb{Z}^+$.

Then, we have $\dfrac{m_1}{2^{n_1}} - \dfrac{m_2}{2^{n_2}} = \dfrac{m_1 2^{n_2} - m_2 2^{n_1}}{2^{n_1} \cdot 2^{n_2}} \in S$ and $\dfrac{m_1}{2^{n_1}} \cdot \dfrac{m_2}{2^{n_2}} = \dfrac{m_1 m_2}{2^{n_1 + n_2}} \in S$ where
$m_1, m_2 \in \mathbb{Z}$ and $n_1, n_2 \in \mathbb{Z}^+$.
Therefore, the subring test is satisfied.

Now, let T be any subring that contains $\dfrac{1}{2}$, then by closure under multiplication, it

contains $\dfrac{1}{2^n}$ and by closure under addition (and subtraction), it contains $\dfrac{m}{2^m}$. Thus, T

contains S. Hence, S is the smallest subring containing $\dfrac{1}{2}$.

Problem 13 Let R be the set of rational numbers that can be written in the form $\dfrac{p}{q}$, where
q is not zero and not divisible by 6. Prove that R is not a subring of $\mathbb{Q}$.

Solution Put $a = \dfrac{1}{2}$ and $b = \dfrac{-1}{3}$. Then, $a, b \in R$ but $a + b = \dfrac{1}{6} \notin R$; therefore, R is not
closed under addition. Hence, R is not a subring of $\mathbb{Q}$. In fact, R is not closed under
multiplication as well.

Problem 14 Let R and S be rings. Show that the subset $\bar{R} = \{(r, 0_S) \mid r \in R\}$ is a subring
of $R \times S$.

Solution Clearly, $\bar{R}$ is a subset of $R \times S$ and $R \times S$ is a ring. (Refer to Problem 10 from
Chapter 2)
   We need to show that $\bar{R} = \{(r, 0_S) \mid r \in R\}$ is a subring of $R \times S$.
   Clearly, $(0, 0_S) \in \bar{R} \Rightarrow \bar{R} \neq \emptyset$.
   Let $a = (r, 0_S)$ and $b = (t, 0_S)$ be arbitrary elements of $\bar{R}$.
Now, we will show that $a - b \in \bar{R}$ and $ab \in \bar{R}$.

Consider, $a - b = (r, 0_S) - (t, 0_S) = (r - t, 0_S - 0_S) = (r - t, 0_S) \in \bar{R}$.

and $ab = (r, 0_S)(t, 0_S) = (rt, 0_S \cdot 0_S) = (rt, 0_S) \in \bar{R}$.

Hence, $\bar{R}$ is a subring of $R \times S$.

Problem 15 If R is a ring, show that $R^* = \{(r, r) \mid r \in R\}$ is a subring of $R \times R$.

Solution Let, R be a ring. We need to show that $R^* = \{(r, r) \mid r \in R\}$ is a subring of
$R \times R$.
   Clearly, $(0, 0) \in R^* \Rightarrow R^* \neq \emptyset$.
   Let, $(r, r), (r', r') \in R^*$. Now, we will show that $(r, r) - (r', r') \in R^*$ and $(r, r)(r', r') \in R^*$.
   Consider, $(r, r) - (r', r') = (r - r', r - r') \in R^*$.
and $(r, r)(r', r') = (rr', rr') \in R^*$.
Hence, R^* is a subring of $R \times R$.

Problem 16 Is $\{1, -1, i, -i\}$ a subring of $\mathbb{C}$?

<u>**Solution**</u> No, since $i + i = 2i \notin \{1, -1, i, -i\}$, thus the given subset is not closed under addition. Hence it is not a subring. Also, $0 = 0_\mathbb{C} \notin \{1, -1, i, -i\}$.

Exercise

(1) Show that the set, $S = \{0, 3\}$ is a subring of the ring of integers modulo 6, $\mathbb{Z}_6$.

(2) Show that $\mathbb{Z}_{12}$ has the proper subrings $\{0, 2, 4, 6, 8, 10\}, \{0, 3, 6, 9\}, \{0, 4, 8\}$, and $\{0, 6\}$.

(3) Determine the smallest subring of Q that contains $\frac{2}{3}$.

(4) Give an example of a subset of a ring that is a subgroup under addition but not a subring.

(5) Show that $\mathbb{Q}[\sqrt{2}]$ is a subring of $\mathbb{R}$.

(6) Show that the set of matrices $\begin{bmatrix} a & b \\ 0 & c \end{bmatrix}$ is a subring of the ring of 2×2 matrices with integral elements.

(7) The set of integers is a subring of the ring of rational numbers. Let, R be the ring of integers. Let, m be any fixed integer and let S be any subset of R such that $S = \{..., -3m, -2m, -m, 0, m, 2m, 3m, ...\}$. Then, show that S is a subring of ring R.

(8) Let $R = M_{2\times2}(\mathbb{Z})$ be the ring of all 2×2 matrices over a ring $\mathbb{Z}$. Let, $S = \left\{ \begin{bmatrix} a & 0 \\ b & c \end{bmatrix} \middle| a, b \in \mathbb{Z} \right\}$. Check whether $(S, +, \cdot)$ is a subring of R.

(9) Give an example of a ring $(R, +, \cdot)$ with identity I_R and a subring $(S, +, \cdot)$ with identity I_S such that $I_R \neq I_S$.

(10) Let R and S be rings. Show that the subset $\bar{S} = \{(0_r, s) \mid s \in S\}$ is a subring of $R \times S$.

(11) Find two proper non-trivial subrings of $\mathbb{Z} \times \mathbb{R}$ (i.e., subrings which are neither zero nor the whole ring).

(12) Does the set of all numbers of the form $a + b\sqrt{3}$ where a and b are rational numbers form a subring of the ring $\mathbb{R}$?

(13) Show that the set of $n \times n$ matrices over the rational numbers is a subring of $n \times n$ matrices over the real numbers under addition and multiplication matrices.

4 INTEGRAL DOMAINS

The word "*Integral Domain*" can be interpreted in language as the *domain of integrity* with the idea that zero divisors are like flaws in the ring as divisibility theory is much more complex in the presence of zero divisors. The term *integral* comes from rings of *algebraic integers*—the study of which motivated the abstraction of many algebraic structures. The fact that integral domain embeds into its field of fractions as integers embed into rational supports the terminology.

In this chapter, the theory of commutative rings with unity is extended to define integral domain, which is further abstracted to fields. The study is elaborated through various examples and theorems.

Before defining integral domain, we define a special class of elements in a ring known as *zero divisor*.

Definition 4.1 (Zero Divisor) Let R be a commutative ring. Let $r(\neq 0) \in R$, then r is said to be a *zero divisor* if there exists a nonzero element $s \in R$ such that $r \cdot s = 0$.

Example 4.2 Consider $\mathbb{Z}_6$, a commutative ring. Then, $2, 3(\neq 0)$ in $\mathbb{Z}_6$ are zero divisors as $2 \cdot 3 = 0$. Also, 4 is a zero divisor in $\mathbb{Z}_6$.

This special class of elements categorizes rings further to define integral domains.

Definition 4.3 (Integral Domain) Let R be a commutative ring with unity. Then, R is said to be an *integral domain* if it has no zero divisor. That is, whenever a, b are nonzero elements in R then $ab \neq 0$.

Example 4.4 The ring of integers, $\mathbb{Z}$ is an integral domain under usual addition and multiplication since it is a commutative ring with unity 1 and for any two integers a, b with $ab = 0$, we have either $a = 0$ or $b = 0$. Thus, it has no zero divisors. Also, recall that the only units of $\mathbb{Z}$ are 1 and -1 (Units are elements with multiplicative inverse).

Example 4.5 The ring of integers modulo n, $\mathbb{Z}_n$ is not an integral domain when n is not prime because if n is not prime, there are integers a and b such that $n = ab$ with $1 < a, b < n$. Then, $ab = 0$ in $\mathbb{Z}_n$, but $a \neq 0$ and $b \neq 0$. So, $\mathbb{Z}_n$ has zero divisors. This means that $\mathbb{Z}_n$ is not an integral domain when n is not prime.

For example, in $\mathbb{Z}_4, 2$ is a zero divisor as $2 \cdot 2 = 0$. And, the set of units of $\mathbb{Z}_n$ is $U(n)$, where $U(n)$ is the set of all numbers relatively prime to n.

Example 4.6 The ring of integers modulo p (where p is a prime), $\mathbb{Z}_p$ is an integral domain. We have $\mathbb{Z}_p$ is a commutative ring with unity 1. For elements $a, b \in \mathbb{Z}_p$ such that $ab \equiv 0$ (mod p), we have $p|ab$, that is, $p|a$ or $p|b$. Hence, we have $a \equiv 0$ (mod p) or $b \equiv 0$ (mod p). Therefore, a and b are not zero divisors.

Example 4.7 The ring $\mathbb{Z}[x]$ of polynomials with integer coefficients is an integral domain.

As product of two nonzero elements in $\mathbb{Z}[x]$ can never be zero. So, it has no zero divisors. And the units in $\mathbb{Z}[x]$ are 1 and -1.

Example 4.8 Given any ring R, the set of polynomials with coefficients in R, $R[x] = \{a_n x^n + a_{n1} x^{n-1} + \cdots + a_1 x + a_0 : a_i \in R\}$ under polynomial addition and multiplication is a commutative ring with unity where unity is the constant polynomial 1. Here, one can observe that $R[x]$ has no (nonzero) zero divisors as the product of two nonzero elements in $R[x]$ can never be zero. Therefore, it is an integral domain. And, the units are the nonzero constant polynomials.

Example 4.9 The ring of Gaussian integers $\mathbb{Z}[i] = \{a + bi \mid a, b \in \mathbb{Z}\}$ is an integral domain. We know that $\mathbb{Z}[i]$ is a commutative ring with unity. We show that $\mathbb{Z}[i]$ has no zero divisors.
Let, $A = a_1 + ib_1$ and $B = a_2 + ib_2$, where $a_1, a_2, b_1, b_2 \in \mathbb{Z}$.
Let, $A \neq 0$, and $AB = 0$
$\Rightarrow (a_1 + ib_1)(a_2 + ib_2) = 0$
$\Rightarrow (a_1 a_2 - b_1 b_2) + (a_1 b_2 + a_2 b_1)i = 0$
So, $a_1 a_2 - b_1 b_2 = 0$ (4.1)
and $a_1 b_2 + b_1 a_2 = 0$.   (4.2)

Now, on multiplying Equation (4.1) with a_1, Equation (4.2) with b_1 and then adding, we obtain

$$a_1^2 a_2 + a_2 b_1^2 = 0 \Rightarrow a_2(a_1^2 + b_1^2) = 0$$

$\Rightarrow a_2 = 0$. $(\because A \neq 0, a_1^2 + b_1^2 \neq 0)$
From (4.1) and (4.2), we obtain $b_1 b_2 = 0$ and $a_1 b_2 = 0$
$\Rightarrow b_2 = 0$ $(\because A \neq 0 \Rightarrow$ both a_1 and b_1 cannot be zero simultaneously.)
Since $a_2 = 0$ and $b_2 = 0$, thus $B = 0$.
Similarly, if $B \neq 0$, then $A = 0$.
Thus, $\mathbb{Z}[i]$ has no zero divisors. Therefore, $\mathbb{Z}[i]$ is an integral domain. And, the set of units of $\mathbb{Z}[i]$ is $\{1, i, -1, -i\}$.

Example 4.10 Let d be an integer. Then, $\mathbb{Z}[\sqrt{d}] = \{a + b\sqrt{d} \mid a, b \in \mathbb{Z}\}$ is an integral domain. We know that $\mathbb{Z}[\sqrt{d}]$ is a commutative ring with unity 1. We show that $\mathbb{Z}[\sqrt{d}]$ has no zero divisors. Let, $A = a_1 + b_1\sqrt{d}$ and $B = a_2 + b_2\sqrt{d}$, where $a_1, a_2 \, b_1, b_2 \in \mathbb{Z}$.
Let, $A \neq 0$, and $AB = 0$
$\Rightarrow (a_1 + b_1\sqrt{d})(a_2 + b_2\sqrt{d}) = 0$
$\Rightarrow (a_1 a_2 + d b_1 b_2) + (a_1 b_2 + b_1 a_2)\sqrt{d} = 0$
So, $a_1 a_2 + d b_1 b_2 = 0$ (4.3)
and $a_1 b_2 + b_1 a_2 = 0$.   (4.4)

Now, on multiplying Equation (4.3) with a_1, Equation (4.4) with $-db_1$ and then adding, we obtain

$a_1^2 a_2 - da_2 b_1^2 = 0 \Rightarrow a_2(a_1^2 - db_1^2) = 0$

$\Rightarrow a_2 = 0.$ $\hspace{3cm} (\because A \neq 0, a_1^2 - db_1^2 \neq 0)$

From (4.3) and (4.4), we obtain $b_1 b_2 = 0$ and $a_1 b_2 = 0$

$\Rightarrow b_2 = 0.$ $\hspace{1cm} (\because A \neq 0 \Rightarrow$ both a_1 and b_1 cannot be zero simultaneously.)

Since $a_2 = 0$ and $b_2 = 0$, therefore $B = 0$.

Similarly, if $B \neq 0$, then $A = 0$.

Thus, $\mathbb{Z}[\sqrt{d}]$ has no zero divisors. Therefore, $\mathbb{Z}[\sqrt{d}]$ is an integral domain.

Example 4.11 The ring $M_2(\mathbb{Z})$ of 2×2 matrices over the integers is not an integral domain.

Let $A = \begin{bmatrix} 0 & 1 \\ 0 & 0 \end{bmatrix}$ and $B = \begin{bmatrix} 1 & 0 \\ 0 & 0 \end{bmatrix}$. Then, both A and B are nonzero. But $A \cdot B =$

$\begin{bmatrix} 0 & 1 \\ 0 & 0 \end{bmatrix} \begin{bmatrix} 1 & 0 \\ 0 & 0 \end{bmatrix} = \begin{bmatrix} 0 & 0 \\ 0 & 0 \end{bmatrix} \Rightarrow A \cdot B = 0$. Thus, the ring $M_2(\mathbb{Z})$ has zero divisors.

Example 4.12 $\mathbb{Z} \oplus \mathbb{Z}$ is not an integral domain. Let us consider $a = (1, 0)$ and $b = (0, 1)$. Then a and b both are nonzero elements, but $ab = (1, 0)(0, 1) = (0, 0)$. Therefore, it has zero divisors.

Although an integral domain does not form a group under multiplication, still it exhibits important group-theoretic properties. One such property is cancellation law.

Theorem 4.13 (Cancellation Law) *Let R be an integral domain and $a, b,$ and c be elements in R. If $a \neq 0$ and $ab = ac$, then $b = c$.*

Proof From $ab = ac$, we have $a(b - c) = 0$. Since $a \neq 0$, we must have $b - c = 0$ $\Rightarrow b = c$. $\hspace{2cm} \square$

Now, we provide subsequent criteria for zero divisors.

Theorem 4.14 *A commutative ring with the cancellation property (under multiplication) has no zero divisors.*

Proof Let $ab = 0$ and $a \neq 0$ for some elements a, b from a commutative ring with the cancellation property. We can write $ab = 0 = a \cdot 0 \Rightarrow ab = a \cdot 0 \Rightarrow b = 0$. Thus, a commutative ring with the cancellation property (under multiplication) has no zero divisors. $\hspace{1cm} \square$

Remark From Theorem 4.14, we conclude that a commutative ring with unity in which the cancellation law (under multiplication) holds is an integral domain.

In the next result, we characterize the elements of a finite commutative ring.

Theorem 4.15 *Let R be a finite commutative ring with unity. Then every nonzero element of R is either a zero divisor or a unit.*

Proof Let $r \in R$ and $r \neq 0$. Let $S = \{xr : x \in R\} \subseteq R$. Then, either $S = R$ or $S \neq R$.

Case 1. $S = R$. This implies that $1 \in S$. Hence, we have $1 = xr$ for some $x \in R$. Since R is commutative, $xr = rx = 1 \Rightarrow r$ has an inverse which implies r is a unit.

Case 2. $S \neq R$. This implies that $S \subset R$. Then, we have $x_i r = x_j r$ for some $x_i \neq x_j$. Thus, we have $(x_i - x_j)r = 0$. Since, $r \neq 0$, $x_i - x_j \neq 0$ and $(x_i - x_j)r = 0$. Hence r is a zero divisor.

Hence, every nonzero element of R is either a zero divisor or a unit. $\qquad\square$

Remark The "finite" condition cannot be relaxed in the above Theorem 4.15. Since, in the ring of integers, $\mathbb{Z}$, 2 is neither a zero divisor nor a unit.

Corollary 4.16 Every nonzero element of $\mathbb{Z}_n$ is a unit or a zero divisor.

Proof From Theorem 4.15.

Alternatively, one can show in the following way.

Let us consider $\mathbb{Z}_n = \{0, 1, \cdots, n-1\}$. Let $k \in \mathbb{Z}_n$ and $k \neq 0$.

Case 1. $\gcd(k, n) = 1$. There exist x, $y \in \mathbb{Z}$ such that $kx + ny = 1$. Then, $kx \oplus_n ny = kx$, that is, $kx = 1$. Hence k has a multiplicative inverse which implies k is a unit of $\mathbb{Z}_n$.

Case 2. $\gcd(k, n) = d(\neq 1)$. Then, $d|k$ and $d|n$ which implies $k = ds$ and $n = dt$, where $s \neq 0$, $t \neq 0 \in \mathbb{Z}$.

Consider, $(k)t = (ds)t = s(dt) = ns = 0 \bmod n$. Hence $kt = 0$ in $\mathbb{Z}_n$. Since $k \neq 0, t \neq 0$ but $kt = 0 \Rightarrow k$ is a zero divisor. $\qquad\square$

Next, we introduce subset of an integral domain, known as *subdomain*. Formally, we define subdomain as follows.

Definition 4.17 (Subdomain) A subset P of an integral domain D is a *subdomain* of D if P is itself an integral domain under the operations of D.

Example 4.18 $\mathbb{Z}$ is a subdomain of $\mathbb{Q}$; $\mathbb{Q}$ is a subdomain of $\mathbb{R}$.

Theorem 4.19 *Let D be an integral domain with unity* 1. *Then,* $P = \{n.1 | n \in \mathbb{Z}\}$, *is the smallest subdomain of D.*

Proof Firstly, we will show that P is a subring. Clearly, $0 \cdot 1 = 0 \in P \Rightarrow P \neq \emptyset$. Let, $X, Y \in P$ where $X = n \cdot 1$ and $Y = m \cdot 1$, where $n, m \in \mathbb{Z}$.

Then, $X - Y = n \cdot 1 - m \cdot 1 = (n - m) \cdot 1 \in P$ ($\because n - m \in \mathbb{Z}$ for $n, m \in \mathbb{Z}$) and $XY = (n \cdot 1)(m \cdot 1) = (nm) \cdot 1 \in P$ ($\because nm \in \mathbb{Z}$ for $n, m \in \mathbb{Z}$).

Hence, P is a subring.

We know that every domain is commutative. Therefore, D is commutative.

Since, $P \subseteq D$ and $n \cdot 1 = n = 1 \cdot n$. Therefore, P is also commutative.

Also, we have $1 \cdot 1 = 1$. Thus, 1 is in P. Therefore, unity exists in P.

Zero divisors: Since, $P \subseteq D$ and D is an integral domain, so it has no zero divisors. Thus, P also does not have zero divisors. Therefore, P is a subdomain of D.

We further show that P is the smallest subdomain of D.

Let, T be any subdomain of D. Then, $1 \in T \Rightarrow 1 + 1 \in T \Rightarrow \underbrace{1 + 1 + \cdots + 1}_{n \text{ times}} \in T$. Thus, we have $n \cdot 1 \in T$ for every positive integer n. Further -1 is in T, therefore $\underbrace{(-1) + (-1) + \ldots + (-1)}_{n \text{ times}} \in T$. This implies $n(-1) = (-n).1 = m.1$ is in T for all negative integers m. Also, $0 = 0.1$ is in T as T is a subdomain. Therefore, n.1 is in T for all integers n. Hence, $P \subseteq T$. Therefore, P is contained in every subdomain of D. Hence, P is the smallest subdomain of D. $\qquad\qquad\qquad\qquad\qquad\qquad\qquad\qquad\qquad\qquad\qquad\qquad\qquad\qquad\square$

Next, the concept of integral domain is abstracted to define and establish useful results of fields.

Definition 4.20 (Field) A commutative ring with unity in which every nonzero element has a multiplicative inverse is called a *field*. In other words, a commutative division ring is called a field.

Example 4.21 The set of rational numbers, $(\mathbb{Q}, +, \cdot)$ is a field.

Example 4.22 The set of real numbers, $(\mathbb{R}, +, \cdot)$ is a field.

Example 4.23 The set of complex numbers, $(\mathbb{C}, +, \cdot)$ is a field.

Example 4.24 For p, a prime number, $\mathbb{Z}_p$ is a field.

Example 4.25 The set of integers, $\mathbb{Z}$ is not a field because the nonzero element 2 of $\mathbb{Z}$ has no multiplicative inverse in $\mathbb{Z}$, i.e., there is no integer m such that $2 \cdot m = 1$.

Example 4.26 $\mathbb{Z}_n$ is not a field when n is not prime because no zero divisor of $\mathbb{Z}_n$ has an inverse in $\mathbb{Z}_n$ w.r.t. multiplication.

Example 4.27 $\mathbb{Z}[x]$ is not a field because no other polynomial except units in $\mathbb{Z}[x]$ has multiplicative inverse.

Example 4.28 $R[x]$ is not a field because the invertible elements of $R[x]$ are just the constant polynomials a_0 with a_0 invertible in R. In particular, $x \in R[x]$ is not invertible.

Example 4.29 $\mathbb{Z}[\sqrt{2}]$ is not a field because the inverse of 2 is $\dfrac{1}{2}$ which does not belong to the ring $\mathbb{Z}[\sqrt{2}]$.

Example 4.30 $\mathbb{Z}_2[i]$, the ring of Gaussian integers modulo 2 is not a field. $\mathbb{Z}_2[i] = \{0, 1, i, 1 + i\}$. Multiplication table for $\mathbb{Z}_2[i]$ is:

$\otimes_2$	0	1	i	$1+i$
0	0	0	0	0
1	0	1	i	$1+i$
i	0	i	1	$1+i$
$1+i$	0	$1+i$	$1+i$	0

Clearly, from the above Cayley table we can conclude that it is not a field as inverse of $1 + i$ does not exist.

Moreover, it is not an integral domain because $(1 + i) \neq 0$, yet $(1 + i)(1 + i) = 0$, which implies $(1 + i)$ is a zero divisor.

Hence, it is neither a field nor an integral domain.

In the next theorem, we will establish that integral domains exist in abundance as every field is an integral domain.

Theorem 4.31 *Every field is an integral domain.*

Proof Let F be a field. Then, F is a commutative ring with unity. Now, we will show that F is an integral domain for which we will show that F has no zero divisors. Let, $ab = 0$ and $a \neq 0$. Since, a is a nonzero element of the field F, therefore, there exists $a^{-1} \in F$ such that $aa^{-1} = a^{-1}a = 1$.

Consider, $ab = 0 \Rightarrow (a^{-1}(ab)) = a^{-1} \cdot 0 = 0 \Rightarrow (a^{-1}a)b = 0 \Rightarrow 1 \cdot b = 0 \Rightarrow b = 0$. Therefore, there are no zero divisors in F. Thus, F is an integral domain. $\qquad\square$

Remark The converse of Theorem 4.31 need not be true which can be concluded with the help of following example.

Example 4.32 $\mathbb{Z}$ is an integral domain but not a field.

The following theorem gives a sufficient condition under which an integral domain is a field.

Theorem 4.33 (Finite Integral Domains Are Fields) *Every finite integral domain is a field.*

Proof Let D be a finite integral domain. Then D will be field if every nonzero element has a multiplicative inverse. Let, $a \in D$, $a \neq 0$.

  Case 1. When $a = 1$. In this case, a is its own inverse, we are done.
  Case 2. When $a \neq 1$. Consider a set $S = \{a, a^2, a^3, \cdots\} \subseteq D$.

Since, D is an integral domain, so $a^i \neq 0$ for any i. Also, D is finite which implies that $a^i = a^j$ for some $i \neq j \Rightarrow a^{i-j} = 1$ (assuming $i > j$), hence we have $a \cdot a^{i-j-1} = 1$. (Here, $i - j > 1$, because $a \neq 1$).

This implies that a has a multiplicative inverse. Thus, every nonzero element of D has a multiplicative inverse. Therefore, D is a field. $\qquad\square$

Corollary 4.34 ($\mathbb{Z}_p$ Is a Field) For every prime p, $\mathbb{Z}_p$, the ring of integers modulo p is a field.

Proof We know that $\mathbb{Z}_p$ is a commutative ring with unity. Since, $\mathbb{Z}_p$ is finite. In order to show that it is a field, we show that it is an integral domain, i.e., it has no zero divisors. Let, $ab = 0$ in $\mathbb{Z}_p$ for some $a, b \in \mathbb{Z}_p$.

$\Rightarrow ab = 0 \bmod p \Rightarrow p|ab \Rightarrow p|a$ or $p|b \Rightarrow a = 0 \bmod p$ or $b = 0 \bmod p$

$\Rightarrow$ either $a = 0$ or $b = 0$ in $\mathbb{Z}_p$. This implies that $\mathbb{Z}_p$ has no zero divisors. Therefore, $\mathbb{Z}_p$ is a finite integral domain. Hence, $\mathbb{Z}_p$ is a field. $\qquad\square$

Definition 4.35 (Subfield) Let F be a field and let K be a non-empty subset of F. We say that K is a *subfield* of F if K is itself a field with respect to the same binary operations as that of F.

Example 4.36 The set of rational numbers, $(\mathbb{Q}, +, \cdot)$ is a subfield of $(\mathbb{R}, +, \cdot)$; the set of real numbers, $(\mathbb{R}, +, \cdot)$ is a subfield of $(\mathbb{C}, +, \cdot)$; $\mathbb{Q}(\sqrt{p})$, p prime, is a subfield of $\mathbb{R}$.

Theorem 4.37 (Subfield Test) *Let F be a field and let K be a subset of F with at least two elements. Then, K is a subfield of F if, for any a, b $(b \neq 0)$ in K, $a - b$ and ab^{-1} belong to K.*

Proof Suppose that S is a subfield of F. Let, $a, b \in S$.
Now, S is a subfield implies that S is a subring of F.
Thus, for any a, $b \in S$, $a - b \in S$.
Again, S is a subfield implies that $S - \{0\}$ is a multiplicative abelian group. Thus, $0 \neq b \in S \Rightarrow b^{-1} \in S$.
Therefore, for any $a \in S$, $b \in S$, $ab^{-1} \in S$.
Conversely, suppose that S is a non-empty subset of the field F satisfying the given conditions, i.e., $a - b \in S$ and $ab^{-1} \in S \ \forall \, a$, $b \in S$. Then, first condition implies that $(S, +)$ is an abelian group and second condition implies that $(S \setminus \{0\}, *)$ is a commutative group.
Lastly, distributivity of multiplication over addition holds in F. Thus, holds in S also.
Hence, $(S, +, *)$ is a subfield. $\qquad\qquad \square$

Observe that, for any element x in $\mathbb{Z}_3$, we have $3x = x \oplus_3 x \oplus_3 x = 0$. Similarly, in $\mathbb{Z}_2[i]$, we have $2x = x \oplus_2 x = 0$. This inspection prompts the following definition.

Definition 4.38 (Characteristic of a Ring) The *characteristic* of a ring R is the least positive integer n such that $nx = 0$ for all x in R. If no such integer exists, we say that R has characteristic 0. The characteristic of R is denoted by char R.

Example 4.39 Rings $\mathbb{Z}, \mathbb{Q}, \mathbb{R}$ have characteristic 0.

Example 4.40 The ring $\mathbb{Z}_n$ has characteristic n as $nx = 0$ for all x in $\mathbb{Z}_n$.

Example 4.41 The ring $\mathbb{Z}_2[x]$ has characteristic 2 as $2 \cdot p(x) = 0$ for all $p(x)$ in $\mathbb{Z}_2[x]$.

Example 4.42 The ring $\mathbb{Z}_3[i]$ has characteristic 3 as $3 \cdot x = 0$ for all x in $\mathbb{Z}_3[i]$.

Theorem 4.43 (Characteristic of a Ring with Unity) *Let R be a ring with unity 1. If 1 has infinite order under addition, then the characteristic of R is 0. If 1 has order n under addition, then the characteristic of R is n.*

Proof Let R be a ring with unity 1.

Case 1. Let the order of 1 under addition be infinite. Then by definition of order of an element, there does not exist any $n > 0$ such that $n \cdot 1 = 0$. This implies, there does not exist any $n > 0$ such that $nx = 0$ for all $x \in R$.
Then, char $R = 0$.

Case 2. Let the order of 1 under addition be n. That is, n is the least positive integer such that $n \cdot 1 = 0$.
Let $x \in R$. Consider, $nx = (n \cdot 1)x = 0 \cdot x = 0 \Rightarrow nx = 0$ for all $x \in R$.
Let $0 < m < n$ such that $mx = 0$ for all $x \in R$.
In particular, $m \cdot 1 = 0$ which is a contradiction to $O(1) = n$. Thus, there does not exist $m < n$ such that $mx = 0$ for all $x \in R$.

Thus, n is the least positive integer such that $nx = 0$ for all $x \in R$. Hence, char $R = n$. $\quad \square$

Theorem 4.44 (Characteristic of an Integral Domain) *The characteristic of an integral domain is either 0 or prime.*

Proof Let, D be an integral domain and $1 \in D$.

Case 1. $O(1) = \infty$ in $(D, +)$. Then, there does not exist any n such that $n \cdot 1 = 0$ which implies that there does not exist any n such that $n \cdot x = 0 \ \forall \ x \in D$. Hence, char $D = 0$.

Case 2. $O(1) = n$ in $(D, +)$. By theorem "characteristic of a ring with unity", we have that char $D = n$.

Claim: n is prime.

Let, if possible, $n = st$ where $1 < s, t < n$. Since, $O(1) = n$, then n is the least positive integer such that $n \cdot 1 = 0 \Rightarrow (st) \cdot 1 = 0 \Rightarrow (s \cdot 1) \cdot (t \cdot 1) = 0 \Rightarrow$ either $s \cdot 1 = 0$ or $t \cdot 1 = 0$
$$(\because D \text{ is an integral domain and } s \cdot 1, \ t \cdot 1 \in D.$$
This is a contradiction to the fact that n is the least positive integer such that $n \cdot 1 = 0$. Hence, our assumption was wrong and there does not exist any s, t such that $n = st$, where $1 < s, \ t < n$. Therefore, n is prime.

Thus, the characteristic of an integral domain is either zero or a prime. $\quad\square$

Theorem 4.45 *Any finite field has order p^k, where p is a prime.*

Proof Let F be any finite field. Then, F is an integral domain which implies that characteristic of F is either 0 or a prime. But, characteristic of F is not 0 because F is finite. This means that characteristic of $F = p$, where p is a prime.

Now, $1 \in F$ and $O(1) = p$ implies that p is the least positive integer such that $p \cdot 1 = 0 \Rightarrow px = 0 \ \forall \ x \in F$. Now, since $O(1) = p$ and $1 \in F$, then by Lagrange's theorem $O(1)$ divides $O(F) \Rightarrow p$ divides $O(F)$ $\hfill (4.5)$

Claim: No other prime divides $|F|$.

Let, if possible, $q \neq p$ be a prime that divides $|F|$.

Since, $(F, +)$ is an abelian group, so $q | O(F)$. By, Cauchy theorem, there exists $a \in F$ such that $O(a) = q \Rightarrow q \cdot a = 0$, also, we have characteristic of $F = p \Rightarrow p \cdot a = 0$. $\hfill (4.6)$

By Equation (4.6), we can conclude that $q | p$ which is a contradiction since $q \neq p$ and both are prime.

Thus, there is no prime other than p that divides $|F|$. Hence, $|F| = p^k$ for some positive integer k. $\quad\square$

Through the following lemma, we establish that every nonzero element has the same additive order in a commutative ring with no zero divisors.

Lemma 4.46 Suppose that R is a commutative ring without zero divisors. Then, all the nonzero elements of R have the same additive order.

Proof Let R be a commutative ring without zero divisors. Let $a \neq 0, b \neq 0 \in R$. Let $O(a) = m$ and $O(b) = n$. We need to show that $m = n$.

Since, $O(a) = m$ therefore, m is the least positive integer such that $ma = 0$ and $O(b) = n$ thus, n is the least positive integer such that $nb = 0$.

If $m < n$, consider $m(ab) = a(mb) = (ma)b \Rightarrow a(mb) = 0$ $\qquad$ ($\because ma = 0 \Rightarrow (ma)b = 0$.)

$\Rightarrow mb = 0$. $\qquad\qquad$ ($\because R$ has no zero divisors and $a \neq 0$.)

This is a contradiction to the fact that n is the least positive integer such that $nb = 0$. Similarly, we can show that $n \not< m \Rightarrow m = n$. Hence, all the nonzero elements have the same order. $\qquad\qquad\square$

Theorem 4.47 *Suppose that R is a commutative ring without zero divisors. Then, the characteristic of R is 0 or prime.*

Proof Let R be a commutative ring without zero divisors. We need to show that characteristic of R is either 0 or a prime.

Case 1. Let the characteristic of R be 0. Then, nothing to prove.

Case 2. Let the characteristic of R be nonzero. Since R is a commutative ring with no zero divisors so by the above Lemma 4.46, the additive order of all the nonzero elements are the same. Let $O(x) = n$ for all $x(\neq 0) \in R$. Thus, we have $nx = 0 \ \forall \ x \in R$.

Then, n is the least positive integer such that $nx = 0 \ \forall \ x \in R$. Thus, characteristic of $R = n$. Now, we need to show that n is prime. Let, if possible, $n = s \cdot t$, where $1 < s, t < n$.

Let $x(\neq 0) \in R$ such that $O(x) = n$. Also, $x \neq 0 \Rightarrow x \cdot x \neq 0 \in R$. ($\because R$ has no zero divisors.) Since, characteristic of $R = n$ and $x \cdot x \neq 0 \in R \Rightarrow n \cdot (x \cdot x) = 0$ $\Rightarrow (s \cdot t)(x \cdot x) = 0 \Rightarrow (sx)(tx) = 0 \Rightarrow$ either $sx = 0$ or $tx = 0$ which is a contradiction to $O(x) = n$ since $1 < s, t < n$. Therefore, n is prime.

$\qquad\qquad\square$

Theorem 4.48 *Let R be a ring with m elements. Then, the characteristic of R divides m.*

Proof Let, R be a ring with m elements, i.e., $|R| = m$. Let characteristic of $R = n$. We need to show that $n|m$.

Since $(R, +)$ is an abelian group and $|R| = m$. Thus, by Lagrange's theorem,

$$O(x)|m \ \ \forall \ x \in R \Rightarrow mx = 0 \text{ for every } x \in R \qquad (4.7)$$

Since, characteristic of $R = n$. That is, n is the least positive integer such that

$$nx = 0 \ \ \forall \ x \in R \qquad (4.8)$$

Thus, by Division Algorithm, there exist integers q and r such that $m = nq + r$ with $0 \leq r < n$. Multiplying both the sides by x, we obtain $mx = (nq + r)x$ for all $x \in R$, that is $mx = nqx + rx \ \forall \ x \in R$. Hence $0 = rx$ $\qquad$ (using (4.7) and (4.8))

$\Rightarrow rx = 0 \ \forall \ x \in R$.

Since, n is the least positive integer such that $nx = 0 \ \forall \ x \in R$, thus we have $r = 0 \Rightarrow m = nq \Rightarrow n|m$. Hence, characteristic of R divides m. $\qquad\square$

Illustrative Problems

Problem 1 Suppose that a and b belong to a commutative ring and ab is a zero divisor. Show that either a or b is a zero divisor.

Solution Let R be a commutative ring and $a, b \in R$. Since, ab is a zero divisor, this implies that there exists $c \neq 0$, $c \in R$ such that $(ab)c = 0$. If $ac = 0$, then, a is a zero divisor. If $ac \neq 0$, then $(ac)b = 0$. Hence, b is a zero divisor.

Problem 2 Let a be a nonzero element of R. Prove that a is a zero divisor of a commutative ring R if and only if $a^2 b = 0$ for some $b \neq 0$.

Solution Suppose that a is a zero divisor and let $ab = 0$ for some $b \neq 0$. Then, $a^2 b = a(ab) = 0$.

   Conversely, assume that $a^2 b = 0$ for some $b \neq 0$. If $ab = 0$, then a is a zero divisor. If $ab \neq 0$, then $a^2 b = a(ab) = 0$, which implies that a is a zero divisor.

Problem 3 Describe all zero divisors and units of $\mathbb{Z} \oplus \mathbb{Q} \oplus \mathbb{Z}$.

Solution Take x, y, z such that $x, z \in \mathbb{Z}$ and $y \in \mathbb{Q}$. The set of zero divisors is $\{(x, y, z) \in \mathbb{Z} \oplus \mathbb{Q} \oplus \mathbb{Z} \mid \text{either } x = 0,\ y = 0 \text{ or } z = 0\}$.
The only two units of $\mathbb{Z}$ are $-1, 1$ while in $\mathbb{Q}$, all nonzero elements are units. The set of units of $\mathbb{Z} \oplus \mathbb{Q} \oplus \mathbb{Z}$ is $\{(x, y, z) \in \mathbb{Z} \oplus \mathbb{Q} \oplus \mathbb{Z} \mid x,\ z \in \{-1, 1\},\ y \neq 0\}$.

Problem 4 Show that every division ring is an integral domain.

Solution Let $ab = 0$ in the division ring R; we must show that either $a = 0$ or $b = 0$. But if $a \neq 0$, then a^{-1} exists, so $b = 1 \cdot b = a^{-1}ab = a^{-1}0 = 0$. Thus, R is an integral domain.

Problem 5 Find two elements a and b in a ring such that both a and b are zero divisors, $a + b \neq 0$, and $a + b$ is not a zero divisor.

Solution In ring $\mathbb{Z}_6$ take $a = 2$ and $b = 3$. Then, 2 and 3 both are zero divisors but $2 + 3 = 5$ is not a zero divisor.

Problem 6 Let a belong to a ring R with unity and suppose that $a^n = 0$ for some positive integer n. (Such an element is called *nilpotent*.) Prove that $1 - a$ has a multiplicative inverse in R.

Solution Since $1, a \in R \Rightarrow 1 + a + a^2 + \cdots + a^{n-1} \in R$.
Take $x = 1 + a + a^2 + \cdots + a^{n-1} \in R$.
Consider, $(1 - a)(1 + a + a^2 + \cdots + a^{n-1}) = 1 - a^n = 1 \ (\because a^n = 0.)$
Now, $x(1 - a) = (1 + a + a^2 + \cdots + a^{n-1})(1 - a) = 1 - a^n = 1$
Thus, $(1 - a)x = x(1 - a) = 1$ implies $1 - a$ has a multiplicative inverse in R.

Problem 7 Show that the nilpotent elements of a commutative ring form a subring.

Solution $S = \{a \in R \mid a^n = 0 \text{ for some } n \in \mathbb{Z}^+\}$.
Since $0^1 = 0$, thus $0 \in S \Rightarrow S \neq \emptyset$.

Let a, $b \in S$, thus, we have $a^n = 0$ and $b^m = 0$ for some n, $m \in \mathbb{Z}^+$. As R is commutative, we have

$$(a-b)^{m+n} = {}^{m+n}C_0 \, a^{m+n} + {}^{m+n}C_1 \, a^{m+n-1} \, b + \cdots + {}^{m+n}C_{m+n} \, a^0 \, b^{m+n} = 0$$

because $a^n = 0$ and $b^m = 0$. Thus, $a - b \in S$.
Also, $(ab)^{mn} = a^{mn} b^{mn} = (a^n)^m \cdot (b^m)^n = 0$. Thus, $ab \in S$. Hence, S is a subring.

Problem 8 Show that 0 is the only nilpotent element in an integral domain.

Solution Let, R be an integral domain. Let, $a \in R$ be nilpotent element i.e., $a^n = 0$, where n is the least such positive integer. Then $a^{n-1} \, a = 0$, we have either $a^{n-1} = 0$ or $a = 0$. But $a^{n-1} \neq 0$ ($\because n$ is the least positive integer such that $a^n = 0$)
Hence, $a = 0$.

Problem 9 A ring element a is called an idempotent if $a^2 = a$. Prove that the only idempotents in an integral domain are 0 and 1.

Solution Let, R be an integral domain. Suppose an element $a \in R$ is an idempotent. Therefore, we have $a^2 = a$, that is $a^2 - a = 0 \Rightarrow a(a-1) = 0$. Hence $a = 0$ or $a - 1 = 0$. Thus, we have $a = 0$ or $a = 1$. Thus, only idempotents in an integral domain are 0 and 1.

Problem 10 Let a and b be idempotents in a commutative ring. Show that each of the following is also an idempotent: ab, $a - ab$, $a + b - ab$, $a + b - 2ab$.

Solution Since a and b are idempotents, thus $a^2 = a$ and $b^2 = b$.

(i) $(ab)^2 = (a^2 b^2) = ab$ Therefore, ab is an idempotent.
(ii) $(a - ab)^2 = a^2 + a^2 b^2 - 2a^2 b = a + ab - 2ab = a - ab$ Thus $a - ab$ is an idempotent.
(iii) $(a + b - ab)^2 = (a + b - ab)(a + b - ab) = a^2 + ab - a^2 b + ba + b^2 - ab^2 - a^2 b - ab^2 + a^2 b^2 = a + ba + b - ab - ab = a + ab + b - ab - ab = a + b - ab \Rightarrow a + b - ab$ is an idempotent.
(iv) Consider,
$$(a + b - 2ab)^2 = (a + b - 2ab)(a + b - 2ab)$$
$$= a^2 + ab - 2a^2 b + ab + b^2 - 2ab^2 - 2a^2 b - 2ab^2 + 4a^2 b^2$$
$$= a + ab - 2ab + ab + b - 2ab - 2ab - 2ab + 4ab$$
$$= a + b + 6ab - 8ab$$
$$= a + b - 2ab.$$
Hence $a + b - 2ab$ is an idempotent.

Problem 11 Show that $\mathbb{Z}_n$ has a nonzero nilpotent element if and only if n is divisible by the square of some prime.

Solution Suppose $\mathbb{Z}_n$ has a nonzero nilpotent element. Let, if possible, n is not divisible by square of any prime. Let $n = p_1 \cdot p_2 \cdots p_k$ where p_i's are distinct primes. Then, $\mathbb{Z}_n \cong \mathbb{Z}_{p_1} \oplus \mathbb{Z}_{p_2} \oplus \cdots \oplus \mathbb{Z}_{p_k}$.
$\mathbb{Z}_n$ has nonzero nilpotent $\Rightarrow \mathbb{Z}_{p_1} \oplus \mathbb{Z}_{p_2} \oplus \cdots \oplus \mathbb{Z}_{p_k}$ has a nonzero nilpotent. Hence $\mathbb{Z}_{p_i}$ has nonzero nilpotent, which is a contradiction.

Conversely, let n be divisible by square of a prime. Then, $n = p^2 m$. Then, $\mathbb{Z}_n \cong \mathbb{Z}_{p^2} \oplus \mathbb{Z}_m$. Then, $(p, 0)$ is a nonzero nilpotent in $\mathbb{Z}_{p^2} \oplus \mathbb{Z}_m$.

Problem 12 Prove that if a is a ring idempotent, then $a^n = a$ for all positive integers n.

Solution We will prove this using Induction.
For $n = 1$, $a^1 = a$. Thus, it is true for $n = 1$.
Now, assuming it is true for $k \in \mathbb{Z}^+$, i.e., $a^k = a$.
For $n = k + 1$, consider, $a^{k+1} = a^k \cdot a = a \cdot a = a^2 = a$ ($\because a$ is a ring idempotent)
Hence, $a^n = a$ is true for all $n \in \mathbb{Z}^+$.

Problem 13 Let R be the set of all real-valued functions defined for all real numbers under function addition and multiplication. Then

   (a) determine all zero divisors of R.
   (b) determine all nilpotent elements of R.
   (c) show that every nonzero element is a zero divisor or a unit.

Solution   (a) A function $f : R \to R$ is a zero divisor of R if and only if there is some $x_0 \in R$ such that $f(x_0) = 0$. In this case, consider

$$g(x) = \begin{cases} 0 & \text{if } x \neq x_0 \\ 1 & \text{if } x = x_0 \end{cases}$$

and notice that $fg = 0$.
   (b) $f(x) = 0$ is the only nilpotent element of R.
   (c) Suppose f is a nonzero function and is not a zero divisor. By part (a), we have $f(x) \neq 0 \ \forall \ x \in R$. Now, we will prove that f is a unit.

Consider the function $g(x) = \dfrac{1}{f(x)}$ (a well-defined real-valued function).

Then, $(fg)(x) = f(x)g(x) = \dfrac{f(x)}{f(x)} = 1$ and hence, g is a multiplicative inverse of f. Thus f is a unit.

Problem 14 Determine all elements of an integral domain that are their own inverses under multiplication.

Solution Let I be an integral domain and $a \in I$ such that $a = a^{-1}$.   (4.9)
Multiplying (4.9) by a, we obtain
$a \cdot a = a^{-1}a \Rightarrow a^2 = 1 \Rightarrow a^2 - 1 = 0 \Rightarrow (a+1)(a-1) = 0 \Rightarrow a = 1, -1$.
Hence, $1, -1$ are the elements of an integral domain I that are their own inverses under multiplication.

Problem 15 Determine all integers $n > 1$ for which $(n-1)!$ is a zero divisor in $\mathbb{Z}_n$.

Solution We know that if n is a prime then $\mathbb{Z}_n$ has no zero divisors. If n is not a prime, we may write $n = ab$ where both a and b are less than n. If $a \neq b$, then $(n-1)!$ includes both a and b among its factors therefore, we have $(n-1)! = 0$. But if $a = b$ and $a > 2$, then $(n-1)! = (a^2-1)(a^2-2)\cdots(a^2-a)\cdots(a^2-2a)\cdots 2 \cdot 1$. Since this product includes $a^2 = n$, it is 0. The only remaining case is $n = 4$ and in this case $3! = 2$ is a zero divisor.

Problem 16 Suppose that a and b belong to an integral domain such that $a^5 = b^5$ and $a^3 = b^3$. Prove that $a = b$.

Solution Suppose I is an integral domain and a and b belong to I such that $a^5 = b^5$ and $a^3 = b^3$. We need to show that $a = b$.
$a^3 = b^3$, that is $(a^3)^2 = (b^3)^2$. Hence, we have $a^6 = b^6 \Rightarrow a \cdot a^5 = b \cdot b^5 \Rightarrow a = b$. ($\because a^5 = b^5$)

Problem 17 Prove that if R is an integral domain, then so is $R[x]$.

Solution The proof is by contradiction. Let $f(x) = a_0 + a_1 x + \cdots a_m x^m$, and $g(x) = b_0 + \cdots + b_n x^n$ be the elements of $R[x]$ such that $f(x)g(x)$ is the zero polynomial. Without loss of generality assume that $a_m \neq 0, b_n \neq 0$ (i.e., $m = \deg f(x)$, $n = \deg g(x)$). Then, $f(x)g(x) = a_0 b_0 + \cdots + a_m b_n x^{m+n}$. Since, R is an integral domain, thus $a_m b_n \neq 0$. Hence $f(x)g(x)$ can't be the zero polynomial, which leads to a contradiction. Thus, we have, either $f(x) = 0$ or $g(x) = 0$. Thus, $R[x]$ has no zero divisors. Hence, $R[x]$ is an integral domain.

Problem 18 Show that the set $S = \{a + b\sqrt{2} + c\sqrt{3} \mid a, b, c \in \mathbb{Q}\}$ is not a field.

Solution The set $S = \{a + b\sqrt{2} + c\sqrt{3} \mid a, b, c \in \mathbb{Q}\}$ is not closed under multiplication. Consider, $\sqrt{6} = \sqrt{2} \cdot \sqrt{3} \notin S$. Let, if possible $\sqrt{6} = a + b\sqrt{2} + c\sqrt{3}$, we have $\sqrt{6} - c\sqrt{3} = a + b\sqrt{2}$. Squaring both sides, we obtain $(6 + 3c^2) - 6c\sqrt{2} = (a^2 + 2b^2) + 2ab\sqrt{2}$. Since $\sqrt{2}$ is irrational, this implies that $2ab = 6c$ and $6 + 3c^2 = a^2 + 2b^2$. Solving the first equation for c we get $c = ab/3$, and then substituting into the second equation, we get $18 + a^2 b^2 = 3a^2 + 6b^2$. Hence, we have $(a^2 - 6)(b^2 - 3) = 0$, which has no rational solutions for a, b. Hence, our assumption was wrong and the set S is not a field.

Problem 19 Show that a field has no zero divisors.

Solution Suppose F is a field and $a \in F$, with $a \neq 0$. Now suppose that $ab = 0$. Then, $0 = a^{-1} \cdot 0 = a^{-1}ab = b$. Thus, a is not a zero divisor.

Problem 20 For any positive integer k and any prime p, determine a necessary and sufficient condition for $\mathbb{Z}_p[\sqrt{k}] = \{a + b\sqrt{k} \mid a, b \in \mathbb{Z}_p\}$ to be a field.

Solution For any positive integer k and any prime p, we need to determine a necessary and sufficient condition for $\mathbb{Z}_p[\sqrt{k}] = \{a + b\sqrt{k} \mid a, b \in \mathbb{Z}_p\}$ to be a field. For this, firstly we will show that $\mathbb{Z}_p[\sqrt{k}]$ is an integral domain.

Case 1. If k has a square root in $\mathbb{Z}_p$, then $\mathbb{Z}_p[\sqrt{k}] = \mathbb{Z}_p$ which is a field.

Case 2. k does not have a square root in $\mathbb{Z}_p$ and $\mathbb{Z}_p[\sqrt{k}]$ is a finite commutative ring with unity.

Let, $A' = a_1' + b_1'\sqrt{k}$ and $B' = a_2' + b_2'\sqrt{k}$ for some $a_1', a_2', b_1', b_2' \in \mathbb{Z}_p$.
Let, $A' \neq 0$. Then, $A'B' = 0 \Rightarrow (a_1' + b_1'\sqrt{k}) \cdot (a_2' + b_2'\sqrt{k}) = 0$
$\Rightarrow a_1' a_2' + a_1' b_2'\sqrt{k} + a_2' b_1'\sqrt{k} + kb_1' b_2' = 0$
$\Rightarrow (a_1' a_2' + kb_1' b_2') + (a_1' b_2' + a_2' b_1')\sqrt{k} = 0$.
So, $a_1' a_2' + kb_1' b_2' = 0$ $\hspace{4cm}$ (4.10)
and $a_1' b_2' + a_2' b_1' = 0$ $\hspace{4cm}$ (4.11)

Now, applying operation $a_1' \times (4.10) - kb_1' \times (4.11)$, we get

$(a_1')^2 a_2' - k a_2' (b_1')^2 = 0 \Rightarrow a_2'((a_1')^2 - k(b_1')^2) = 0 \Rightarrow a_2' = 0$ (since, $A' \neq 0 \Rightarrow (a_1')^2 - k(b_1')^2 \neq 0$).

From (4.10) and (4.11), we obtain $b_1' b_2' = 0$ and $a_1' b_2' = 0$, which implies $b_2' = 0$ ($\because A \neq 0 \Rightarrow$ both a_1' and b_1' cannot be zero simultaneously.)

Now, since $a_2' = 0$ and $b_2' = 0 \Rightarrow B' = 0$.

Similarly, if $B' \neq 0$, then $A' = 0$.

Hence, $\mathbb{Z}_p[\sqrt{k}]$ has no zero divisors. Therefore, $\mathbb{Z}_p[\sqrt{k}]$ is an integral domain.

Now, we need to find the conditions under which an element a of $\mathbb{Z}_p[\sqrt{k}]$ has a multiplicative inverse. So, in $\mathbb{Z}_p[\sqrt{k}]$, consider

$$(a + b\sqrt{k})^{-1} = \frac{1}{a + b\sqrt{k}} = \frac{1}{a + b\sqrt{k}} \cdot \frac{a - b\sqrt{k}}{a - b\sqrt{k}} = \frac{a - b\sqrt{k}}{a^2 - kb^2}.$$

So, $\dfrac{a - b\sqrt{k}}{a^2 - kb^2}$ exists if and only if $a^2 - b^2 k \neq 0$, that is $a \neq 0$ and $b \neq 0$ simultaneously. Therefore, $\mathbb{Z}_p[\sqrt{k}]$ is a field when $a, b \neq 0$ simultaneously, i.e., the necessary and the sufficient condition for $\mathbb{Z}_p[\sqrt{k}] = \{a + b\sqrt{k} \mid a, b \in \mathbb{Z}\}$ to be a field is that $a^2 - b^2 k \neq 0$, that is, $a \neq 0$ and $b \neq 0$ simultaneously for some $k > 0$ and prime p.

Problem 21 If D is an integral domain with $na = 0$ for some $a \neq 0$ in D and some integer $n \neq 0$. Prove that D is of finite characteristic.

Solution Let $b \in D$. Consider $nab = (na)b = 0$. Thus, we have $nab = 0$. On the other hand, $nab = nba$ as integral domain is a commutative ring. Therefore, $0 = nab = nba$. But $a \neq 0$ implies $nb = 0$ for all $b \in D$.

Problem 22 If D is an integral domain with finite characteristic. Then prove that the characteristic of D is a prime number.

Solution Let a be any nonzero element of D. Then $a^2 \neq 0$ as D is an integral domain. Let D be of positive characteristic q, then for all $a \neq D$, we have $qa^2 = 0$. If q is a composite number, let p_1 be a prime number dividing q and we have $q = p_1 q_1$ for some q_1. Now, $qa^2 = p_1 q_1 a^2 = p_1 a q_1 a = 0$. Since D is integral domain, then either $p_1 a = 0$ or $q_1 a = 0$. Either of these equations gives a contradiction to the assumption that q is the smallest positive integer such that $qx = 0$ for all $x \in D$. Thus q is not composite, it is a prime.

Problem 23 Let x and y belong to a commutative ring R with prime characteristic p.

    (a) Show that $(x + y)^p = x^p + y^p$.

    (b) Show that, for all positive integers n, $(x + y)^{p^n} = x^{p^n} + y^{p^n}$.

    (c) Find elements x and y in a ring of characteristic 4 such that $(x + y)^4 \neq x^4 + y^4$.

Solution   **(a)** We need to show that $(x + y)^p = x^p + y^p$. By binomial expansion, we have

$$(x + y)^p = x^p + {}^p C_1 x^{p-1} y + {}^p C_2 x^{p-2} y^2 + \cdots + {}^p C_{p-1} x y^{p-1} + y^p$$

$$= x^p + p x^{p-1} y + \frac{p(p-1)}{2} x^{p-2} y^2 + \cdots + p x y^{p-1} + y^p$$

$$= x^p + 0 + \cdots + 0 + y^p \qquad\qquad (\because \text{char}(R) = p \Rightarrow px = 0 \ \forall \ x \in D.)$$

$$= x^p + y^p.$$

    (b) We need to show that, for all positive integers n, $(x + y)^{p^n} = x^{p^n} + y^{p^n}$. We will prove this using principle of mathematical induction.

For $n = 1$, we have $(x + y)^p = x^p + y^p$ (by part a.). The result is true for $n = 1$.
Assume that the result is true for $n = k$. i.e., $(x + y)^{p^k} = x^{p^k} + y^{p^k}$.
We have to show that the result is true for $n = k + 1$. Consider,
$(x + y)^{p^{k+1}} = ((x + y)^{p^k})^p = (x^{p^k} + y^{p^k})^p = x^{p^{k+1}} + y^{p^{k+1}}$.   (Using part a.)
Thus, the result is true for $n = k + 1$.
Therefore by principle of mathematical induction, the result is true for all n. i.e.,
$(x + y)^{p^n} = x^{p^n} + y^{p^n}$.

(c) We need to find elements x and y in a ring of characteristic 4 such that $(x + y)^4 \neq x^4 + y^4$. In the ring $\mathbb{Z}_4$, consider $x = 1, y = 1$. So, $(1 + 1)^4 = 2^4 = 16 = 0$ in $\mathbb{Z}_4$ but $1^4 + 1^4 = 1 + 1 = 2 \neq 0$ in $\mathbb{Z}_4$.

Problem 24 Let R be a commutative ring with unity 1 and prime characteristic. If $a \in R$ is a nilpotent, prove that there is a positive integer k such that $(1 + a)^k = 1$.

Solution Let R be a commutative ring with unity 1 and characteristic of $R = p$, a prime.
Let, $a \in R$ be a nilpotent, therefore $a^n = 0$ for some $n > 0$.
Let $k = p^n$. Consider, $(1 + a)^k = (1 + a)^{p^n} = 1^{p^n} + a^{p^n}$ (By Problem 23)
$= 1 + 0 = 1$.

Problem 25 Explain why a finite ring must have a nonzero characteristic.

Solution Let R be a finite ring. Let, if possible, characteristic of $R = 0$ which implies that there does not exist any $n > 0$ such that $nx = 0$ for all $x \in R$. Then there does not exist any $n > 0$ for some $x \in R$ such that $nx = 0 \Rightarrow O(x)$ is infinite. By Lagrange's theorem, we have $O(x)|O(R)$ but $O(R)$ is finite which contradict the fact that $O(x)$ is infinite. Therefore, our assumption was wrong and hence characteristic of $R \neq 0$.

Problem 26 Let F be a finite field with n elements. Prove that $x^{n-1} = 1$ for all nonzero x in F.

Solution Let F be a finite field with n elements, i.e., $O(F) = n$.
Let, $F^* = F \backslash \{0\} \Rightarrow |F^*| = n - 1$. Since, F is a field. So, $(F^*, \cdot)$ is an abelian group with unity 1. Let $x \in F^*$. Then, by Lagrange's theorem, we have
$O(x)|O(F^*) \Rightarrow O(x)|(n - 1) \Rightarrow x^{n-1} = 1$ for all $x (\neq 0) \in F$.

Problem 27 Let F be a field of order 2^n. Prove that char $F = 2$.

Solution Since every field is an integral domain. Therefore, F is an integral domain, and it is known that the characteristic of an integral domain is either 0 or prime. Since, $O(F)$ is finite, thus char $F \neq 0 \Rightarrow$ char $F = p$ (prime).
Hence, $O(1) = p$. Then, by Lagrange's theorem, $O(1)|O(F) \Rightarrow p|2^n \Rightarrow p = 2 \Rightarrow$ char $F = 2$.

Exercise

(1) Show that a commutative ring D is an integral domain if and only if for $a, b, c \in D$ with $a \neq 0$ the relation $ab = ac$ implies that $b = c$.

(2) Find a nonzero element in a ring that is neither a zero divisor nor a unit. List all zero divisors in $\mathbb{Z}_{20}$. Can you see a relationship between the zero divisors of $\mathbb{Z}_{20}$ and the units of $\mathbb{Z}_{20}$?

(3) Find elements a, b, and c in the ring $\mathbb{Z} \oplus \mathbb{Z} \oplus \mathbb{Z}$ such that ab, ac, and bc are zero divisors but abc is not a zero divisor.

(4) Give an example of a commutative ring without zero divisors that is not an integral domain.

(5) Suppose that a and b belong to an integral domain. If $a^m = b^m$ and $a^n = b^n$, where m and n are positive integers that are relatively prime, prove that $a = b$.

(6) Let R be the ring of real-valued continuous functions on $[-1, 1]$. Show that R has zero divisors.

(7) Find a zero divisor in $\mathbb{Z}_5[i] = \{a + bi \mid a,\ b \in \mathbb{Z}_5\}$.

(8) Find an idempotent in $\mathbb{Z}_5[i] = \{a + bi \mid a,\ b \in \mathbb{Z}_5\}$.

(9) Find all units, zero divisors, idempotents, and nilpotent elements in $\mathbb{Z}_3 \oplus \mathbb{Z}_6$.

(10) Show that $\mathbb{Z}(\sqrt{2}) = \{n + m\sqrt{2} \mid n, m \in \mathbb{Z}\}$ is a subring of $\mathbb{C}$ and find any four units (in fact there are infinitely many).

(11) Prove that there is no integral domain with exactly six elements. Can your argument be adapted to show that there is no integral domain with exactly four elements? What about 15 elements? Use these observations to guess a general result about the number of elements in a finite integral domain.

(12) Find an example of an integral domain and distinct positive integers m and n such that $a^m = b^m$ and $a^n = b^n$, but $a \neq b$.

(13) Show that a finite commutative ring with no zero divisors and having at least two elements, has a unity.

(14) Suppose that D is an integral domain and that ϕ is a nonconstant function from D to the nonnegative integers such that $\phi(xy) = \phi(x)\phi(y)$. If x is a unit in D, show that $\phi(x) = 1$.

(15) Let $R = \{0, 2, 4, 6, 8\}$ under addition and multiplication modulo 10. Prove that R is a field.

(16) Show that $\mathbb{Z}_7[\sqrt{3}] = \{a + b\sqrt{3} \mid a, b \in \mathbb{Z}_7\}$ is a field.

(17) Describe the smallest subfield of the field of real numbers that contains $\sqrt{2}$.

(18) Let R be a finite commutative ring with more than one element and no zero divisors. Show that R is a field.

(19) Let R be the field of real numbers and let F be the set of all 2×2 matrices of the form $\begin{bmatrix} a & b \\ -3b & a \end{bmatrix}$, where $a, b \in R$. Show that F is a field under the usual matrix operations.

(20) Let R be a commutative ring with identity. Show that R is an integral domain if and only if R is a subring of a field.

(21) Suppose that F is a field with 27 elements. Show that for every element $a \in F$, $5a = -a$.

(22) Let F be a field of order 32. Show that the only subfields of F are F itself and $\{0, 1\}$.

(23) Find a finite field in which $a^2 + b^2 = 0$ implies that $a = b = 0$, and find another in which this is not true.

(24) Show that $\{a + b\sqrt{3} \mid a, b \in \mathbb{Q}\}$ and $\{a + b\sqrt{2} + c\sqrt{5} + d\sqrt{10} \mid a, b, c, d \in \mathbb{Q}$ are subfields of R.

(25) Show that $\{a + b\sqrt{2} \mid a, b \in \mathbb{Q}\}$ is not a subfield of R.

(26) Let $R = \left\{ \begin{bmatrix} a & -b \\ b & a \end{bmatrix} \,\middle|\, a, b \in \mathbb{Z}_7 \right\}$ with the usual matrix addition and multiplication. Prove that R is a commutative ring. How many elements are in R? Is R a field? What happens when $\mathbb{Z}_7$ is replaced by $\mathbb{Z}_5$?

(27) Characterize those integral domains for which 1 is the only element that is its own multiplicative inverse.

(28) If D is an integral domain and if $na = 0$ for some $a \neq 0$ in D and some integer $n \neq 0$, prove that D is of finite characteristic.

(29) Let F be a field of characteristic 2 with more than two elements. Show that $(x + y)^3 \neq x^3 + y^3$ for some x and y in F.

(30) Let R be a ring and let $M_2(R)$ be the ring of 2×2 matrices with entries from R. Explain why these two rings have the same characteristics.

(31) Find the characteristic of $\mathbb{Z}_4 \oplus 4\mathbb{Z}$.

(32) In a commutative ring of characteristic 2, prove that the idempotents form a subring.

5 IDEALS AND FACTOR RINGS

Ideals, in modern algebra, are subrings of a mathematical ring with certain absorption properties. The concept of an ideal was first defined and developed by German mathematician *Richard Dedekind* in 1871. In particular, he used ideals to translate ordinary properties of arithmetic into properties of sets.

The origin of the notion of ideals in a ring lies in the idea of "ideal numbers", numbers which are missing but are really ought to be there. *Ernst Kummer* invented the concept of ideal numbers to serve as the "missing" factors in number rings in which unique factorization fails; here the word "ideal" is in the sense of existing in imagination only.

In this chapter, the abstraction of ideals is explored through various examples. The study is examined through various problems to enable students to apprehend the notion of the ideals.

The ring of integers $(\mathbb{Z}, +, \cdot)$ has subrings of the form $n\mathbb{Z}$, for $n \in \mathbb{Z}$. In particular $2\mathbb{Z}$ is a subring of $\mathbb{Z}$ and for any $k \in \mathbb{Z}, 2m \in 2\mathbb{Z}$, the element $k(2m) = 2(km) \in 2\mathbb{Z}$. Also, $(2m) \cdot k = 2(mk) \in 2\mathbb{Z}$ so we can say that $2\mathbb{Z}$ absorbs the elements of $\mathbb{Z}$ under multiplication, i.e., $2\mathbb{Z}$ is an ideal of $\mathbb{Z}$. In fact $2\mathbb{Z}$ is a two sided ideal of $\mathbb{Z}$. In general, suppose that $(R, +, \cdot)$ is a ring. Then, a subset U of R is said to be a left ideal of a ring R if it is an additive subgroup of $(R, +)$ that absorbs multiplication from the left by elements of R. Thus, we define left ideal of a ring as follows.

Definition 5.1 (Left Ideal) A non-empty subset U of a ring R is called a *left ideal* if

(i) $a - b \in U$ for all $a, b \in U$.
(ii) $a \in U, r \in R \Rightarrow ra \in U$.

Example 5.2 For $a \in R$ let $Ra = \{xa \mid x \in R\}$. Then, Ra is a left ideal of R. Since, $0 = 0.a \in Ra$, thus Ra is non-empty. Suppose $x, y \in Ra$, where $x = r_1 a$ and $y = r_2 a$ for some $r_1, r_2 \in R$. Then, we have $x - y = r_1 a - r_2 a = (r_1 - r_2)a = r_3 a$, where $r_3 \in R$. Thus $x - y \in Ra$.

Now, for some $x \in Ra$ and $r \in R$, where $x = r_1 a$ where $r_1 \in R$, we have $rx = r(r_1 a) = (rr_1)a = r_4 a$ for some $r_4 \in R$. Thus $rx \in Ra \,\forall\, x \in Ra$ and $r \in R$. Hence, Ra is a left ideal of R.

Example 5.3 In the ring M of 2×2 matrices over integers, the set

$$U = \left\{ \begin{pmatrix} a & 0 \\ b & 0 \end{pmatrix} \;\middle|\; a, b \in \mathbb{Z} \right\} \text{ is a left ideal of } M.$$

Since, $\begin{pmatrix} 0 & 0 \\ 0 & 0 \end{pmatrix} \in U$, therefore, $U \neq \emptyset$.

Let $A = \begin{pmatrix} a & 0 \\ b & 0 \end{pmatrix}$, $B = \begin{pmatrix} c & 0 \\ d & 0 \end{pmatrix} \in U$, where $a, b, c, d \in \mathbb{Z}$. Then,

$$A - B = \begin{pmatrix} a-c & 0 \\ b-d & 0 \end{pmatrix} \in U. \hspace{3cm} (\because a-c, b-d \in \mathbb{Z})$$

Also, for $P = \begin{pmatrix} \alpha & \beta \\ \gamma & \delta \end{pmatrix} \in M$ and $A = \begin{pmatrix} a & 0 \\ b & 0 \end{pmatrix} \in U$, where $\alpha, \beta, \gamma, \delta, a, b \in \mathbb{Z}$, we have

$$PA = \begin{pmatrix} \alpha & \beta \\ \gamma & \delta \end{pmatrix} \begin{pmatrix} a & 0 \\ b & 0 \end{pmatrix} = \begin{pmatrix} \alpha a + \beta b & 0 \\ \gamma a + \delta b & 0 \end{pmatrix} \in U. \hspace{1cm} (\because \alpha a + \beta b, \gamma a + \delta b \in \mathbb{Z})$$

Hence, U is a left ideal of M.

Similarly, a subset U of a ring R is said to be a *right ideal* of R if it is an additive subgroup of $(R, +)$ that absorbs multiplication from the right by elements of R. Thus, right ideal of a ring is defined as follows.

Definition 5.4 (Right Ideal) A non-empty subset U of a ring R is called a *right ideal* if

    (i) $a - b \in U$ for all $a, b \in U$.
    (ii) $a \in U, r \in R \Rightarrow ar \in U$.

Example 5.5 For $a \in R$ let $aR = \{ax \mid x \in R\}$. Then, aR is a right ideal of R. Since, $0 \in aR$, thus aR is non-empty. Suppose $x, y \in aR$, where $x = ar_1$ and $y = ar_2$ for some $r_1, r_2 \in R$. Then, we have $x - y = ar_1 - ar_2 = a(r_1 - r_2) = ar_3$ where $r_3 \in R$. Thus $x - y \in aR$.
Now, for some $x \in aR$ and $r \in R$, where $x = ar_1$ for some $r_1 \in R$, we have $xr = (ar_1)r = a(r_1 r) = ar_4$, where $r_4 \in R$. Thus $xr \in aR$ and hence, aR is a right ideal of R.

Example 5.6 In the ring M of 2×2 matrices over integers, the set

$$U = \left\{ \begin{pmatrix} a & b \\ 0 & 0 \end{pmatrix} \;\middle|\; a, b \in \mathbb{Z} \right\} \text{ is a right ideal of } M.$$

Since, $\begin{pmatrix} 0 & 0 \\ 0 & 0 \end{pmatrix} \in U$, therefore $U \neq \emptyset$.

Let $A = \begin{pmatrix} a & b \\ 0 & 0 \end{pmatrix}$, $B = \begin{pmatrix} c & d \\ 0 & 0 \end{pmatrix} \in U$, where $a, b, c, d \in \mathbb{Z}$, then

$$A - B = \begin{pmatrix} a - c & b - d \\ 0 & 0 \end{pmatrix} \in U. \qquad\qquad (\because a - c, b - d \in \mathbb{Z})$$

Also, for $P = \begin{pmatrix} \alpha & \beta \\ \gamma & \delta \end{pmatrix} \in M$ and $A = \begin{pmatrix} a & b \\ 0 & 0 \end{pmatrix} \in U$, where $\alpha, \beta, \gamma, \delta, a, b \in \mathbb{Z}$, we have

$$AP = \begin{pmatrix} a & b \\ 0 & 0 \end{pmatrix}\begin{pmatrix} \alpha & \beta \\ \gamma & \delta \end{pmatrix} = \begin{pmatrix} a\alpha + b\gamma & a\beta + b\delta \\ 0 & 0 \end{pmatrix} \in U. \; (\because a\alpha + b\gamma, a\beta + b\delta \in \mathbb{Z})$$

Hence, U is a right ideal of M.

Accordingly, a two-sided ideal is a left ideal that is also a right ideal, and is sometimes simply called an ideal which is mentioned in the next definition.

Definition 5.7 (Ideal) A non-empty subset U of a ring R is called an *ideal* (*two-sided ideal*) of R if

(i) $a, b \in U \Rightarrow a - b \in U$, and
(ii) $a \in U, r \in R \Rightarrow ar \in U$ and $ra \in U$.

Remark An ideal U is said to be a *proper ideal* of R if U is a proper subset of R.

Example 5.8 For any ring R, $\{0\}$ and R both are ideals of R and are known as *trivial ideals*.

Example 5.9 For any positive integer n, the set $n\mathbb{Z} = \{0, \pm n, \pm 2n, ...\}$ is an ideal of $\mathbb{Z}$.
Let, $I = n\mathbb{Z}$. Clearly, $0 = 0 \cdot n \Rightarrow 0 \in n\mathbb{Z} = I$, thus $0 \in I \Rightarrow I \neq \emptyset$.
Let, a, b be arbitrary elements of I such that $a = nx$ and $b = ny$ for some $x, y \in \mathbb{Z}$. Then, $a - b = nx - ny = n(x - y) \in n\mathbb{Z} = I$.
Also, for arbitrary element $a \in I$ such that $a = nx$ for some $x \in \mathbb{Z}$ and $k \in \mathbb{Z}$, we have $ka = ak = (nx)k = n(xk) \in n\mathbb{Z} = I$. Hence, $n\mathbb{Z}$ is an ideal of $\mathbb{Z}$.

The definition of an ideal implies every ideal is a subring. However, the converse need not be true, that is, a subring of a ring R need not necessarily be an ideal of R. The ring $\mathbb{Z} \oplus \mathbb{Z}$ has a subring $S = \{(x, x) \mid x \in \mathbb{Z}\}$ which is not an ideal of $\mathbb{Z} \oplus \mathbb{Z}$. We have $\mathbb{Z} \oplus \mathbb{Z} = \{(m, n) \mid m, n \in \mathbb{Z}\}$. Let $S = \{(x, x) \mid x \in \mathbb{Z}\}$. Let, $x_1, x_2 \in \mathbb{Z}$. Then, $(x_1, x_1) - (x_2, x_2) = (x_1 - x_2, x_1 - x_2) \in S \; \forall \, x_1, x_2 \in \mathbb{Z}$. Also, we have $(x_1, x_1) \cdot (x_2, x_2) = (x_1 x_2, x_1 x_2) \in S \; \forall \, x_1, x_2 \in \mathbb{Z}$.
Hence, S is a subring. But S is not an ideal. Since $(1, 1) \in S$ and $(1, 2) \in \mathbb{Z} \oplus \mathbb{Z}$, however $(1, 2)(1, 1) = (1, 2) \notin S$. Hence, S is a subring of $\mathbb{Z} \oplus \mathbb{Z}$ but not an ideal of $\mathbb{Z} \oplus \mathbb{Z}$.

Suppose R is a ring with unity then any ideal of R that contains unity of R is the ring itself, which is proved in the following proposition.

Proposition 5.10 *If R is a ring with unity and U is an ideal of R such that $1 \in U$, then $U = R$.*

Proof Let, R be a ring with unity and U be an ideal of R containing the unity. Since, U is an ideal of R, thus $U \subseteq R$ $\qquad\qquad(5.1)$
Let $x \in R$. Now $x \in R, 1 \in U \Rightarrow x \cdot 1 = x \in U$. $\qquad$ (since U is an ideal of R)
Therefore, $R \subseteq U$. $\qquad\qquad(5.2)$
From Equations (5.1) and (5.2), we have $U = R$. $\qquad\qquad\square$

One can observe that the ring of reals, $\mathbb{R}$ is a field and the only ideals of $\mathbb{R}$ are $\{0\}$ and $\mathbb{R}$ itself. This observation is generalised for all the fields which is stated in the following theorem.

Theorem 5.11 *A field has no proper ideals.*

Proof Let F be a field and U be an ideal of R. Then we will prove that either $U = \{0\}$ or $U = F$.

From the definition of an ideal, we have $U \subseteq F$.   (5.3)

Let if possible, $U \neq \{0\}$, thus there exists some $a(\neq 0) \in U$.

Therefore, $a \in F$ $(\because U \in F)$

$\Rightarrow a^{-1} \in F$. $(\because F$ is a field$)$

Now $a \in U, a^{-1} \in F$, hence $aa^{-1} = 1 \in U$.

Let $x \in F$ and, $1 \in U \Rightarrow x \cdot 1 \in U$ $(\because U$ is an ideal$)$

$\Rightarrow x \in U$. Thus, $F \subseteq U$ (5.4)

From (5.3) and (5.4), we have $U = F$.

Therefore, $\{0\}$ and F are the only ideals of F. $\square$

Note that for any ring R, $N(R)$, the set of nilpotent elements of R, is not necessarily an ideal of R since it does not form a subring in general as the class of nilpotent elements of R need not be closed. For example, consider the ring $R = M_2(\mathbb{C})$, and let $A = \begin{pmatrix} 0 & 1 \\ 0 & 0 \end{pmatrix}$

and $B = \begin{pmatrix} 0 & 0 \\ 1 & 0 \end{pmatrix}$. Then $A^2 = B^2 = \mathbf{0}$. On the other hand, consider

$A + B = \begin{pmatrix} 0 & 1 \\ 1 & 0 \end{pmatrix} \neq \begin{pmatrix} 0 & 0 \\ 0 & 0 \end{pmatrix}$ here, we have $(A + B)^2 = I_2$, thus $(A + B)^n \neq O$ for all

$n \in \mathbb{N}$. So $A + B \notin N(R)$.

Next theorem gives the sufficient condition under which the set of nilpotent elements of a ring R is an ideal of R.

Theorem 5.12 *Let R be a commutative ring. Then, $N(R)$ is an ideal of R.*

Proof Let $a, b \in N(R)$ where $a^k = 0$ and $b^n = 0$ for some $n, k \in \mathbb{N}$. Then

$$(ab)^{nk} = a^{nk} b^{nk} = 0.$$

Similarly, $(ra)^k = r^k a^k = 0$, since R is commutative. It remains to check that $N(R)$ is closed under addition. By the binomial theorem,

$$(a + b)^{k+n} = {}^{k+n}C_0 a^{k+n} + {}^{k+n}C_1 a^{k+n-1} b + \dots + {}^{k+n}C_n a^k b^n + {}^{k+n}C_{n+1} a^{k-1} b^{n+1} + \dots +$$
$${}^{k+n}C_{n+k-1} ab^{n+k-1} + {}^{k+n}C_{n+k} b^{n+k}$$
$$= {}^{k+n}C_0 a^k \cdot a^n + {}^{k+n}C_1 a^k \cdot a^{n-1} b + \dots + {}^{k+n}C_n a^k b^n + {}^{k+n}C_{n+1} a^{k-1} b^n \cdot b + \dots +$$
$${}^{k+n}C_{k+n-1} ab^n \cdot b^{k-1} + {}^{k+n}C_{n+k} b^n \cdot b^k$$
$$= 0 \qquad\qquad (\because a^k = 0 \text{ and } b^n = 0 \text{ for some } n, k \in \mathbb{N})$$

This implies that $a + b$ belongs to $N(R)$. Hence, $N(R)$ is an ideal of R. $\square$

For any positive integer n, the set $n\mathbb{Z} = \{0, \pm n, \pm 2n, ...\}$ is a proper ideal of $\mathbb{Z}$. Also, we can notice that $\mathbb{Z}$ is a commutative ring but not a field. It is possible to classify any commutative ring as a field depending upon the existence of proper ideal. A commutative ring with unity is a field if it has no proper ideal, which gives us a sufficient condition for a ring to be field that is proved in the succeeding theorem.

Theorem 5.13 *A commutative ring with unity is a field if it has no proper ideal.*

Proof Let R be a commutative ring with unity such that R has no proper ideal. In order to prove that every nonzero element in R has a multiplicative inverse, let $a(\neq 0) \in R$, then the set $Ra = \{ra : r \in R\}$ is an ideal of R.

By the given hypothesis, we have, $1 \in R$. Therefore $1 \cdot a = a \in Ra$ i.e., $a(\neq 0) \in Ra$, thus Ra is not a zero ideal.

As R has no proper ideal and $Ra \neq \{0\}$, therefore, we have $Ra = R$.

Now, $1 \in R \Rightarrow 1 \in Ra$ which implies that there exists an element $b \in R$ such that $1 = ba$. As R is commutative, we have $1 = ba = ab$. Thus, b is the inverse of a.

Thus, every nonzero element of R has a multiplicative inverse in R. Hence, R is a field. $\square$

Just like subrings, ideals can also be analyzed under the set-theoretic operations like intersection, union, sum and product of ideals.

Following theorems will define the algebra of ideals of a ring R.

Theorem 5.14 *The intersection of two ideals of a ring R is an ideal of R.*

Proof Let U_1 and U_2 be two ideals of a ring R. Then, $0 \in U_1, 0 \in U_2$.

Then, we have $0 \in U_1 \cap U_2 \Rightarrow U_1 \cap U_2 \neq \emptyset$.

Let, $a, b \in U_1 \cap U_2$, and $r \in R$. Then, $a, b \in U_1$ and $a, b \in U_2$.

Thus, we have $a - b, ar, ra \in U_1$ and, $a - b, ar, ra \in U_2$ ($\because U_1$ and U_2 are ideals of R)

Therefore, $a, b \in U_1 \cap U_2, r \in R \Rightarrow a - b, ar, ra \in U_1 \cap U_2$.

Hence, the intersection of two ideals of a ring R is an ideal of R. $\square$

The above result can be generalized to arbitrary collection of ideals of a ring R which is provided in the next theorem.

Theorem 5.15 *The arbitrary intersection of ideals of a ring is an ideal.*

Proof Suppose $I_1, I_2, ...$ be ideals of R. Let $I = \bigcap_{j \in J} I_j$.

We need to show that $I = \bigcap_{j \in J} I_j$ is an ideal.

Since, $0 \in I_1, I_2, ...$, So, it must be contained in the intersection of all ideals I_j. Thus, $0 \in \bigcap_{j \in J} I_j$. Therefore, $\bigcap_{j \in J} I_j \neq \emptyset$.

Now, let $x, y \in I$, thus $x, y \in I_j$ and I_j is an ideal. So, $x - y \in I_j \ \forall \ j \in J$. Therefore, $x - y \in \bigcap_{j \in J} I_j = I$.

Again, let $x \in I$ and $r \in R$; therefore $x \in I_j$ and $r \in R$ for all $j \in J$. So, $rx \in I_j \ \forall \ j \in J$. Therefore, $rx \in \bigcap_{j \in J} I_j = I$. Similarly, we have $xr \in \bigcap_{j \in J} I_j = I$.

Hence, $I = \bigcap_{j \in J} I_j$ is an ideal. Thus, the arbitrary intersection of ideals of a ring is again an ideal. $\qquad\square$

Note that since the intersection of ideals of a ring R is again an ideal, so intersection of left ideals will also be a left ideal and it holds true for right ideals too. However, the intersection of a left ideal and a right ideal of R need not be a left or a right ideal. We substantiate our statement with the following example.

Example 5.16 Consider the ring $M_2(\mathbb{Z}) = \left\{ \begin{pmatrix} a & b \\ c & d \end{pmatrix} \mid a, b, c, d \in \mathbb{Z} \right\}.$

We define $U_1 = \left\{ \begin{pmatrix} a & 0 \\ b & 0 \end{pmatrix} \mid a, b \in \mathbb{Z} \right\}$ and $U_2 = \left\{ \begin{pmatrix} a & b \\ 0 & 0 \end{pmatrix} \mid a, b \in \mathbb{Z} \right\}.$

Then, by Example 5.3 and Example 5.6, we know that U_1 is a left ideal of $M_2(\mathbb{Z})$ and U_2 is a right ideal of $M_2(\mathbb{Z})$. Then, $U_1 \cap U_2 = \left\{ \begin{pmatrix} a & 0 \\ 0 & 0 \end{pmatrix} \mid a \in \mathbb{Z} \right\}.$

Clearly, $U_1 \cap U_2$ is not a left ideal as for $\begin{pmatrix} 1 & 0 \\ 0 & 0 \end{pmatrix} \in U_1 \cap U_2$ and $\begin{pmatrix} 1 & 1 \\ 1 & 1 \end{pmatrix} \in M_2(\mathbb{Z})$,

we have $\begin{pmatrix} 1 & 1 \\ 1 & 1 \end{pmatrix} \begin{pmatrix} 1 & 0 \\ 0 & 0 \end{pmatrix} = \begin{pmatrix} 1 & 0 \\ 1 & 0 \end{pmatrix} \notin U_1 \cap U_2.$

Similarly $U_1 \cap U_2$ is not a right-ideal as for $\begin{pmatrix} 1 & 0 \\ 0 & 0 \end{pmatrix} \in U_1 \cap U_2$ and $\begin{pmatrix} 1 & 1 \\ 1 & 1 \end{pmatrix} \in M_2(\mathbb{Z})$, we have $\begin{pmatrix} 1 & 0 \\ 0 & 0 \end{pmatrix} \in U_1 \cap U_2 \begin{pmatrix} 1 & 1 \\ 1 & 1 \end{pmatrix} = \begin{pmatrix} 1 & 1 \\ 0 & 0 \end{pmatrix} \notin U_1 \cap U_2.$ Therefore, $U_1 \cap U_2$ is neither a left ideal nor a right ideal.

Remark Union of two ideals of a ring R need not be an ideal. For example, in the ring of integers, $\mathbb{Z}$, $2\mathbb{Z}$ and $3\mathbb{Z}$ are ideals of $\mathbb{Z}$. However, $2\mathbb{Z} \cup 3\mathbb{Z}$ is not an ideal of $\mathbb{Z}$ since for $2, 3 \in 2\mathbb{Z} \cup 3\mathbb{Z}$, but $3 - 2 = 1 \notin 2\mathbb{Z} \cup 3\mathbb{Z}$. Therefore, $2\mathbb{Z} \cup 3\mathbb{Z}$ is not an ideal of $\mathbb{Z}$.

The following theorem states the necessary and sufficient condition under which the union of ideals will be an ideal.

Theorem 5.17 *Let U_1, U_2 be ideals of a ring R. Then, union of these ideals is an ideal of R if and only if either $U_1 \subseteq U_2$ or $U_2 \subseteq U_1$.*

Proof Let $(R, +, \cdot)$ be a ring. Let U_1 and U_2 be its two ideals. If $U_1 \subseteq U_2$ or $U_2 \subseteq U_1$ then $U_1 \cup U_2 = U_2$ or $U_1 \cup U_2 = U_1$. This implies, $U_1 \cup U_2$ is an ideal.

Conversely, suppose that U_1 and U_2 are two ideals such that $U_1 \cup U_2$ is also an ideal. Now, we need to show that either $U_1 \subseteq U_2$ or $U_2 \subseteq U_1$.

Let if possible, $U_1 \not\subseteq U_2$ and $U_2 \not\subseteq U_1$. Then, there exists $a \in U_1$ such that $a \notin U_2$ and there exists $b \in U_2$ such that $b \notin U_1$.

Now, $a \in U_1, b \in U_2$ implies that $a, b \in U_1 \cup U_2$ which is an ideal. Thus, $ab, a - b \in U_1 \cup U_2$.

Now, $a - b \in U_1 \cup U_2 \Rightarrow a - b \in U_1$ or $a - b \in U_2$.

Case 1. If $a - b \in U_1$

Also, $a \in U_1$. Then, $a - (a - b) \in U_1$. Thus, $b \in U_1$, which is a contradiction.

Case 2. If $a - b \in U_2$

Also, $b \in U_2$. Then, $b + (a - b) \in U_2$. Thus, $a \in U_2$, which is a contradiction.

Hence, in both the cases we arrive at a contradiction. Thus, our assumption was wrong. Therefore, either $U_1 \subseteq U_2$ or $U_2 \subseteq U_1$.

Hence, union of two ideals is an ideal if and only if one of them is contained in another. $\square$

Next we define the sum of two ideals of a ring.

Definition 5.18 (Sum of Ideals) Let U_1 and U_2 be two ideals of a ring R, then the set $U_1 + U_2 = \{a + b : a \in U_1, b \in U_2\}$ is called the *sum of ideals U_1 and U_2*. In subrings, we have seen that the sum of two subrings need not be a subring. However, the sum of two ideals is an ideal which is given by the following theorem.

Theorem 5.19 *If U_1 and U_2 are any two ideals of a ring R, then $U_1 + U_2$ is an ideal containing both U_1 and U_2.*

Proof Clearly $0 = 0 + 0 \in U_1 + U_2$; therefore, $U_1 + U_2 \neq \emptyset$.

Let, $\alpha, \beta \in U_1 + U_2$, where $\alpha = a_1 + b_1$, $\beta = a_2 + b_2$ for some $a_1, a_2 \in U_1$, and $b_1, b_2 \in U_2$.

Then, we have $\alpha - \beta = (a_1 - a_2) + (b_1 - b_2) \in U_1 + U_2$ (since $a_1 - a_2 \in U_1; b_1 - b_2 \in U_2$).

Let, $\alpha \in U_1 + U_2$ and $r \in R$, where $\alpha = a_1 + b_1$, with $a_1 \in U_1$ and $b_1 \in U_2$

Then, $\alpha r = a_1 r + b_1 r \in U_1 + U_2$ and $r\alpha = ra_1 + rb_1 \in U_1 + U_2$ since $a_1 r, ra_1 \in U_1$, and $b_1 r, rb_1 \in U_2$. $(\because U_1, U_2 \text{ are ideals of } R)$

Hence, $U_1 + U_2$ is an ideal of R.

Now, consider $a \in U_1 \Rightarrow a = a + 0 \in U_1 + U_2$ $(\because 0 \in U_2)$

$\Rightarrow U_1 \subseteq U_1 + U_2$

Consider $b \in U_2 \Rightarrow b = 0 + b \in U_1 + U_2$ $(\because 0 \in U_1)$

$\Rightarrow U_2 \subseteq U_1 + U_2$. $\square$

In the next theorem, we show that the product of two ideals is again an ideal.

Theorem 5.20 *If U and V are ideals of R, let UV be the set of all elements that can be written as finite sums of elements of the form uv where $u \in U$ and $v \in V$. Then, UV is an ideal of R.*

Proof Suppose U and V are ideals of R and

$$UV = \{\sum_{i \in \Lambda} u_i v_i \mid u_i \in U \text{ and } v_i \in V \text{ and } \Lambda \text{ being some finite index set}\}.$$

So we need to show $UV = I$ (say) is an ideal of R.

Suppose $x, y \in I$, therefore $x = \sum_{i \in \Lambda} u_i v_i$ and $y = \sum_{i \in \Gamma} u_i' v_i'$ for $u_i, u_i' \in U$ and $v_i, v_i' \in V$ for all $i \in \Lambda \cup \Gamma$, where Λ, Γ are some finite index sets.

Now, consider $x - y = \sum_{i \in \Lambda} u_i v_i - \sum_{i \in \Gamma} u_i' v_i' = \sum_{i \in \Lambda} u_i v_i + \sum_{i \in \Gamma} (-u_i') v_i'$. Thus, $x - y \in I$.

Also for some $x \in I$ and $r \in R$, where $x = \sum_{i \in \Delta} u_i v_i$ for all $i \in \Delta$, where Δ is some finite index set, we have

$$xr = \left(\sum_{i \in \Delta} u_i v_i \right) r = \sum_{i \in \Delta} u_i v_i r = \sum_{i \in \Delta} u_i v_i' \text{ for some } v_i' \in V. \text{ Thus, } xr \in I \text{ for all } x \in I \text{ and}$$

$r \in R$.

Similarly, $rx \in I$ for all $x \in I$ and $r \in R$.

Therefore, I is an ideal of R. $\square$

Definition 5.21 Let S be any subset of a ring R and U be an ideal of R. Then U is said to be generated by S if

(i) $S \subseteq U$ and
(ii) for any ideal V of R, such that $S \subseteq V$, we have $U \subseteq V$.

If the ideal U is generated by a subset S of a ring R, then we denote U by the symbol $\langle S \rangle$.

Remark The set $\langle S \rangle$ is the intersection of all the ideals of R which contain S.

Next theorem shows that the sum of two ideals is equal to the set generated by the union of both the ideals.

Theorem 5.22 *If U_1 and U_2 are any two ideals of a ring R, then $U_1 + U_2 = \langle U_1 \cup U_2 \rangle$.*

Proof If U_1, U_2 are ideals of R, then by Theorem 5.19, $U_1 + U_2$ is also an ideal of R, such that $U_1 \subseteq U_1 + U_2$ and $U_2 \subseteq U_1 + U_2$.

We have $U_1 \subseteq U_1 + U_2, U_2 \subseteq U_1 + U_2$, then $U_1 \cup U_2 \subseteq U_1 + U_2$.

Let V be any ideal of R such that $U_1 \cup U_2 \subseteq V$.

If $x \in U_1 + U_2$, then $x = a + b, a \in U_1, b \in U_2$.

Now, $a \in U_1 \cup U_2, b \in U_1 \cup U_2 \Rightarrow a, b \in V \Rightarrow a + b \in V \Rightarrow x \in V$.

Hence, $U_1 + U_2 \subseteq V$, consequently, by definition $U_1 + U_2 = \langle U_1 \cup U_2 \rangle$. $\square$

The following theorem shows that the set generated by the product of two ideals is contained in the intersection of both the ideals.

Theorem 5.23 *If U and V are ideals of a ring R, then $\langle UV \rangle \subseteq U \cap V$.*

Proof If U and V are ideals of a ring R, then by Theorem 5.20, UV is also an ideal of R.

Now, suppose some $x \in \langle UV \rangle$, therefore $x = \sum_{i \in \Gamma} u_i v_i$ for $u_i \in U$ and $v_i \in V$ for all $i \in \Gamma$, where Γ is some finite index set. But $u_i v_i \in U \ \forall \ i \in \Gamma$ as U is an ideal of R. Therefore,

$\sum_{i \in \Gamma} u_i v_i \in U$.

Similarly, $\sum_{i \in \Gamma} u_i v_i \in V$ as V too is an ideal of R. Thus $x \in U$ and $x \in V$.

Therefore $x \in U \cap V$. Hence, $\langle UV \rangle \subseteq U \cap V$. $\square$

We note that the set of $n \times n$ matrices, $M_n(\mathbb{R})$ is a ring, which is not a division ring as the element $\begin{pmatrix} 1 & 0 \\ 0 & 0 \end{pmatrix} \in M_n(\mathbb{R})$ is not invertible.

In the next theorem, we provide a sufficient condition for a ring to be a division ring.

Proposition 5.24 *If R is a ring with unity and R has no right ideals except R and $\{0\}$. Then, R is a division ring.*

Proof Let $x \neq 0 \in R$. Consider $xR = \{xr : r \in R\}$.
Since $x \in R$, therefore $x = x \cdot 1 \in xR$, so $xR \neq \emptyset$ and xR is nonzero.
Also, $xy - xz = x(y - z) \in xR \ \forall \ xy, xz \in xR$
and for any $s \in R$, we have $(xr)s = x(rs) \in xR$.
Thus, xR is a right ideal of R. By the given hypothesis, we have $xR = R$.
Since, R is a ring with unity, there exists $y \in R$ such that $xy = 1$.
Thus, $R \backslash \{0\}$ is a semigroup with unity and every nonzero element of $R \backslash \{0\}$ is right invertible. Hence, $R \backslash \{0\}$ is a group under multiplication. Hence, R is a division ring. $\square$

We know that $n\mathbb{Z}$ is an ideal of $\mathbb{Z}$. Notice that $2\mathbb{Z} = \langle 2 \rangle$, $3\mathbb{Z} = \langle 3 \rangle$ and so on where $\langle 2 \rangle, \langle 3 \rangle \dots$ are the sets generated by a single element $2, 3, \dots$ respectively. This gives us a motivation to study about the ideals generated by a single element which is referred to as a *principal ideal*. Thus, a principal ideal is an ideal I in a ring R that is generated by a single element a of R through multiplication by every element of R. It can be understood more precisely by the following definition and some other results related to the principal ideals.

Definition 5.25 (Principal Ideal Generated by a) Let R be a commutative ring with unity and let $a \in R$ be any element. Then, the smallest ideal of R containing a is called the *principal ideal generated by a*. It is denoted by the set $\langle a \rangle = \{ra \mid r \in R\}$.

Example 5.26 The set $(\mathbb{Z}, +, \cdot)$ is a commutative ring with unity. The set of even integers, $2\mathbb{Z} = \langle 2 \rangle = \{2n : n \in \mathbb{Z}\}$ is an ideal of $\mathbb{Z}$ generated by 2. Thus, $2\mathbb{Z}$ is a principal ideal of $\mathbb{Z}$.

Now, a curiosity comes to check whether a principal ideal generated by a single element a is also an ideal of a ring R that will be cleared by the succeeding theorem.

Theorem 5.27 *Let R be a commutative ring with unity and let $a \in R$. Then, $\langle a \rangle$, the principal ideal generated by a, is an ideal of R.*

Proof Let R be a commutative ring with unity. Let, $a \in R$.
Let $\langle a \rangle = \{ra \mid r \in R\}$. Since $1 \in R$, $1 \cdot a \in \langle a \rangle \Rightarrow a \in \langle a \rangle \Rightarrow \langle a \rangle \neq \emptyset$.
Let $x, y \in \langle a \rangle$ such that $x = r_1 a, y = r_2 a$ for some $r_1, r_2 \in R$.
Consider, $x - y = r_1 a - r_2 a = (r_1 - r_2)a \in \langle a \rangle$ for some $r_1, r_2 \in R$.
Now, let $r \in R$. Then, $rx \in \langle a \rangle$. Since, R is commutative, so $rx = xr \Rightarrow rx, xr \in \langle a \rangle$ for all $r \in R$.
Hence, $\langle a \rangle$ is an ideal of R. $\square$

In the following result, we discuss about the structure of the principal ideal of a ring.

Theorem 5.28 *If R is a commutative ring and $a \in R$, then the principal ideal $\langle a \rangle$ is equal to the set $\{ar + na : r \in R, n \in \mathbb{Z}\}$.*

Proof Let $U = \{ar + na : r \in R, n \in \mathbb{Z}\}$. We shall prove that U is the ideal generated by a.
Clearly $a = a0 + 1a \in U$ so that $U \neq \emptyset$.
Let $ar + na, as + ma \in U$, where $r, s \in R, n, m \in \mathbb{Z}$. Then, $(ar + na) - (as + ma) = a(r - s) + (n - m)a \in U$.
Further, if $s \in R$, then $(ar + na)s = a(rs) + a(ns) = a(rs + ns) + 0a \in U$
which shows that U is a right ideal of R. Since R is commutative, U is also a left ideal of R.
Hence, U is an ideal of R such that $a \in U$.
Let V be another ideal of R, containing a. Since $a \in V$ and V is an ideal of $R, ar \in V, na \in V$
for all $r \in R$ and $n \in \mathbb{Z}$.
Therefore, $ar + na \in V$ for all $r \in R$ and for all $n \in \mathbb{Z}$.
Hence, $U \subseteq V$. Consequently, $U = \langle a \rangle$. $\qquad \square$

Definition 5.29 (Principal Ideal Ring) A ring for which every ideal is a principal ideal
is called a *principal ideal ring*.

Example 5.30 The ring $(\mathbb{Z}_5, \oplus_5, \otimes_5)$ is a principal ideal ring.

Theorem 5.31 *Every field is a principal ideal ring.*

Proof Let F be a field. We know, that a field has no proper ideal, i.e., $\{0\}$ and F are the
only ideals of a field F.
Since, the ideal $U = \{0\}$ is generated by 0, i.e., $U = \{0\} = \langle 0 \rangle$ and the ideal F is generated
by 1, i.e., $F = \langle 1 \rangle$. Therefore $U = \langle 0 \rangle$ and F are the principal ideals of F. i.e., every ideal
of F is a principal ideal.
Hence, F is a principal ideal ring. $\qquad \square$

Consider noncommutative ring $R = \left\{ \begin{pmatrix} a & b \\ c & d \end{pmatrix} \,\middle|\, a, b, c, d \in \mathbb{Z} \right\}$.

Let $a = \begin{pmatrix} 1 & 1 \\ 0 & 0 \end{pmatrix}$. Consider, $aR = \left\{ \begin{pmatrix} a & b \\ 0 & 0 \end{pmatrix} \,\middle|\, a, b \in \mathbb{Z} \right\}$.

Clearly, aR is not an ideal as for $\begin{pmatrix} 1 & 1 \\ 0 & 0 \end{pmatrix} \in aR$ and $\begin{pmatrix} 1 & 1 \\ 1 & 1 \end{pmatrix} \in R$, we have

$\begin{pmatrix} 1 & 1 \\ 1 & 1 \end{pmatrix} \begin{pmatrix} 1 & 1 \\ 0 & 0 \end{pmatrix} = \begin{pmatrix} 1 & 1 \\ 1 & 1 \end{pmatrix} \notin aR$. Thus, aR, in general, is not an ideal of a ring.

The next theorem provides a sufficient condition for aR to be an ideal of a ring R.

Theorem 5.32 *Let R be a commutative ring and $a \in R$. Then, $aR = \{ar \mid r \in R\}$ is an
ideal of R.*

Proof Let R be a commutative ring and $a \in R$.
We need to show that $aR = \{ar \mid r \in R\}$ is an ideal of R.
Since $0 \in aR \Rightarrow aR \neq \emptyset$.
Suppose $x, y \in aR$, where $x = ar_1$ and $y = ar_2$ for some $r_1, r_2 \in R$. Then,
we have $x - y = ar_1 - ar_2 = a(r_1 - r_2) = ar_3$ for some $r_3 \in R \Rightarrow x - y \in aR$.
Also, for some $x \in aR$ where $x = ar_1$ and $r \in R$, we have $rx = xr = ar_1 r = a(r_1 r) = ar_3$

for some $r_3 \in R$ as R is commutative. Thus, for all $x \in aR$ and $r \in R$, we have $rx, xr \in aR$. Hence, aR is an ideal of R. $\square$

Proposition 5.33 *If R is a commutative ring with unity, then the ideal Ra is the smallest ideal containing a.*

Proof Let $\langle a \rangle = \cap \{U : U$ is an ideal of R and $a \in U\}$.
Clearly, $\langle a \rangle$ is the smallest ideal which contains a. We shall show that $\langle a \rangle = Ra$. Since R is a ring with unity, $a = 1.a$ belongs to Ra. Therefore $\langle a \rangle$ is contained in Ra as Ra is an ideal.

Let U be any ideal of R such that $a \in U$. For any $r \in R$, we have $ra \in U$ by the definition of an ideal. Hence, $Ra \subset U$. Since U is an arbitrary ideal containing a it follows that $Ra \subseteq \cap \{U : U$ is an ideal and $a \in U\}$, i.e., $Ra \subseteq \langle a \rangle$. Hence, $Ra = \langle a \rangle$. $\square$

Earlier we have shown that the ring $\mathbb{Z} \oplus \mathbb{Z}$ has a subring $S = \{(x, x) \mid x \in \mathbb{Z}\}$ which is not an ideal of $\mathbb{Z} \oplus \mathbb{Z}$. However, if the ring is cyclic under addition, then every subring is an ideal.

Theorem 5.34 *Every subring of a cyclic ring is an ideal.*

Proof Let R be a cyclic ring and r be a generator of $(R, +)$, where $(R, +)$ is the additive group of R. Since R is a ring, then $r^2 \in R$. Since R is cyclic under addition, therefore there exist $k \in \mathbb{Z}$ such that $r^2 = kr$. Let S be a subring of R. Then S is a cyclic ring. Let s be a generator of $(S, +)$, where $(S, +)$ is the additive group of S. Since S is a subring of R, then $s \in R$. Thus, there exists $z \in \mathbb{Z}$ with $s = zr$.

Let $t \in R$ and $u \in S$. Then there exist $a, b \in \mathbb{Z}$ with $t = ar$ and $u = bs$. Since $tu = (ar)(bs) = (ar)[b(zr)] = (ar)[(bz)r] = (abz)r^2 = (abz)(kr) = (abkz)r = (abk)(zr) = (abk)s \in S$ and multiplication is commutative in R, it follows that S is an ideal of R. $\square$

The idea of a coset in ring theory is basically the same as in group theory, except with ideals replacing subgroups. Let R be a ring, and I be an ideal in R. First we define an equivalence relation on R as follows: elements a, b of R are equivalent, written $a \sim b$, whenever $a - b$ is in I. This partitions our ring into disjoint subsets, called *cosets*. One of the cosets is just I itself: this is the equivalence class of 0, since $a \sim 0$ exactly when $a = a - 0 \in I$. The other cosets look similar to I. The collection of all the cosets of I is denoted by R/I. For a familiar example, take R to be $\mathbb{Z}$, and I to be the ideal $\langle 3 \rangle$. Then the cosets are $\langle 3 \rangle$, $1+ <3>$ and $2+ <3>$.

Now that we've defined the cosets, we want a way to turn the set of cosets R/I into a ring itself, and this is where the fact that I is an ideal comes into play. To fix notation, let R be a ring, I be an ideal in R, and if r is an element of R, we write $r + I$ for the coset containing r. Another way of writing it is $r + I = \{r + i \mid i \in I\}$, namely the elements of I all shifted by r. The notion of coset of ring leads to another important class of rings, known as *factor rings*.

Definition 5.35 (Factor Rings) Let R be a ring and I be an ideal of R. Then, $R/I = \{r + I \mid r \in R\}$ is called a *factor ring* with respect to the following operations

  (i) $(r + I) + (s + I) = (r + s) + I$
  (ii) $(r + I) \cdot (s + I) = rs + I$.

Example 5.36 For $R = \mathbb{Z}, I = 4\mathbb{Z}, \dfrac{R}{I} = \dfrac{\mathbb{Z}}{4\mathbb{Z}} = \{k + 4\mathbb{Z} \mid k \in \mathbb{Z}\} = \{0 + 4\mathbb{Z}, 1 + 4\mathbb{Z}, 2 + 4\mathbb{Z}, 3 + 4\mathbb{Z}\}$ is a factor ring.

Example 5.37 $\dfrac{2\mathbb{Z}}{6\mathbb{Z}} = \{0 + 6\mathbb{Z}, 2 + 6\mathbb{Z}, 4 + 6\mathbb{Z}\}$ is a factor ring.

Example 5.38 For $R = \left\{ \begin{bmatrix} a_1 & a_2 \\ a_3 & a_4 \end{bmatrix} \middle| a_i \in \mathbb{Z} \right\}$ and $I = \left\{ \begin{bmatrix} b_1 & b_2 \\ b_3 & b_4 \end{bmatrix} \middle| b_i \text{ is even} \right\} \subseteq R$ be the ideal of R, $\dfrac{R}{I} = \left\{ \begin{bmatrix} a_1 & a_2 \\ a_3 & a_4 \end{bmatrix} + I \middle| a_i \in \{0, 1\} \right\}$ is a factor ring having 16 elements.

The next theorem shows that the definition of factor rings works as long as I is an ideal of R.

Theorem 5.39 (Existence of Factor Rings) *Let R be a ring and let A be a subring of R. The set of cosets $\{r + A \mid r \in R\}$ is a ring under the operations $(s + A) + (t + A) = (s + t) + A$ and $(s + A)(t + A) = st + A$ if and only if A is an ideal of R.*

Proof Let R be a ring and let A be a subring of R. Let A be an ideal of R. We need to show that $\dfrac{R}{A} = \{r + A \mid r \in A\}$ is a ring.

Since R is ring, therefore $\dfrac{R}{A} \neq \emptyset$.

Let $r_1 + A, r_2 + A, r_3 + A \in \dfrac{R}{A}$, where $r_1, r_2, r_3 \in R$.

(i) Consider $(r_1 + A) + (r_2 + A) = (r_1 + r_2) + A \in \dfrac{R}{A}$. $\hspace{1em} (\because r_1 + r_2 \in R.)$

(ii) $(r_1 + A) + (r_2 + A) = (r_1 + r_2) + A = (r_2 + r_1) + A \hspace{1em} (\because (R, +) \text{ is commutative.})$
$\Rightarrow (r_1 + A) + (r_2 + A) = (r_2 + A) + (r_1 + A)$.

(iii) $((r_1 + A) + (r_2 + A)) + (r_3 + A) = ((r_1 + r_2) + A) + (r_3 + A) =$
$((r_1 + r_2) + r_3) + A = (r_1 + (r_2 + r_3)) + A \hspace{1em} (\because (R, +) \text{ is associative.})$
$= (r_1 + A) + ((r_2 + r_3) + A) = (r_1 + A) + ((r_2 + A) + (r_3 + A))$.

(iv) We have $(r_1 + A) + (0 + A) = (r_1 + 0) + A = r_1 + A$.
Similarly, $(0 + A) + (r_1 + A) = r_1 + A$ $\hspace{1em} (\because 0 \text{ is zero of } R.)$
$\Rightarrow 0 + A = A$ is zero of $\dfrac{R}{A}$.

(v) Since $r_1 \in R$ and R is a ring, therefore $-r_1 \in R$.
consider $(r_1 + A) + ((-r_1) + A) = 0 + A = A \Rightarrow (-r_1) + A$ is negative of
$r_1 + A \in \dfrac{R}{A}$.

(vi) $(r_1 + A) \cdot (r_2 + A) = (r_1 \cdot r_2) + A \in \dfrac{R}{A}$.

(vii) Similarly, associativity also holds for multiplication.

(viii) $(r_1 + A)((r_2 + A) + (r_3 + A)) = (r_1 + A)((r_2 + r_3) + A) = r_1(r_2 + r_3) + A =$
$(r_1 r_2 + r_1 r_3) + A \hspace{1em} (\because R \text{ is a ring.})$
$= (r_1 r_2 + A) + (r_1 r_3 + A) = (r_1 + A)(r_2 + A) + (r_1 + A)(r_3 + A)$.

(ix) Similarly, $((r_1 + A) + (r_2 + A)) \cdot (r_3 + A) = (r_1 + A)(r_3 + A) + (r_2 + A)(r_3 + A)$.
Therefore, R/A forms a ring.

Conversely, suppose that the set of cosets $\{r + A \mid r \in R\}$ is a ring under the operations $(s + A) + (t + A) = (s + t) + A$ and $(s + A)(t + A) = st + A$. We need to show that A is an ideal of R.

Given that A is a subring of R. For showing A is ideal it is enough to show $ar, ra \in A$ whenever $a \in A, r \in R$.

Let, $a \in A, r \in R \Rightarrow a + A = A$ $\hfill (\because a \in A)$
and $r + A \in \dfrac{R}{A}$.

Now, $ar + A = (a + A)(r + A) = (A)(r + A) = (0 + A)(r + A) = 0 + A = A$, zero of $\dfrac{R}{A}$.

Thus, we have $ar + A = A \Rightarrow ar \in A$.
Similarly, $ra \in A \Rightarrow A$ is ideal of R. $\hfill \square$

Next we discuss an important class of principal ideal called as the principal ideal domain. A principal ideal domain is an integral domain in which every proper ideal can be generated by a single element. The term "principal ideal domain" is often abbreviated P.I.D. Examples of P.I.D.s include the integers, the Gaussian integers, and the set of polynomials in one variable with real coefficients.

Definition 5.40 (Principal Ideal Domain) An integral domain D is called a *principal ideal domain* if every ideal of D is principal ideal, that is, every ideal of D is of the form $\langle a \rangle = \{ad \mid d \in D\}$ for some a in D.

Example 5.41 The ring of integers, $\mathbb{Z}$ is a principal ideal domain. We know that $\mathbb{Z}$ is an integral domain. In order to show that $\mathbb{Z}$ is a principal ideal domain, we show that any ideal of $\mathbb{Z}$ is a principal ideal. Let, I be any ideal of $\mathbb{Z}$.

> **Case 1.** $I = \{0\} \Rightarrow I = \langle 0 \rangle \Rightarrow I$ is a principal ideal and we are done.
> **Case 2.** $I \neq \{0\}$. Let m be the least positive integer in I. (Note that by the well-ordering principle, such an m will exist, as I contains both positive and negative integers, as I is an ideal.)
> **Claim:** $I = \langle m \rangle$.

Let, $x \in \langle m \rangle \Rightarrow x = mk$ for some $k \in \mathbb{Z}$.
Since, $m \in I, k \in \mathbb{Z} \Rightarrow mk \in I \Rightarrow x \in I$ $\hfill (\because I$ is an ideal of $\mathbb{Z}.)$
But, x was an arbitrary element of $\langle m \rangle \Rightarrow \langle m \rangle \subseteq I$. $\hfill (5.5)$
Let, $n \in I$. By division algorithm there exist $q, r \in \mathbb{Z}$ such that $n = mq + r$ with $0 \leq r < m \Rightarrow r = n - mq$.
Since, $n \in I, mq \in I$ (because m belongs to I, q belongs to $\mathbb{Z}$ and I is an ideal) $\Rightarrow n - mq \in I \Rightarrow r \in I$.
But, m is the least positive integer in I and $0 \leq r < m \Rightarrow r = 0 \Rightarrow n = mq \Rightarrow n \in \langle m \rangle$.
But, n was an arbitrary element of $I \Rightarrow I \subseteq \langle m \rangle$. $\hfill (5.6)$
From Equations (5.5) and (5.6), we conclude that $I = \langle m \rangle$.
Therefore, I is a principal ideal. But, I was an arbitrary ideal of $\mathbb{Z}$. Thus, every ideal of $\mathbb{Z}$ is a principal ideal. Hence, $\mathbb{Z}$ is a principal ideal domain.

Example 5.42 Every ideal of the polynomial ring $K[x]$, where K is a field, is a principal ideal.

Example 5.43 $\mathbb{Z}[x]$ is not a principal ideal domain: the ideal $\langle x, 2 \rangle = \{a(x)\,x + 2\,b(x) \mid a(x), b(x) \in \mathbb{Z}[x]\}$ is not principal. First note that $\langle x, 2 \rangle \neq \mathbb{Z}[x]$ as $1 \notin \langle x, 2 \rangle$ because if it were then $1 = xf(x) + 2g(x)$ for $f(x), g(x) \in \mathbb{Z}[x]$ but $xf(x) + 2g(x)$ has even constant term.

Now, suppose that $\langle x, 2 \rangle = \langle p(x) \rangle$ for some $p(x) \in \mathbb{Z}[x]$. Then, we must have $x = p(x)f(x)$ and $2 = p(x)g(x)$ for some $f(x), g(x) \in \mathbb{Z}[x]$. But the second implies that $p(x)$ must be a constant polynomial, specifically $p(x) = -2, -1, 1$ or 2. We can't have $p(x) = \pm 1$ because then $\langle p(x) \rangle = \mathbb{Z}[x]$ so $p(x) = \pm 2$. But then $x = \pm 2f(x)$, a contradiction since $\pm 2f(x)$ has even coefficients and coefficient of x is 1 on LHS.

If an ideal exhibits some additional properties then the ideal can be categorized into prime ideals and maximal ideals.

The notion of a prime ideal is a generalization of "prime" in $\mathbb{Z}$. Suppose $n \in \mathbb{Z}^+ \backslash \{1\}$ such that n divides ab. In this case, n is guaranteed to divide either a or b when n is prime. This idea of primes has been extended to define prime ideals in a ring.

Definition 5.44 (Prime Ideal) A *prime ideal* A of a commutative ring R is a proper ideal of R such that if $ab \in A$ we have either $a \in A$ or $b \in A$.

Example 5.45 In a field F, the only prime ideal of F is $\{0\}$.

Example 5.46 Let $n\mathbb{Z}$ be a proper ideal in $\mathbb{Z}$ with $n > 1$ and suppose $ab \in \mathbb{Z}$ for $a, b \in \mathbb{Z}$. Then, in $\mathbb{Z}$, the ideal $n\mathbb{Z}$ is prime if and only if n is prime. That is, the nonzero prime ideals of $\mathbb{Z}$ are of the form $p\mathbb{Z}$, where p is prime. $\{0\}$ is also a prime ideal of $\mathbb{Z}$.

Example 5.47 The ideal $\langle 3 \rangle = \{3n : n \in \mathbb{Z}\}$ is a prime ideal in $\mathbb{Z}$, since $\langle 3 \rangle$ is a proper ideal of $\mathbb{Z}$ and for elements $a, b \in \mathbb{Z}$ and $ab \in \langle 3 \rangle$, we have $3|ab \Rightarrow 3|a$ or $3|b \Rightarrow a \in \langle 3 \rangle$ or $b \in \langle 3 \rangle$.

Example 5.48 The whole ring R is (by definition) never a prime ideal as it is not a proper ideal of R.

Example 5.49 Let $R = \mathbb{Z}[x]$ and $I = \langle x^2 + 1 \rangle$. Then, I is a prime ideal of R. Suppose that $p(x) \cdot q(x) \in I$. Then, $(x^2 + 1)|p(x) \cdot q(x)$. Reduce $p(x), q(x)$ modulo $x^2 + 1$ to get $p(x) \equiv (ax + b)\bmod(x^2 + 1)$ and $q(x) = (cx + d)\bmod(x^2 + 1)$. So now $(x^2 + 1)|(ax + b)(cx + d)$. This is not possible unless $a = b = 0$ or $c = d = 0$. In the first case, $(x^2 + 1)|p(x)$, and in the second case, $(x^2 + 1)|q(x)$. That is, either $p(x) \in I$ or $q(x) \in I$. Thus, I is a prime ideal of R.

Example 5.50 The ideal $\langle 6 \rangle$ is not a prime ideal in $\mathbb{Z}$ since $2 \times 3 \in \langle 6 \rangle$ although neither 2 nor 3 belongs to $\langle 6 \rangle$.

We see that the ideals, $4\mathbb{Z}, 8\mathbb{Z}$ are not prime ideals of $\mathbb{Z}$. Also, one can observe that the rings $\mathbb{Z}/4\mathbb{Z}, \mathbb{Z}/8\mathbb{Z}$ are not integral domains. The next theorem provides a necessary and sufficient condition for a factor ring to be an integral domain.

Theorem 5.51 (R/A Is an Integral Domain If and Only If A Is Prime) *Let R be a commutative ring with unity and let A be an ideal of R. Then R/A is an integral domain if and only if A is prime.*

Proof Let $\dfrac{R}{A}$ be an integal domain with A as the zero element and $1 + A$ as unity.

Now, we need to show that A is a prime.

Let $ab \in A$. Then $ab + A = A \Rightarrow (a + A)(b + A) = A \Rightarrow a + A = A$ or $b + A = A$ ($\because R/A$ is an integral domain. Therefore, it has no zero divisors.)

that is, $a \in A$ or $b \in A$, thus A is a prime ideal.

Now, suppose that A is a prime ideal of R.

We need to show that $\dfrac{R}{A}$ is an integral domain.

Since, R is a commutative ring with unity 1. Then, $\dfrac{R}{A}$ is a commutative ring with unity $1 + A$.

So, we need to show that $\dfrac{R}{A}$ does not have zero divisors.

Let if possible $(a + A)(b + A) = A$. Then, $ab + A = A \Rightarrow ab \in A \Rightarrow a \in A$ or $b \in A \Rightarrow a + A = A$ or $b + A = A$. So, $\dfrac{R}{A}$ has no zero divisors.

Therefore, $\dfrac{R}{A}$ is an integral domain. $\qquad\qquad\square$

In ring theory, a maximal ideal is an ideal that is maximal (with respect to set inclusion) amongst all proper ideals. In other words, I is a maximal ideal of a ring R if there are no other ideals between I and R.

Maximal ideals are important because the quotients of rings by maximal ideals are simple rings, and in the special case of unital commutative rings they are also fields.

Definition 5.52 (Maximal Ideal) A *maximal ideal A* of a commutative ring R is a proper ideal of R such that, whenever B is an ideal of R and $A \subseteq B \subseteq R$, then $B = A$ or $B = R$.

Example 5.53 The ideal $\langle x^2 + 1 \rangle$ is maximal in $\mathbb{R}[x]$.

Example 5.54 Let $R = \mathbb{Z}[x]$ and $I = \langle x^2 + 1 \rangle$. Then, the ideal I is not a maximal ideal in R. Consider $J = \langle x^2 + 1, 2 \rangle$. Since $2 \in J$ but not belongs to I and $1 \in R$ but not belongs to J, we have $I \subset J \subset R$. Thus, I is not a maximal ideal of R.

The prime ideals of $\mathbb{Z}$ are $\langle 0 \rangle, \langle 2 \rangle, \langle 3 \rangle, \langle 5 \rangle, \ldots$. All these are also maximal ideals except $\langle 0 \rangle$. This observation is extended to give the next result.

Theorem 5.55 *An ideal of ring of integers $\mathbb{Z}$ is a maximal ideal if and only if it is generated by some prime number.*

Proof Let M be an ideal of $\mathbb{Z}$ generated by a prime number p, that is, $M = \langle p \rangle = \{pn : n \in \mathbb{Z}\}$.

Let U be any ideal of $\mathbb{Z}$, such that $M \subseteq U \subseteq \mathbb{Z}$. Since, every ideal of $\mathbb{Z}$ is a principal ideal. Thus, we have $U = \langle q \rangle$, where q is an integer.

Now $M \subseteq U \subseteq \mathbb{Z} \Rightarrow \langle p \rangle \subseteq \langle q \rangle \subseteq \mathbb{Z} \Rightarrow p \in \langle q \rangle$ and hence $p = qm$ for some $m \in \mathbb{Z}$.

But p is prime which implies either $q = 1$ or $m = 1$.

If $m = 1$ then $p = q \Rightarrow \langle p \rangle = \langle q \rangle$ and hence $M = U$.

And if $q = 1 \Rightarrow \langle q \rangle = \mathbb{Z} \Rightarrow U = \mathbb{Z}$.

Hence, M is maximal ideal.

Conversely, let M be a maximal ideal in $\mathbb{Z}$ and let $M = \langle p \rangle$, where p is a positive integer (Note that such a p will exist as $\mathbb{Z}$ is a PID).

Let us assume that p is not prime, Then, there exist integers a and b such that $p = ab$ where $1 < a, b < p$.

Let $U = \langle a \rangle$ then $U \supset M$.

Thus, $M \subset U \subset \mathbb{Z}$ but M is a maximal ideal.

Hence, $M = U$ or $U = Z$.

Case 1. Let $U = \mathbb{Z}$, we have $U = \langle a \rangle = \langle 1 \rangle \Rightarrow a = 1$, which is a contradiction.

Case 2. Let $U = M$ we have $U = \langle a \rangle = M \Rightarrow a \in M \Rightarrow a = rp$ for some $r \in \mathbb{Z}$.

Therefore, $p = ab = (rp)b = p(rb) \Rightarrow 1 = rb$. Hence, we get b is a unit, which is a contradiction.

Hence, M is generated by a prime p. $\qquad\qquad\qquad\qquad\qquad\qquad\qquad\qquad\qquad\qquad$ $\square$

Note that $\langle 0 \rangle$ is a prime ideal of $\mathbb{Z}$ but not a maximal ideal. Also, if $A = \mathbb{Z}[x]$, the polynomial ring in one variable over $\mathbb{Z}$ and p is a prime number, then $\langle 0 \rangle, \langle p \rangle, \langle x \rangle$, and $\langle p, x \rangle = \{ap + bx \mid a, b \in A\}$ are all prime ideals of A. Of these, only $\langle p, x \rangle$ is maximal. Thus, we observe that a prime ideal need not be a maximal ideal. In fact, there are some ideals that are neither prime nor maximal. For example, the ideals $4\mathbb{Z}, 8\mathbb{Z}$ are neither prime nor maximal ideals of $\mathbb{Z}$. In the next result, we talk about the necessary and sufficient condition under which an ideal is a maximal ideal.

Theorem 5.56 (R/A Is a Field If and Only If A Is Maximal) *Let R be a commutative ring with unity and A be an ideal of R. Then R/A is a field if and only if A is maximal.*

Proof Let, $\dfrac{R}{A}$ be a field. We need to show that A is a maximal ideal.

Let, B be any ideal such that $A \subseteq B \subseteq R$.

Let, if possible, $A \neq B$, so $A \subset B$, therefore there exists $b \in B$ such that $b \notin A$.

So, $b + A$ is a nonzero element of $\dfrac{R}{A}$ and $\dfrac{R}{A}$ is a field. Thus, $b + A$ has the multiplicative inverse in $\dfrac{R}{A}$. Let, $c + A$ be the inverse of $b + A$ for some $c \in R$.

$\Rightarrow (b + A)(c + A) = 1 + A \Rightarrow bc + A = 1 + A \Rightarrow 1 - bc \in A \subset B \Rightarrow 1 - bc \in B$ $\qquad$ (5.7)

Since, $b \in B, c \in R \Rightarrow bc \in B$ ($\because B$ is an ideal.) $\qquad\qquad\qquad\qquad\qquad\qquad\qquad$ (5.8)

From Equations (5.7) and (5.8), we have $1 - bc + bc \in B \Rightarrow 1 \in B$.

Since, $1 \in B, r \in R$ (arbitrary) $\Rightarrow 1 \cdot r \in B$ ($\because B$ is an ideal.)

$\Rightarrow r \in B$. So, $R \subseteq B \subseteq R$ Therefore, $B = R$. Hence, A is a maximal ideal.

Conversely, let A be a maximal ideal of R.

We need to show that $\dfrac{R}{A}$ is a field.

Since, R is a commutative ring with unity 1 which implies that $\dfrac{R}{A}$ is a commutative ring with unity $1 + A$. In order to show that $\dfrac{R}{A}$ is a field, we need to show that every nonzero element of $\dfrac{R}{A}$ has the multiplicative inverse.

Let, $b + A$ be any nonzero element of $\dfrac{R}{A}$; therefore, $b + A \neq A$, which implies $b \notin A$.

Claim: $b + A$ has a multiplicative inverse.

We construct $B = \{br + a : r \in R, a \in A\}$. We first show that B is an ideal of R. Clearly, $0 = 0 \cdot b + 0 \in B \Rightarrow B \neq \emptyset$.

Let, $x, y \in B$ and $s \in R$ where $x = br_1 + a_1$ and $y = br_2 + a_2$ for some $r_1, r_2 \in R$ and $a_1, a_2 \in A$.

Now, $x - y = br_1 + a_1 - (br_2 - a_2) = b(r_1 - r_2) + (a_1 - a_2) \in B$ $\quad (\because \ r_1 - r_2 \in R$ and $a_1 - a_2 \in A)$

Consider $sx = s(br_1 + a_1) = b(sr_1) + (sa_1) \in B$ $(\because sr_1 \in R$ and $sa_1 \in A$ as $s \in R, a \in A$ and A is an ideal.)

It is given that R is commutative $\Rightarrow xs \in B \Rightarrow B$ is an ideal of R.

We now show that $A \subset B$. Since, $b \cdot 0 + a \in B \ \forall \ a \in A \Rightarrow a \in B \ \forall \ a \in A \Rightarrow A \subseteq B$.

Also, $b = b \cdot 1 + 0 \in B$ and $b \notin A \Rightarrow A \neq B$. Therefore, $A \subset B \subseteq R$.

But, A is a maximal ideal of R. Thus, $B = R$. Since, $1 \in R \Rightarrow 1 \in B \Rightarrow 1 = bc + a$ for some $c \in R, a \in A$.

Then, $1 + A = (bc + a) + A = bc + A$ $\hfill (\because a + A = A \Rightarrow a \in A)$

$\Rightarrow 1 + A = (b + A)(c + A) \Rightarrow b + A$ has a multiplicative inverse, which is $c + A$. Hence, $\dfrac{R}{A}$ is a field. $(\because b + A$ is an arbitrary element of R/A and we have shown that every nonzero element has a multiplicative inverse.) $\hfill \square$

Next, we check that whether every maximal ideal is a prime ideal in a ring R through an example. Consider a commutative ring $R = 2\mathbb{Z}$ without unity, the ideal $I = 4\mathbb{Z}$ is maximal but not prime because $2 \cdot 6 \in I$ while neither 2 nor 6 belongs to I.

The next result provides a sufficient condition under which every maximal ideal is a prime ideal.

Theorem 5.57 *In a commutative ring with unity, every maximal ideal is a prime ideal.*

Proof Let I be a maximal ideal in a commutative ring R with unity.

By Theorem 5.56, we know that R/I is a field if and only if I is a maximal ideal. Thus, R/I is a field.

Since every field is an integral domain, therefore R/I is an integral domain.

Thus, by Theorem 5.51, we know that R/I is an integral domain if and only if I is a prime ideal. $\hfill \square$

The above theorem can be applied to show that $\dfrac{\mathbb{Z}_3[x]}{\langle x^2 + x + 1 \rangle}$ is not a field.

In order to show that $\dfrac{\mathbb{Z}_3[x]}{\langle x^2 + x + 1 \rangle}$ is not a field, we need to show that $\langle x^2 + x + 1 \rangle$ is not a maximal ideal.

Here, $x + 2 \in \mathbb{Z}_3[x]$ and $(x + 2)(x + 2) = x^2 + 4x + 4 = x^2 + x + 1 \in \mathbb{Z}_3[x]$.

Since, $x^2 + x + 1 \in \langle x^2 + x + 1 \rangle \Rightarrow (x + 2)(x + 2) \in \langle x^2 + x + 1 \rangle$.

But $x + 2 \notin \langle x^2 + x + 1 \rangle \Rightarrow \langle x^2 + x + 1 \rangle$ is not a prime ideal. Thus, $\langle x^2 + x + 1 \rangle$ is not a maximal ideal. Hence, $\dfrac{\mathbb{Z}_3[x]}{\langle x^2 + x + 1 \rangle}$ is not a field.

Above we have discussed some examples of ideals that are prime but not maximal and also some ideals that are neither prime nor maximal. So, it is clear that a prime ideal need not be a maximal ideal.

Illustrative Problems

Problem 1 Find the principal ideals of the ring $(\mathbb{Z}_6, \oplus_6, \otimes_6)$.

Solution We have $Z_6 = \{0, 1, 2, 3, 4, 5\}$.
Clearly $(\mathbb{Z}_6, \oplus_6, \otimes_6)$ is a commutative ring with unity and it is not a field. (Since it has zero divisor.)
Now, we have
$\langle 0 \rangle = \{0\}$ the null ideal is a principal ideal of $\mathbb{Z}_6$
$\langle 1 \rangle = \mathbb{Z}_6$, the unit ideal is a principal ideal of the ring
$\langle 2 \rangle = \{0, 2, 4\}$ is a principal ideal of the ring
$\langle 3 \rangle = \{0, 3\}$ is a principal ideal of the ring
$\langle 4 \rangle = \langle 2 \rangle$ is a principal ideal
$\langle 5 \rangle = \langle 1 \rangle$ is a principal ideal.
Hence, the principal ideals of $(\mathbb{Z}_6, \oplus_6, \otimes_6)$ are $\langle 0 \rangle, \langle 1 \rangle, \langle 2 \rangle$ and $\langle 3 \rangle$.

Problem 2 If U is an ideal of R, let $r(U) = \{x \in R \mid xu = 0 \text{ for all } u \in U\}$. Prove that $r(U)$ is an ideal of R.

Solution We have $0.u = 0$ for all $u \in U$; therefore, $U \neq \emptyset$. Let $x_1, x_2 \in r(U)$ and let $u \in U$. Then $(x_1 - x_2)u = x_1 u - x_2 u = 0$ as $x_1 u = 0$ and $x_2 u = 0$. Hence $x_1 - x_2 \in r(U)$. Now let $r \in R$ and $x \in r(U)$. Then $(rx)u = r(xu) = 0 \implies (xr)u = x(ru) = 0$ as U is an ideal, $ru \in U$. Hence $r(U)$ is an ideal of R.

Problem 3 If U is an ideal of R, let $[R:U] = \{x \in R \mid rx \in U \text{ for all } r \in R\}$. Prove that $[R:U]$ is an ideal of R containing U.

Solution Consider, $[R:U] = \{x \in R \mid rx \in U \text{ for all } r \in R\}$. Clearly, $0 \in [R:U]$ so $[R:U] \neq \emptyset$. Let $x, y \in [R:U]$ therefore, for $r \in R, rx \in U$ and $ry \in U$. Then $r(x - y) = rx - ry \in U$ as U is an ideal. Hence, $x - y \in [R:U]$. Now for $s \in R$, and $x \in [R:U]$, $r(sx) \in U$ as $sx \in U$ and U is an ideal. Thus, $sx \in [R:U]$. Similarly $(rx)s = r(xs) \in U$ for any $r \in R$, i.e. $xs \in [R:U]$ for any $s \in R$.

Problem 4 If A and B are ideals in a ring R such that $A \cap B = \langle 0 \rangle$, prove that $ab = 0$, for $a \in A$ and $b \in B$.

Solution Let $a \in A$ and $b \in B$. Then, $ab \in A$ and $ab \in B$ as A and B both are ideals. Hence, $ab \in A \cap B = \{0\}$.

Problem 5 Let R be a ring containing $\mathbb{Z}$ as a subring. Prove that if integers m, n are contained in a proper ideal of R, then they have a common factor > 1.

Solution Let R be a ring containing $\mathbb{Z}$ as a subring. Let I be a proper ideal of R and $m, n \in I$. Let, if possible, that m and n do not have a common factor greater than 1. Then they are relatively prime.

Hence there exist $x, y \in \mathbb{Z}$ such that $xm + ny = 1$. Since $n, m \in I$ we have xm and $ny \in I$. But this implies $1 \in I$, which is a contradiction, since I is a proper ideal.

Problem 6 Prove that any nonzero ideal in the Gaussian integers $\mathbb{Z}[i]$ must contain some positive integer.

Solution Let I be a nonzero ideal in $\mathbb{Z}[i]$ and let $m + ni \in I$ for some $m, n \in \mathbb{Z}$. Since I is an ideal, therefore $(m + ni)(m - ni) \in I$. Hence, $m^2 + n^2 \in I$ where $m^2 + n^2 \in \mathbb{Z}$ is positive as either m or n is nonzero.

Problem 7 Let $R[x]$ denote the set of all polynomials with real coefficients and let A denote the subset of all polynomials with constant term 0. Show that A is an ideal of $R[x]$ and $A = \langle x \rangle$.

Solution Here, $R[x] = \{a_0 + a_1 x + a_2 x^2 + \ldots \mid a_i \in R\}$ and
$A = \{a_1 x + a_2 x^2 + \ldots \mid a_i \in R\}$.
Clearly, $0 = 0 \cdot x + 0 \cdot x^2 + \ldots = 0 \in A \Rightarrow A \neq \emptyset$.
Let, $A_1 = a_1 x + a_2 x^2 + \ldots$, where $a_i \in R$ and
$A_2 = b_1 x + b_2 x^2 + \ldots$, where $b_i \in R$.
Consider, $A_1 - A_2 = a_1 x + a_2 x^2 + \ldots - (b_1 x + b_2 x^2 + \ldots)$
$= (a_1 - b_1)x + (a_2 - b_2)x^2 + \ldots \in A$.
Also, for $r \in R[x]$ and $A = \{a_1 x + a_2 x^2 + \ldots \mid a_i \in R\}$, we have
$rA = r(a_1 x + a_2 x^2 + \ldots) = (ra_1)x + (ra_2)x^2 + \ldots \in A$
and $Ar = (a_1 x + a_2 x^2 + \ldots)r = \{(a_1 r)x + (a_2 r)x^2 + \ldots \in A\}$.
Hence, A is an ideal of $R[x]$.
$A = \{a_1 x + a_2 x^2 + \ldots \mid a_i \in R\}$ and $\langle x \rangle = \{rx \mid r \in R[x]\}$.
Now, we show that $A \subseteq \langle x \rangle$ and $\langle x \rangle \subseteq A$. Let, $a_1 x + a_2 x^2 + \ldots \in A$.
Consider, $a_1 x + a_2 x^2 + \ldots = x(a_1 + a_2 x + \ldots) = xr = rx \in \langle x \rangle$ where $r = a_1 + a_2 x + \ldots$
$\in R[x]$.
Since, $a_1 x + a_2 x^2 + \ldots$ was any arbitrary element of A. Thus, $A \subseteq \langle x \rangle$.
Let, $rx \in \langle x \rangle$ where, $r \in R[x]$.
Take $r = a_0 + a_1 x + \ldots \Rightarrow rx = x(a_0 + a_1 x + \ldots) = a_0 x + a_1 x^2 + \ldots \in A$.
Since, rx was any arbitrary element of $\langle x \rangle$. Thus, $\langle x \rangle \subseteq A$.
Hence, $A = \langle x \rangle$.

Problem 8 Let R be a commutative ring with unity and let $a_1, a_2, \ldots, a_n$ belong to R. Let, $I = \langle a_1, a_2, \ldots, a_n \rangle = \{r_1 a_1 + r_2 a_2 + \ldots + r_n a_n \mid r_i \in R\}$. Prove that I is an ideal of R.

Solution Clearly, $0 = 0 \cdot a_1 + 0 \cdot a_2 + \ldots + 0 \cdot a_n \in I \Rightarrow I \neq \emptyset$.
Let, $X, Y \in I$ and $r \in R$ where
$X = r_1 a_1 + r_2 a_2 + \ldots + r_n a_n$ and $Y = s_1 a_1 + s_2 a_2 + \ldots + s_n a_n$ for all $r_i, s_i \in R$.
Consider, $X - Y = (r_1 a_1 + r_2 a_2 + \ldots + r_n a_n) - (s_1 a_1 + s_2 a_2 + \ldots + s_n a_n) = (r_1 - s_1)$
$a_1 + (r_2 - s_2)a_2 + \ldots + (r_n - s_n)a_n \in I$. ($\because r_i - s_i \in R$ for all i as $r_i, s_i \in R$ and R is a ring.)
For $r \in R$, consider $rX = r(r_1 a_1 + r_2 a_2 + \ldots + r_n a_n) = (rr_1)a_1 + \ldots + (rr_n)a_n \in I$.
($\because rr_i \in R \ \forall \ i$.)
Now, since R is commutative, therefore $rX = Xr \Rightarrow Xr \in I$.
Hence, I is an ideal of R.

Problem 9 Let $\mathbb{Z}[x]$ denote the ring of all polynomials with integer coefficients and let I be the subset of $\mathbb{Z}[x]$ of all polynomials with even constant term. Show that I is an ideal of $\mathbb{Z}[x]$ and $I = \langle x, 2 \rangle$.

Solution Let $\mathbb{Z}[x] = \{a_0 + a_1 x + a_2 x^2 + ... \mid a_i \in \mathbb{Z}\}$ and
$I = \{a_0 + a_1 x + a_2 x^2 + ... \mid a_i \in \mathbb{Z}, a_0 \in 2\mathbb{Z}\} = \{2m + a_1 x + a_2 x^2 + ... \mid a_i, m \in \mathbb{Z}\}$.
Clearly, $0 = 2 \cdot 0 + 0 \cdot x + 0 \cdot x^2 + ... \in I$, therefore $I \neq \emptyset$.
Let, $X, Y \in I$ and $r \in \mathbb{Z}[x]$ where
$X = 2m + a_1 x + a_2 x^2 + ...,\quad Y = 2n + b_1 x + b_2 x^2 + ...$ and $r = c_0 + c_1 x + c_2 x^2 + ...$ for
some $m, n, a_i, b_i, c_i \in \mathbb{Z}$.
Consider, $X - Y = (2m + a_1 x + a_2 x^2 + ...) - (2n + b_1 x + b_2 x^2 + ...) = 2(m - n) + (a_1 - b_1)x + (a_2 - b_2)x^2 + ... \in I$
Now $r \cdot X = (c_0 + c_1 x + c_2 x^2 ...) \cdot (2m + a_1 x + a_2 x^2 + ...)$
$= (c_0 \cdot 2m) + (c_0 \cdot a_1)x + (c_0 \cdot a_2)x^2 + ... + (c_1 \cdot 2m)x + ...$
$= (c_0 \cdot 2m) + (c_0 \cdot a_1 + c_1 \cdot 2m + ...)x + ... \in I$ for $m, c_i, a_i \in \mathbb{Z}$.
Similarly, $X \cdot r \in I$.

Hence, I is an ideal.

We now show that $I = \langle x, 2 \rangle$. Here, $\langle x, 2 \rangle = \{2 \cdot r + s \cdot x \mid r, s \in \mathbb{Z}[x]\}$.
Let, $X \in I \Rightarrow X = 2m + a_1 x + a_2 x^2 + ...,$ where $m, a_i \in \mathbb{Z}$
$= 2(m) + x(a_1 + a_2 x + ...) \in \langle x, 2 \rangle$ ($\because m \in \mathbb{Z}[x]$ and $a_1 + a_2 x + ... \in \mathbb{Z}[x]$).
But X was an arbitrary element of $I \Rightarrow I \subseteq \langle x, 2 \rangle$.   (5 9)
Let, $Y \in \langle x, 2 \rangle \Rightarrow Y = rx + 2s$ where $r, s \in \mathbb{Z}[x]$ such that
$r = a_0 + a_1 x + ..., s = b_0 + b_1 x + ...$ for $a_i, b_i \in \mathbb{Z}$.
Then, $Y = a_0 x + a_1 x^2 + ... + 2b_0 + 2b_1 x + ... = 2b_0 + (a_0 + 2b_1)x + (a_1 + 2b_2)x^2 + ... \in I \Rightarrow \langle x, 2 \rangle \subseteq I$.
Hence, $I = \langle x, 2 \rangle$.

Problem 10 In the ring of integers, find a positive integer a such that $\langle a \rangle = \langle 2 \rangle + \langle 3 \rangle$.

Solution We have $\langle 2 \rangle = \{..., -6, -4, -2, 0, 2, 4, 6, ...\}$
$\langle 3 \rangle = \{..., -9, -6, -3, 0, 3, 6, 9, ...\}$
Since $\gcd(2, 3) = 1$, this implies that there exist $m, n \in \mathbb{Z}$ such that
$1 = 2m + 3n \in \langle 2 \rangle + \langle 3 \rangle \Rightarrow \langle 2 \rangle + \langle 3 \rangle = \mathbb{Z} = \langle 1 \rangle$.
Therefore, $a = 1$.

Problem 11 Find a positive integer a such that $\langle a \rangle = \langle 3 \rangle \langle 4 \rangle$.

Solution Since, every element of $\langle 3 \rangle$ has the form $3t$ for some $t \in \mathbb{Z}$ and that of $\langle 4 \rangle$ has the form $4k$ for some $k \in \mathbb{Z}$. Therefore, every element of $\langle 3 \rangle \langle 4 \rangle$ has the form $3t4k$ for some $t, k \in \mathbb{Z}$, we have $12s \in \langle 12 \rangle$ (where $s = t\,k$).
So, $\langle 3 \rangle \langle 4 \rangle \subseteq \langle 12 \rangle$.
Conversely, since $12 \in \langle 3 \rangle \langle 4 \rangle$, so we have $\langle 12 \rangle \subseteq \langle 3 \rangle \langle 4 \rangle$.
Therefore, we conclude that $\langle 3 \rangle \langle 4 \rangle = \langle 12 \rangle \Rightarrow \langle a \rangle = \langle 12 \rangle \Rightarrow a = 12$.

Problem 12 If A and B are ideals of a commutative ring R with unity and $A + B = R$, show that $A \cap B = AB$.

Solution Let, A and B be ideals of a commutative ring R with unity such that $A + B = R$.

In order to show that $A \cap B = AB$, the set A and B should satisfy these two conditions, $A \cap B \subseteq AB$ and $AB \subseteq A \cap B$.

Let, $x \in A \cap B \Rightarrow x \in A$ and $x \in B$.

It is given that $A + B = R$ and $1 \in R \Rightarrow 1 \in A + B$. Let, $1 = a + b$ for some $a \in A$ and $b \in B$.

Now $x = x.1 = x(a + b) = xa + xb \in AB$ (as $x \in B$, $a \in A$ implies $xa \in AB$ and $x \in A$, $b \in B$ implies $xb \in AB$)

For $ax, a \in A, x \in B$ and $xb, x \in A$ and $b \in B$, we have $ax + xb = 1 \in B$.

But x was an arbitrary element of $A \cap B \Rightarrow A \cap B \subseteq AB$. $\hfill (5.10)$

Conversely, since A and B are ideals, then for any $y \in B$ and $z \in A$, the product $yz \in AB$.

Now, for $x \in AB$, let $x = \sum_{i=1}^{n} a_i b_i$.

Here $a_i \in A$ and $b_i \in B \Rightarrow a_i b_i \in A$ as A is an ideal. Similarly, $a_i b_i \in B$.

Thus, $x = \sum_{i=1}^{n} a_i b_i \in A \cap B$.

But x was an arbitrary element of $AB \Rightarrow AB \subseteq A \cap B$. $\hfill (5.11)$

From Equations (5.10) and (5.11), we have $A \cap B = AB$.

Problem 13 Consider the factor ring of the Gaussian integers $R = \mathbb{Z}[i]$. Let, $I = \langle 2 - i \rangle$. Describe the elements of $\dfrac{R}{I}$.

Solution We know that $I = \langle 2 - i \rangle$ is an ideal of $\mathbb{Z}[i]$, then $\dfrac{R}{I}$ is defined as

$\dfrac{R}{I} = \{a + ib + I \mid a, b \in \mathbb{Z}\}$.

Since, $I = \langle 2 - i \rangle \Rightarrow 2 - i \in I \Rightarrow 2 - i + I = I = 0 + I \Rightarrow 2 - i = 0$ in coset representation $\Rightarrow i = 2$ in coset representation.

Therefore, $\dfrac{R}{I} = \{a + 2b + I \mid a, b \in \mathbb{Z}\} = \{m + I \mid m \in \mathbb{Z}\}$.

So, $i = 2$ in coset representation

$\Rightarrow i^2 = 4$ in coset representation

$\Rightarrow -1 = 4$ in coset representation

$\Rightarrow 5 = 0$ in coset representation

$\Rightarrow \dfrac{R}{I} = \{I, 1 + I, 2 + I, 3 + I, 4 + I\}$.

Claim: $O(1+I)=5$.

Consider, $5(1 + I) = 5 + I = 0 + I = I \Rightarrow O(1 + I)$ divides $5 \Rightarrow O(1 + I) = 1$ or 5.

If $O(1 + I) = 1 \Rightarrow 1 \cdot (1 + I) = I \Rightarrow 1 + I = I \Rightarrow 1 \in I = \langle 2 - i \rangle = \{r(2 - i) \mid r \in \mathbb{Z}[i]\}$

$\Rightarrow 1 = (a + ib)(2 - i)$ for some $a, b \in \mathbb{Z}$

$\Rightarrow 1 = (2a + b) + i(2b - a)$.

Comparing real and imaginary parts, $1 = 2a + b \Rightarrow 0 = 2b - a$. On solving these, we obtain $a = \dfrac{2}{5}, b = \dfrac{1}{5}$ which is a contradiction as $a, b \in \mathbb{Z} \Rightarrow O(1 + I) \neq 1 \Rightarrow O(1 + I) = 5$

$\Rightarrow$ no further simplification is possible.

Therefore, $\dfrac{R}{I} = \{I, 1 + I, 2 + I, 3 + I, 4 + I\}$.

Problem 14 Let $R[x]$ be the ring of polynomials with real coefficients. Let, $I = \langle x^2 + 1 \rangle$. Describe the elements of $\dfrac{R}{I}$.

Solution Since, I is an ideal, $\dfrac{R}{I}$ is defined as $\dfrac{R}{I} = \{f(x) + I \mid f(x) \in R[x]\}$.
Applying division algorithm to $f(x)$ and $(x^2 + 1)$, we get polynomial $q(x)$ and $r(x)$ such that $f(x) = q(x)(x^2 + 1) + r(x)$ with $0 \leq \deg(r(x)) < \deg(x^2 + 1)$
$f(x) = q(x)(x^2 + 1) + ax + b$ where $a, b \in R$
$f(x) + I = q(x)(x^2 + 1) + ax + b + I = ax + b + I$ ($\because q(x)(x^2 + 1) \in I$ as $q(x) \in R[x], x^2 + 1 \in I$ and I is an ideal.)
Therefore, $\dfrac{R}{I} = \{ax + b + I \mid a, b \in R\}$.

Problem 15 In $\mathbb{Z}_5[x]$, let $I = \langle x^2 + x + 2 \rangle$. Find the multiplicative inverse of $2x + 3 + I$ in $\dfrac{\mathbb{Z}_5[x]}{I}$.

Solution We have $\frac{\mathbb{Z}_5[x]}{I} = \{ax + b + I \mid a, b \in \mathbb{Z}_5\}$. Let, $ax + b + I$ be the multiplicative inverse of $2x + 3 + I$. Therefore,
$(2x + 3)(ax + b) + I = 1 + I$
$\Rightarrow 2ax^2 + 2bx + 3ax + 3b + I = 1 + I$
$\Rightarrow 2ax^2 + 2ax + 4a - 2ax - 4a + 2bx + 3ax + 3b + I = 1 + I$
$2a(x^2 + x + 2) + (a + 2b)x - 4a + 3b + I = 1 + I$
$(a + 2b)x - 4a + 3b + I = 1 + I$
($\because 2a(x^2 + x + 2) + I = I$ as $2a(x^2 + x + 2) \in I$ since $x^2 + x + 2 \in I, 2a \in \mathbb{Z}_5[x]$ and I is an ideal.)
Comparing both the sides, we obtain $a + 2b = 0$ and $-4a + 3b = 1$.
On solving, we obtain $a = \dfrac{-2}{11}, b = \dfrac{1}{11}$.
So, $b = \dfrac{1}{11} = 1 \cdot 11^{-1} = 1^{-1} = 1$ and
$a = \dfrac{-2}{11} = (-2)(11^{-1}) = (-2) \cdot 1 = -2 = 3$. (3 is the additive inverse of 2 in $\mathbb{Z}_5[x]$.)
Hence, multiplicative inverse of $2x + 3 + I$ in $\dfrac{R}{I}$ is $3x + 1 + I$.

Problem 16 If n is an integer greater than 1, show that $\langle n \rangle = n\mathbb{Z}$ is a prime ideal of $\mathbb{Z}$ if and only if n is prime.

Solution Let, $n\mathbb{Z}$ be a prime ideal. Let, if possible $n = s \cdot t$ where $1 < s, t < n$
$n \in n\mathbb{Z} \Rightarrow s \cdot t \in n\mathbb{Z} \Rightarrow s \in n\mathbb{Z}$ or $t \in n\mathbb{Z}$ $\hfill (\because n\mathbb{Z}$ is prime ideal.)
which is a contradiction to $1 < s, t < n$.
Hence, our assumption was wrong and n is prime.
Conversely, let n be prime. We need to show that $n\mathbb{Z}$ is a prime ideal.
Let, $ab \in n\mathbb{Z} \Rightarrow n|ab \Rightarrow n|a$ or $n|b \Rightarrow a \in n\mathbb{Z}$ or $b \in n\mathbb{Z}$. Hence, $n\mathbb{Z}$ is a prime ideal.

Problem 17 In $\mathbb{Z}[x]$, the ring of polynomials with integer coefficients, let $I = \{f(x) \in \mathbb{Z}[x] \mid f(0) = 0\}$. Prove that I is an ideal and $I = \langle x \rangle$.

<u>Solution</u> Let, $I = \{f(x) \in \mathbb{Z}[x] \mid f(0) = 0\}$
$= \{f(x) = a_0 + a_1x + a_2x^2 + ... \mid f(0) = 0, a_i \in \mathbb{Z}\}$
$= \{a_1x + a_2x^2 + ... \mid a_i \in \mathbb{Z}\}$. Clearly, $0 = 0 \cdot x + 0 \cdot x^2 + 0 \cdot x^3 + ... \in I \Rightarrow I \neq \emptyset$.
Let, $f(x) = a_1x + a_2x^2 + ...$ and $g(x) = b_1x + b_2x^2 + ...$ for $f(x), g(x) \in I$.
Let, $h(x) = c_0 + c_1x + c_2x^2 + ... \in \mathbb{Z}[x]$.
Now, we need to prove that I is an ideal.
Consider, $f(x) - g(x) = (a_1x + a_2x^2 + ...) - (b_1x + b_2x^2 + ...) = (a_1 - b_1)x + (a_2 - b_2)x^2$
$+ ... \in I$
Also, $h(x)f(x) = (c_0 + c_1x + c_2x^2 + ...)(a_1x + a_2x^2 + ...) = (c_0a_1)x + (c_0a_2 + c_1a_1)x^2$
$+ ... \in I$
And $f(x)h(x) = (a_1c_0)x + (a_2c_0 + a_1c_1)x^2 + ... \in I \; \forall \; c_i, a_i \in \mathbb{Z}$.
Hence, I is an ideal.
Now, we need to prove that $I = \langle x \rangle$.
We have $\langle x \rangle = \{rx \mid r \in \mathbb{Z}[x]\}$.
Let, $f(x) \in I$ such that $f(x) = a_1x + a_2x^2 + ... = x(a_1 + a_2x + ...) \in \langle x \rangle$
$(\because a_1 + a_2x + ... \in \mathbb{Z}[x])$
$\Rightarrow I \subseteq \langle x \rangle$ $\hfill (5.12)$
Let, $h(x) \in \langle x \rangle \Rightarrow h(x) = r(x) \cdot x$ for some $r(x) \in \mathbb{Z}[x]$ where $r(x) = a_0 + a_1x + a_2x^2 + ...$
$\Rightarrow h(x) = (a_0 + a_1x + a_2x^2 + ...)x = a_0x + a_1x^2 + a_2x^3 + ... \in I$ $\hfill (\because h(0) = 0.)$
$\Rightarrow \langle x \rangle \subseteq I$ $\hfill (5.13)$
From Equations (5.12) and (5.13), we conclude that $I = \langle x \rangle$.

Exercise

(1) Let, $I = \langle a_1, a_2, ..., a_n \rangle = \{r_1a_1 + r_2a_2 + ... + r_na_n \mid r_i \in R \; \forall i\}$ is an ideal of R. If J is any ideal of R that contains $a_1, a_2, ..., a_n$, then show that $I \subseteq J$. (Hence, $\langle a_1, a_2, ..., a_n \rangle$ is the smallest ideal of R that contains $a_1, a_2, ..., a_n$.)

(2) Let R be the ring of all real-valued functions of a real variable. Show that the subset S of all differentiable functions is a subring of R but not an ideal of R.

(3) In the ring of integers, find a positive integer a such that
  (a) $\langle a \rangle = \langle 6 \rangle + \langle 8 \rangle$.
  (b) $\langle a \rangle = \langle m \rangle + \langle n \rangle$.

(4) Find a positive integer a such that
  (a) $\langle a \rangle = \langle 6 \rangle \langle 8 \rangle$.
  (b) $\langle a \rangle = \langle m \rangle \langle n \rangle$.

(5) Let A and B be ideals of a ring. Prove that $AB \subseteq A \cap B$.

(6) How many elements are there in $\dfrac{\mathbb{Z}[i]}{\langle 3 + i \rangle}$?

(7) Show that $I = \langle x^2 + 1 \rangle$ is not prime ideal in $\mathbb{Z}_2[x]$.

(8) Find all maximal ideals in
  (a) $\mathbb{Z}_8$.
  (b) $\mathbb{Z}_{10}$.
  (c) $\mathbb{Z}_{12}$.
  (d) $\mathbb{Z}_{36}$.
  (e) $\mathbb{Z}_n$.

(9) Prove that $I = k\langle 2 + 2i\rangle$ is not a prime ideal of $\mathbb{Z}[i]$. How many elements are in $\mathbb{Z}[i]/I$? What is the characteristic of $\mathbb{Z}[i]/I$?

(10) Let R be a subring of a field F such that for each x in F either $x \in R$ or $x^{-1} \in R$. Prove that if I and J are two ideals of R, then either $I \subseteq J$ or $J \subseteq I$.

(11) Let R be any ring with identity, and n be any positive integer. If $M_n(R)$ denotes the ring of $n \times n$ matrices with entries in R, prove that $M_n(I)$ is an ideal of $M_n(R)$ whenever I is an ideal of R, and that every ideal of $M_n(R)$ has this form.

(12) Let $\mathbb{Q}$ be the field of rational numbers and $D = \{a + b\sqrt{2} \mid a, b \in \mathbb{Q}\}$.

(a) Show that D is a subring of the field of real numbers.

(b) Show that D is a principal ideal domain (PID).

(c) Show that $\sqrt{3}$ is not an element of D.

(13) **(a)** Let R be a commutative ring with 1. Show that if M is a maximal ideal of R then M is a prime ideal of R.

(b) Give an example of a nonzero prime ideal in a ring R that is not a maximal ideal.

(14) Let R be a ring with unity. Show that every proper ideal of R is contained in a maximal ideal.

(15) Let R be a commutative ring with 1 and P a prime ideal of R. Show that if I and J are ideals of R such that $I \cap J \subseteq P$ and $J \not\subseteq P$, then $I \subseteq P$.

(16) Let R be a commutative ring with 1 such that for every x in R there is an integer $n > 1$ (depending on x) such that $x^n = x$. Show that every prime ideal of R is maximal

(17) Let R be a commutative ring with 1 in which every ideal is a prime ideal. Prove that R is a field.

(18) Let R_1 and R_2 be commutative rings with identities and let $R = R_1 \times R_2$. Show that every ideal I of R is of the form $I = I_1 \times I_2$ with I_i an ideal of R_i for $i = 1, 2$.

(19) Suppose that I and J are ideals in a ring R. Assume that $I \cup J$ is an ideal of R. Prove that $I \subseteq J$ or $J \subseteq I$.

(20) Let R be a ring of rational 2×2 matrices. Prove that this ring has no ideals other than $\{0\}$ and the ring itself.

6 RING HOMOMORPHISM

Understanding the intricacies of ring homomorphisms is a fundamental journey into the realm of abstract algebra. Ring homomorphisms serve as essential bridges between different algebraic structures, providing a means to explore the relationships and connections within the realm of rings. A ring homomorphism is a function between two rings that preserves the ring structure, capturing the algebraic essence of the source ring in the target ring. In this journey, we encounter concepts such as kernel and image, crucial aspects that delineate the behavior of these mappings.

Exploring the properties of ring homomorphisms, we encounter concepts like injective, surjective, and bijective homomorphisms, each shedding light on different facets of the relationship between the source and target rings. Injective homomorphisms preserve distinctiveness, surjective homomorphisms cover the entire target ring, and bijective homomorphisms embody a perfect one-to-one correspondence.

The study of ring homomorphisms extends its influence beyond mere theoretical exploration. It finds applications in various branches of mathematics, including algebraic geometry, functional analysis, and coding theory. Furthermore, the universal properties associated with ring homomorphisms offer a powerful toolkit for constructing and understanding mathematical structures.

This chapter aims to elucidate the foundations, properties, and applications of ring homomorphisms, equipping readers with the tools to unravel the algebraic tapestry woven by these transformative mappings. The subsequent sections delve into specific aspects, properties, and applications, unraveling the layers of abstraction that make ring homomorphisms a cornerstone in the study of rings and their interconnections.

Let us explore how the rings interact with each other through ring homomorphisms. The definition of a ring homomorphism is provided underneath.

Definition 6.1 (Ring Homomorphism) A *ring homomorphism* ϕ from a ring R to a ring S is a mapping from R to S that preserves the two ring operations; that is, for all a, b in R,

(i) $\phi(a + b) = \phi(a) + \phi(b)$
(ii) $\phi(a \cdot b) = \phi(a) \cdot \phi(b)$.

Note that the operations '+' and '·' occurring on the left hand sides of the equations in the above definition are defined on ring R while those occurring on the right hand sides are defined on ring S.

Based on certain other properties satisfied by the function, ring homomorphism can be further classified as mentioned below:

Definition 6.2 (Ring Monomorphism) Recall that a function $\phi\colon R \to S$ is called *one-one* if $\phi(a) = \phi(b)$ implies $a = b$. In other words, we can say that whenever $a \neq b$, we have $\phi(a) \neq \phi(b)$ for all a, $b \in R$. If ϕ is one-one and a ring homomorphism, then it is called a *ring monomorphism*.

Definition 6.3 (Ring Epimorphism) A function $\phi\colon R \to S$ is called *onto* if for every $y \in S$, there exists an $x \in R$ such that $\phi(x) = y$.

If ϕ is onto and ring homomorphism, then it is called a *ring epimorphism*.

Definition 6.4 (Ring Isomorphism) A mapping $\phi\colon R \to S$ is called a *ring isomorphism* if it is one-one, onto and a ring homomorphism.

Definition 6.5 (Ring Endomorphism) A homomorphism $\phi\colon R \to R$ is called *ring endomorphism*.

Now, to understand the concept of ring homomorphism and isomorphism in a better way, take a look at the examples discussed below.

Example 6.6 Let R be a ring and let $\phi\colon R \to R$ be a map defined as $\phi(x) = 0$ for all $x \in R$. Then, it can be easily verified that ϕ is a ring homomorphism called the *trivial homomorphism*.

$(\phi(x + y) = 0 = 0 + 0 = \phi(x) + \phi(y)$ and $\phi(xy) = 0 = 0 \cdot 0 = \phi(x) \cdot \phi(y).)$

Example 6.7 Let R be a ring. Then, the mapping $I\colon R \to R$ defined as $I(x) = x$ for all $x \in R$ is a ring homomorphism known as the *identity homomorphism*.

For x, $y \in R$, consider $I(x + y) = x + y = I(x) + I(y)$ and $I(xy) = xy = I(x)I(y)$.

Example 6.8 For any positive integer n, the mapping $k \longmapsto k \bmod n$ is a ring homomorphism from $\mathbb{Z}$ onto $\mathbb{Z}_n$ called *natural homomorphism* from $\mathbb{Z}$ to $\mathbb{Z}_n$.

Let us define a map $\phi\colon \mathbb{Z} \to \mathbb{Z}_n$ such that $\phi(k) = k \bmod n$.

Clearly, ϕ is well-defined. Let k_1, $k_2 \in \mathbb{Z}$, consider

$$\phi(k_1 + k_2) = (k_1 + k_2) \bmod n$$
$$= k_1 \bmod n + k_2 \bmod n$$
$$= \phi(k_1) \oplus_n \phi(k_2).$$

Also,

$$\phi(k_1 \cdot k_2) = k_1 \cdot k_2 \bmod n$$
$$= k_1 \bmod n \cdot k_2 \bmod n$$
$$= \phi(k_1) \odot_n \phi(k_2).$$

Hence, ϕ is a ring homomorphism.

The map defined in Example 6.8 is a ring homomorphism but not an isomorphism as ϕ is not one-one. Let $x \in \mathbb{Z}$ such that $x = 2n + 1$ for $n \in \mathbb{Z}$. Thus, $\phi(2n + 1) = (2n + 1) \bmod n = (2n \bmod n) + (1 \bmod n) = 0 + 1 = 1$. Now, let $y \in \mathbb{Z}$ such that $y = n + 1$. Then, $\phi(n + 1) = (n + 1) \bmod n = (n \bmod n) + (1 \bmod n) = 0 + 1 = 1$.

Therefore, $\phi(2n + 1) = \phi(n + 1)$ but $2n + 1 \neq n + 1$ in $\mathbb{Z}$. Thus, ϕ is not one-one. Hence, ϕ is not an isomorphism.

However, ϕ is onto and therefore an epimorphism.

Example 6.9 The mapping $\phi \colon \mathbb{C} \to \mathbb{C}$ defined by

$$\phi(a + ib) = (a - ib)$$

is a ring isomorphism.

First, we show that the mapping ϕ is well-defined. Let, $a + ib = c + id$ therefore, we have $(a - c) + i(b - d) = 0$. Thus $(a - c) + i(b - d) = 0 + 0i$. On comparing real and imaginary parts, we get $a - c = 0$ and $b - d = 0$. Hence, $a = c$ and $b = d$, thus $a - ib = c - id$. Therefore, $\phi(a + ib) = \phi(c + id)$. Thus, ϕ is well defined.

Now, we show that ϕ is one-one. Let, $\phi(a + ib) = \phi(c + id)$, therefore $a - ib = c - id$ which implies $a = c$, $b = d$ on comparing the real and imaginary parts. Thus, $a + ib = c + id$. Hence, ϕ is one-one.

Further, we show that ϕ is onto. For each $x + iy \in \mathbb{C}$, there exists $x - iy \in \mathbb{C}$ such that $\phi(x - iy) = x + iy$. Therefore, ϕ is onto.

Finally, we show that ϕ is a ring homomorphism.

Let $a + ib, c + id \in \mathbb{C}$. Then,

$$\phi((a + ib) + (c + id)) = \phi((a + c) + i(b + d))$$
$$= (a + c) - i(b + d)$$
$$= (a - ib) + (c - id)$$
$$= \phi(a + ib) + \phi(c + id).$$

Also,

$$\phi((a + ib)(c + id)) = \phi(ac + iad + ibc - bd)$$
$$= \phi((ac - bd) + i(ad + bc))$$
$$= (ac - bd) - i(ad + bc)$$
$$= ac - bd - iad - ibc$$
$$= ac - ibc - bd - iad$$
$$= (a - ib)c - (ia + b)d$$

$$= (a - ib)c - i(a - ib)d$$
$$= (a - ib)(c - id)$$
$$= \phi(a + ib)\phi(c + id).$$

Hence, ϕ is a ring isomorphism.

Example 6.10 The mapping $\phi\colon \mathbb{Z}_4 \to \mathbb{Z}_{20}$ defined as $\phi(x) = 5x$ is a ring homomorphism.
   Firstly, we will show that the mapping ϕ is well-defined. For this
   Let, for $x,\ y \in \mathbb{Z}$ such that $x = y$. Thus, $x - y = 0$ in $\mathbb{Z}_4$.
$\Rightarrow x - y = 4q$ for some $q \in \mathbb{Z}$
$\Rightarrow 5x - 5y = 20q \Rightarrow 5x - 5y = 0$ in $\mathbb{Z}_{20}$
$\Rightarrow 5x = 5y$ in $\mathbb{Z}_{20}$
$\Rightarrow \phi(x) = \phi(y)$. Thus, ϕ is well-defined.
Now, we show that $\phi(x + y) = \phi(x) + \phi(y)$ for all $x,\ y \in \mathbb{Z}_4$. By applying division algorithm
on $x + y$ and 4, there exist $q,\ r \in \mathbb{Z}$ such that
$x + y = 4q + r$ with $0 \le r < 4$
Thus $x + y = r$ in $\mathbb{Z}_4$
Consider $\phi(x + y) = \phi(r) = 5r = 5(x + y - 4q) = 5x + 5y - 20q = 5x + 5y$ in $\mathbb{Z}_{20}$
Hence, $\phi(x + y) = \phi(x) + \phi(y)$.
Now, we will show that $\phi(xy) = \phi(x)\phi(y)$ for all $x,\ y \in \mathbb{Z}_4$
By division algorithm for xy and 4, there exist $q,\ r \in \mathbb{Z}$ such that
$xy = 4q + r$ with $0 \le r < 4$
Therefore, $xy = r$ in $\mathbb{Z}_4$.
Now, consider
   $\phi(xy) = \phi(r) = 5r = 5(xy - 4q) = 5xy - 20q = (5 \cdot 5)(xy)$ in $\mathbb{Z}_{20}$ (as $5 \cdot 5 = 5$ in $\mathbb{Z}_{20}$).
Thus, $\phi(xy) = (5 \cdot 5)xy = 5x \cdot 5y = \phi(x) \cdot \phi(y)$.
Hence, ϕ is a ring homomorphism.

Example 6.11 The mapping $\phi\colon \mathbb{Z}_5 \to \mathbb{Z}_{10}$ defined by $\phi(x) = 5x$ is not a ring
homomorphism since it does not preserve addition.
   Let us choose, $x = 2,\ y = 4$ in $\mathbb{Z}_5$. Then, $\phi(2 \oplus_5 4) = \phi(6 \bmod 5) = \phi(1) = 5(1) = 5$
in $\mathbb{Z}_{10}$.
   Further, $\phi(2) = 5(2) = 10 = 0$ in $\mathbb{Z}_{10}$ and $\phi(4) = 5(4) = 20 = 0$ in $\mathbb{Z}_{10}$. Therefore,
$\phi(2) \oplus_{10} \phi(4) = 0 \oplus_{10} 0 = 0$. Thus, $\phi(2 \oplus_5 4) \neq \phi(2) \oplus_{10} \phi(4)$.

Example 6.12 The mapping $\phi\colon \mathbb{Z}_4 \to \mathbb{Z}_{12}$ defined by $\phi(x) = 3x$ is not a ring
homomorphism as it does not preserve multiplication.
   Let us choose, $x = 1,\ y = 3$ in $\mathbb{Z}_4$. Then,
   $\phi(x \odot_4 y) = \phi(1 \odot_4 3) = \phi(3 \bmod 4) = \phi(3) = 3(3) = 9$ in $\mathbb{Z}_{12}$.
   Also, $\phi(x) = \phi(1) = 3(1) = 3$ in $\mathbb{Z}_{12}$ and $\phi(y) = \phi(3) = 3(3) = 9$ in $\mathbb{Z}_{12}$. Therefore,
$\phi(x) \odot_{12} \phi(y) = 3 \odot_{12} 9 = 27 \bmod 12 = 3$. Thus, $\phi(x \odot_4 y) \neq \phi(x) \odot_{12} \phi(y)$.

Example 6.13 Let $\mathbb{R}[x]$ denote the ring of all polynomials with real coefficients. Consider
the mapping $\phi\colon \mathbb{R}[x] \to \mathbb{R}$, defined as $\phi(f(x)) = f(1)$.
   Here, the mapping ϕ is well defined as for $f(x),\ g(x) \in \mathbb{R}[x]$. Suppose we have that
$f(x) = g(x)$. Then, $f(1) = g(1)$, therefore $\phi(f(x)) = \phi(g(x))$. Hence, ϕ is well-defined.

Next, we show that ϕ is onto. Let, $y \in \mathbb{R}$. Take $f(x) = y$ for all $x \in \mathbb{R}$.
Then, $\phi(f(x)) = f(1) = y$. Therefore, ϕ is onto.
Further, we show that ϕ is a ring homomorphism.
  Let $f(x),\, g(x) \in \mathbb{R}[x]$.

$$\begin{aligned}
\phi(f(x) + g(x)) &= \phi((f+g)(x)) \\
&= (f+g)(1) \\
&= f(1) + g(1) \\
&= \phi(f(x)) + \phi(g(x)).
\end{aligned}$$

Also,

$$\begin{aligned}
\phi(f(x) \cdot g(x)) &= \phi((f \cdot g)(x)) \\
&= (f \cdot g)(1) \\
&= f(1) \cdot g(1) \\
&= \phi(f(x)) \cdot \phi(g(x)).
\end{aligned}$$

Therefore, ϕ is a ring homomorphism.
However, ϕ is not one-one as for $f(x) = 2 + 3x + 4x^2$ and $g(x) = 1 + 4x + 4x^2$, we have
$\phi(f(x)) = f(1) = 2 + 3(1) + 4(1)^2 = 2 + 3 + 4 = 9$ and $\phi(g(x)) = g(1) = 1 + 4(1) + 4(1)^2$
$= 1 + 4 + 4 = 9$ which implies that $\phi(f(x)) = \phi(g(x))$. But $f(x) \neq g(x)$. Therefore, ϕ is
not one-one. Hence, ϕ is an epimorphism.

Example 6.14 Consider the function $\phi\colon \mathbb{R}[x] \to \mathbb{R}$ defined by $\phi(p(x)) = p(0)$ for any
polynomial $p(x) \in \mathbb{R}[x]$ with real coefficients.
  Then, ϕ is a surjective homomorphism. Let $p(x),\, q(x) \in \mathbb{R}[x]$ be polynomials. Then,
$\phi(p(x) + q(x)) = \phi((p+q)(x)) = (p+q)(0) = p(0) + q(0) = \phi(p(x)) + \phi(q(x))$. Also, $\phi(p(x)$
$q(x)) = \phi((pq)(x)) = (pq)(0) = p(0)q(0) = \phi(p(x))\phi(q(x))$. Thus, ϕ is a homomorphism.
Further, let $a \in \mathbb{R}$. Then, $\phi(x + a) = a$, thus ϕ is surjective.

Example 6.15 Let $F[x]$ represents the ring of polynomials with coefficients in a field F,
and $F(a)$ represents the field of rational functions in a with coefficients in a field F.
  Then, the mapping $\alpha\colon F[x] \to F(a)$ defined by $\alpha(h(x)) = h(a)$ for $h(x) \in F(x)$ is a
homomorphism.
  Let $f(x),\, g(x)$ be two elements in $F[x]$. Then,

$$\begin{aligned}
\alpha(f(x) + g(x)) &= \alpha(f+g)(x) \\
&= (f+g)(a) \\
&= f(a) + g(a) \\
&= \alpha(f(x)) + \alpha(g(x)).
\end{aligned}$$

Moreover,

$$\alpha(f(x)g(x)) = \alpha(fg)(x)$$
$$= (fg)(a)$$
$$= f(a)g(a)$$
$$= \alpha(f(x))\alpha(g(x)).$$

Hence, α is a homomorphism.

Example 6.16 Consider the ring $C[0,1]$, the space of all real-valued continuous functions defined on the closed interval $[0,1]$. Then, the map $\phi : C[0,1] \to \mathbb{R}$ defined as $\phi(f) = f(1/2)$ is an epimorphism.

Let $f,\ g \in C[0,1]$. Then the operations '+' and '$\cdot$' are defined by $(f+g)(x) = f(x) + g(x)$ and $(f \cdot g)(x) = f(x)g(x)$ for all $x \in [0,1]$.

Now, consider $\phi(f+g) = (f+g)(1/2) = f(1/2) + g(1/2) = \phi(f) + \phi(g)$ also $\phi(fg) = (fg)(1/2) = f(1/2)g(1/2) = \phi(f)\phi(g)$.

Thus, ϕ is a ring homomorphism.

Now, ϕ is onto because for any $r \in \mathbb{R}$, consider the constant function $g : [0,1] \to \mathbb{R}$ such that $g(x) = r$. We already know that g is continuous on $[0,1]$. Also $g(1/2) = r$. Thus, $\phi(g) = r$. Hence, ϕ is an epimorphism.

However, ϕ is not one-one. For this, let us consider, $f : [0,1] \to \mathbb{R}$ such that $f(x) = x - \dfrac{1}{2}$ and g is the zero function, that is $g(x) = 0$ for all $x \in [0,1]$. Then $\phi(f) = \phi(g)$, but $f \neq g$.

Example 6.17 The mapping from $M_2(\mathbb{Z})$ into $\mathbb{Z}$ given by

$$\begin{bmatrix} a & b \\ c & d \end{bmatrix} \longmapsto a$$

is not a ring homomorphism as for elements

$$\begin{bmatrix} 1 & 1 \\ 1 & 2 \end{bmatrix} \text{ and } \begin{bmatrix} 2 & 1 \\ 1 & 3 \end{bmatrix} \in M_2(\mathbb{Z}),$$

$$\phi\left(\begin{bmatrix} 1 & 1 \\ 1 & 2 \end{bmatrix}\begin{bmatrix} 2 & 1 \\ 1 & 3 \end{bmatrix}\right) = \phi\left(\begin{bmatrix} 3 & 4 \\ 4 & 7 \end{bmatrix}\right)$$
$$= 3 \neq 2$$
$$= \phi\left(\begin{bmatrix} 1 & 1 \\ 1 & 2 \end{bmatrix}\right) \cdot \phi\left(\begin{bmatrix} 2 & 1 \\ 1 & 3 \end{bmatrix}\right).$$

Thus, multiplication is not preserved under ϕ. Hence, the mapping $\begin{bmatrix} a & b \\ c & d \end{bmatrix} \mapsto a$ is not a ring homomorphism.

Example 6.18 Consider the map ϕ from $\mathbb{Z} \to \mathbb{Z}$ defined by $\phi(n) = n^k$, where k is a fixed integer. Then the map ϕ is not a ring homomorphism unless $k = 1$ because if $k = 2$, then since $2 = 1^2 + 1^2 \neq (1+1)^2 = 4$, therefore it is not preserved under addition. When $k = 1$, the map becomes the identity map, and the conditions for a ring homomorphism are satisfied.

Example 6.19 Let $S = \left\{ \begin{bmatrix} a & b \\ -b & a \end{bmatrix} \,\middle|\, a, b \in \mathbb{R} \right\}$. Then, the mapping $\phi : \mathbb{C} \to S$ given by

$$\phi(a + ib) = \begin{bmatrix} a & b \\ -b & a \end{bmatrix} \text{ is a ring isomorphism.}$$

First, we see that ϕ is a bijection map since it has a two-sided inverse; namely, the map ϕ^{-1} given as

$$\phi^{-1}\left(\begin{bmatrix} a & b \\ -b & a \end{bmatrix} \right) = a + ib.$$

Furthermore, for any two complex numbers $a + ib$ and $c + id$,

$$\phi((a + ib) + (c + id)) = \phi((a + c) + i(b + d))$$
$$= \begin{bmatrix} a + c & b + d \\ -(b + d) & a + c \end{bmatrix}$$
$$= \begin{bmatrix} a & b \\ -b & a \end{bmatrix} + \begin{bmatrix} c & d \\ -d & c \end{bmatrix}$$
$$= \phi(a + ib) + \phi(c + id).$$

Therefore, addition is preserved.

Similarly, we have

$$\phi((a + ib)(c + id)) = \phi((ac - bd) + i(ad + bc))$$
$$= \begin{bmatrix} ac - bd & ad + bc \\ -(ad + bc) & ac - bd \end{bmatrix}$$
$$= \begin{bmatrix} a & b \\ -b & a \end{bmatrix} \cdot \begin{bmatrix} c & d \\ -d & c \end{bmatrix}$$
$$= \phi(a + ib) \cdot \phi(c + id).$$

Therefore, multiplication is also preserved. Thus, ϕ is a ring homomorphism.

Since, ϕ is one-one, onto and ring homomorphism, hence, ϕ is a ring isomorphism.

Example 6.20 Let $\mathbb{Z}[\sqrt{2}] = \{a + b\sqrt{2} \mid a, b \in \mathbb{Z}\}$ and $H = \left\{ \begin{bmatrix} a & 2b \\ b & a \end{bmatrix} \,\middle|\, a, b \in \mathbb{Z} \right\}$.

Then, the rings $\mathbb{Z}[\sqrt{2}]$ and H are isomorphic.

Define a mapping $\phi : H \to \mathbb{Z}[\sqrt{2}]$ by $\phi\left(\begin{bmatrix} a & 2b \\ b & a \end{bmatrix}\right) = a + b\sqrt{2}$.

Then, for $\begin{bmatrix} a & 2b \\ b & a \end{bmatrix}, \begin{bmatrix} c & 2d \\ d & c \end{bmatrix} \in H$, we have

$$\phi\left(\begin{bmatrix} a & 2b \\ b & a \end{bmatrix} + \begin{bmatrix} c & 2d \\ d & c \end{bmatrix}\right) = \phi\left(\begin{bmatrix} a+c & 2b+2d \\ b+d & a+c \end{bmatrix}\right) = (a+c) + (b+d)\sqrt{2} =$$

$$(a + b\sqrt{2}) + (c + d\sqrt{2}) = \phi\left(\begin{bmatrix} a & 2b \\ b & a \end{bmatrix}\right) + \phi\left(\begin{bmatrix} c & 2d \\ d & c \end{bmatrix}\right).$$

And $\phi\left(\begin{bmatrix} a & 2b \\ b & a \end{bmatrix}\begin{bmatrix} c & 2d \\ d & c \end{bmatrix}\right) = \phi\left(\begin{bmatrix} ac+2bd & 2ad+2bc \\ ad+bc & ac+2bd \end{bmatrix}\right) = (ac + 2bd) +$

$(ad + bc)\sqrt{2} = (a + b\sqrt{2})(c + d\sqrt{2}) = \phi\left(\begin{bmatrix} a & 2b \\ b & a \end{bmatrix}\right) \cdot \phi\left(\begin{bmatrix} c & 2d \\ d & c \end{bmatrix}\right).$

Hence, ϕ is a ring homomorphism.

Now, ϕ is surjective since, given any $x \in \mathbb{Z}[\sqrt{2}]$, we have that $x = a + b\sqrt{2}$ for some $a, b \in \mathbb{Z}$,

so $\begin{bmatrix} a & 2b \\ b & a \end{bmatrix} \in H$ such that $\phi\left(\begin{bmatrix} a & 2b \\ b & a \end{bmatrix}\right) = a + b\sqrt{2} = x$.

Finally, to see that ϕ is injective, suppose that $\begin{bmatrix} a & 2b \\ b & a \end{bmatrix}, \begin{bmatrix} c & 2d \\ d & c \end{bmatrix} \in H$ and $\phi\left(\begin{bmatrix} a & 2b \\ b & a \end{bmatrix}\right) =$

$\phi\left(\begin{bmatrix} c & 2d \\ d & c \end{bmatrix}\right)$. This means that $a + b\sqrt{2} = c + d\sqrt{2}$. If $b \neq d$ (so that $d - b \neq 0$), then this

would imply that $\sqrt{2} = \dfrac{a - c}{d - b}$ is rational, a contradiction. Hence, $b = d$. Then, $a + b\sqrt{2} =$

$c + d\sqrt{2} = c + b\sqrt{2}$.

Subtracting $b\sqrt{2}$ from both sides of the above equation shows that $a = c$. Thus, $\begin{bmatrix} a & 2b \\ b & a \end{bmatrix} =$

$\begin{bmatrix} c & 2d \\ d & c \end{bmatrix}$, so it follows that ϕ is injective. Hence, ϕ is a ring isomorphism.

Example 6.21 The ring $2\mathbb{Z}$ is not isomorphic to the ring $3\mathbb{Z}$. Let, if possible, $2\mathbb{Z} \cong 3\mathbb{Z}$ then there exists an isomorphism $\phi : 2\mathbb{Z} \to 3\mathbb{Z}$.

Let, $\phi(2) = a$, $a \in 3\mathbb{Z}$. Since, ϕ is a ring homomorphism. So, $\phi(4) = \phi(2 + 2) = \phi(2) + \phi(2) = a + a = 2a$.

Also, $\phi(4) = \phi(2 \cdot 2) = \phi(2) \cdot \phi(2) = a \cdot a = a^2$.

So, $a^2 = 2a \Rightarrow a(a - 2) = 0 \Rightarrow a = 0$ or $(a - 2) = 0$.

If $a = 0$, then $\phi(2) = 0$. Also, $\phi(4) = 0$ which contradicts the assumption that ϕ is an isomorphism and hence one-one. If $a - 2 = 0$, then $a = 2$, which is a contradiction to $a \in 3\mathbb{Z}$. Hence, $2\mathbb{Z} \ncong 3\mathbb{Z}$.

Example 6.22 The mapping $f : \mathbb{Z} \to \mathbb{Z}$ defined as $f(x) = x$ is a ring homomorphism. For $m, n \in \mathbb{Z}$, $m = n \Rightarrow f(m) = f(n)$.

Further, $f(m+n) = m+n = f(m) + f(n)$ and $f(m \cdot n) = m \cdot n = f(m) \cdot f(n)$ for m, $n \in \mathbb{Z}$. Hence, f is a ring homomorphism.

However, the mapping $h(x) = -f(x) = -x$ is not a ring homomorphism since for $m, n \in \mathbb{Z}$, $h(m \cdot n) = -(m \cdot n)$ while $h(m) \cdot h(n) = (-m) \cdot (-n) = m \cdot n$. Thus, multiplication is not preserved.

Example 6.23 Consider the map $g : \mathbb{Z} \to \mathbb{Z}$ such that $g(x) = 2x$. Then, g is not a ring homomorphism. For $m, n \in \mathbb{Z}$, take $m = 1$ and $n = 1$. Then, $g(m \cdot n) = g(1 \cdot 1) = g(1) = 2 \cdot 1 = 2$ and $g(m) \cdot g(n) = g(1) \cdot g(1) = (2 \cdot 1)(2 \cdot 1) = 2 \cdot 2 = 4$.
Since, $g(1 \cdot 1) \neq g(1) \cdot g(1)$. Thus, multiplication is not preserved. Hence, g is not a ring homomorphism.

In ring theory, understanding the properties of ring homomorphisms is crucial for exploring the relationships between different rings. In the next theorem, we will examine some properties of ring homomorphisms.

Theorem 6.24 *Let $\phi : R \to S$ be a ring homomorphism. Then*

    (i) $\phi(0_R) = 0_S$.
   (ii) $\phi(-r) = -\phi(r)$ *for all* $r \in R$.
  (iii) *If* $r \in U(R)$ *then* $\phi(r) \in U(S)$ *and* $\phi(r^{-1}) = \phi(r)^{-1}$.
  (iv) *If* $R' \subset R$ *is a subring, then* $\phi(R')$ *is a subring of* S.

Proof   **(i)** Since $0_R + 0_R = 0_R$, $\phi(0_R) + \phi(0_R) = \phi(0_R)$. This implies $\phi(0_R) + \phi(0_R) = \phi(0_R) + 0_S$, which implies $\phi(0_R) = 0_S$ (by left cancellation).

  (ii) Let $r \in R$. Since $r + -r = -r + r = 0_R$, we have $\phi(r) + \phi(-r) = \phi(-r) + \phi(r) = \phi(0_R) = 0_S$, where the last equality comes from (i). Thus $\phi(-r) = -\phi(r)$ as additive inverses are unique.

 (iii) Suppose that $r \in U(R)$. Then there exists $r^{-1} \in R$ such that $r \cdot r^{-1} = r^{-1} \cdot r = 1_R$. Then since ϕ is a ring homomorphism we have $\phi(r) \cdot \phi(r^{-1}) = \phi(r^{-1})\phi(r) = \phi(r^{-1}.r) = \phi(1_R) = 1_S$. Thus $\phi(r)$ has a multiplicative inverse and it is $\phi(r^{-1})$.

 (iv) Let $R' \subset R$ be a subring. To show that $\phi(R')$ is a subring we must show that $0_S \in \phi(R')$ and for all $s_1, s_2 \in \phi(R')$, $s_1 - s_2$ and $s_1 s_2$ are also in $\phi(R')$. Clearly, $\phi(0_R) \in \phi(R)$ as $0_R \in R'$. This implies $0_S \in \phi(R)$ (using (i)). Now, since $s_1, s_2 \in \phi(R')$, there exists $r_1, r_2 \in R'$ such that $\phi(r_1) = s_1$ and $\phi(r_2) = s_2$. Thus, $s_1 - s_2 = \phi(r_1) - \phi(r_2) = \phi(r_1) + \phi(-r_2) = \phi(r_1 - r_2)$, and $s_1 s_2 = \phi(r_1)\phi(r_2) = \phi(r_1 r_2)$. Since R' is a subring, $r_1 - r_2$ and $r_1 r_2$ are contained in R'. Hence, $s_1 - s_2$ and $s_1 s_2$ are in $\phi(R')$. Therefore, $\phi(R')$ is a subring of S.

$\square$

In the consecutive theorem, fundamental properties of ring homomorphism have been discussed.

Theorem 6.25 (Properties of Ring Homomorphisms) *Let ϕ be a ring homomor-phism from a ring R to a ring S. Let A be a subring of R.*

  (i) *For any $r \in R$ and any positive integer n, $\phi(nr) = n\phi(r)$ and $\phi(r^n) = (\phi(r))^n$.*
 (ii) *$\phi(A) = \{\phi(a) \mid a \in A\}$ is a subring of S, i.e., homomorphic image of a subring is a subring.*

(iii) *If A is an ideal and ϕ is onto, then $\phi(A)$ is an ideal.*

(iv) *If R is commutative, then $\phi(R)$ is commutative.*

 (v) *If R has a unity 1, $S \neq \{0\}$, and ϕ is onto, then $\phi(1)$ is the unity of S.*

(vi) *If ϕ is an isomorphism from R onto S, then ϕ^{-1} is an isomorphism from S onto R.*

Proof   **(i)** For any $r \in R$ and any positive integer n, consider
$$\phi(nr) = \phi\underbrace{(r + r + ... + r)}_{n \text{ times}} = \underbrace{\phi(r) + \phi(r) + ... + \phi(r)}_{n \text{ times}} = n\phi(r).$$
Also, $\phi(r^n) = \phi\underbrace{(r \cdot r \cdot r ... \cdot r)}_{n \text{ times}} = \underbrace{\phi(r) \cdot \phi(r) \cdot ... \cdot \phi(r)}_{n \text{ times}} = (\phi(r))^n.$

(ii) Since A is a subring of R, then A is a ring in itself. So, $0 \in A \Rightarrow \phi(0) \in \phi(A) \Rightarrow$ $\phi(A) \neq \emptyset$. Clearly, $\phi(A) \subseteq S$.

Now, suppose that $\phi(a), \phi(b) \in \phi(A)$, where $a, b \in A$. We need to show that $\phi(a) - \phi(b) \in \phi(A)$ and $\phi(a)\phi(b) \in \phi(A)$.

Since $a, b \in A$ and A is a subring of R, so $a - b \in A$ and $ab \in A$ which implies that $\phi(a - b) \in \phi(A)$ and $\phi(ab) \in \phi(A)$. Therefore, $\phi(a) - \phi(b) \in \phi(A)$ and $\phi(a)\phi(b)$ $\in \phi(A)$ ($\because \phi$ is a ring homomorphism, so $\phi(ab) = \phi(a)\phi(b)$ and $\phi(a - b) = \phi(a + (-b))$ $= \phi(a) + \phi(-b) = \phi(a) - \phi(b)$.)

Hence, $\phi(A)$ is a subring of S.

(iii) Let A be an ideal and ϕ be onto. We show that $\phi(A)$ is an ideal of S.

From property (ii), $\phi(A)$ is a subring of S.

Now, suppose that $\phi(a) \in \phi(A)$. Let, $s \in S$. We need to show that $\phi(a) \cdot s \in \phi(A)$ and $s\phi(a) \in \phi(A)$.

Since $\phi : R \to S$ is onto thus for $s \in S$, there exists $r \in R$ such that $\phi(r) = s$.

Consider, $\phi(a) \cdot s = \phi(a) \cdot \phi(r) = \phi(ar) \in \phi(A)$ ($\because a \in A$, $r \in R$ and A is an ideal of $R \Rightarrow ar \in A$.)

Similarly, $s \cdot \phi(a) \in \phi(A)$. Hence, $\phi(A)$ is an ideal of S.

(iv) We have $\phi(R) = \{\phi(x) \mid x \in R\}$. Let $\phi(a), \phi(b) \in \phi(R)$, where $a, b \in R$. We show that $\phi(a) \cdot \phi(b) = \phi(b) \cdot \phi(a)$.

Consider $\phi(a) \cdot \phi(b) = \phi(ab) = \phi(ba) = \phi(b) \cdot \phi(a)$ for some $a, b \in R$
$$(\because R \text{ is commutative and } \phi \text{ is homomorphism.})$$
Therefore, $\phi(a) \cdot \phi(b) = \phi(b) \cdot \phi(a)$. Hence, $\phi(R)$ is commutative.

 (v) The mapping $\phi : R \to S$ is onto which implies that for $s \in S$, there exists $r \in R$ such that $\phi(r) = s$. We show that $\phi(1)$ is unity of S, i.e., $\phi(1) \cdot s = s \cdot \phi(1) = s$, where $s \in S$, that is we need to show that $\phi(1)\phi(r) = \phi(r)\phi(1) = \phi(r)$ ($\because \phi(r) = s$.)

Consider $\phi(1) \cdot \phi(r) = \phi(1 \cdot r) = \phi(r) = s$ ($\because \phi$ is homomorphism and since $1 \in R$ is unity $\Rightarrow 1 \cdot r = r \cdot 1 = r \ \forall \ r \in R$.)

And $\phi(r) \cdot \phi(1) = \phi(r \cdot 1) = \phi(r) = s$.   ($\because \phi$ is homomorphism and since $1 \in R$ is unity $\Rightarrow 1 \cdot r = r \cdot 1 = r \ \forall \ r \in R$.)

Thus, we have $\phi(1) \cdot \phi(r) = \phi(r) \cdot \phi(1) = \phi(r) \Rightarrow \phi(1) \cdot s = s \cdot \phi(1) = s \Rightarrow \phi(1)$ is the unity of S.

(vi) Since the mapping $\phi : R \to S$ is an isomorphism, so ϕ is one-one, onto and homomorphism. Since ϕ is one-one and onto, the inverse of ϕ exists, and the inverse is also one-one and onto. Therefore, ϕ^{-1} is one-one and onto. Now, we show that

ϕ^{-1} is homomorphism. i.e., for $x, \ y \in S$, $\phi^{-1}(x+y) = \phi^{-1}(x) + \phi^{-1}(y)$ and $\phi^{-1}(xy) = \phi^{-1}(x) \cdot \phi^{-1}(y)$.

Let $\phi^{-1}(x+y) = a \Rightarrow x+y = \phi(a)$, $\phi^{-1}(x) = b \Rightarrow x = \phi(b)$ and $\phi^{-1}(y) = c \Rightarrow y = \phi(c)$.

Then, consider $x+y = \phi(a) \Rightarrow \phi(b) + \phi(c) = \phi(a) \Rightarrow \phi(b+c) = \phi(a)$

$(\because \phi$ is homomorphism.$)$

$\Rightarrow b + c = a$ $\hfill (\because \phi$ is one-one.$)$

$\Rightarrow \phi^{-1}(x) + \phi^{-1}(y) = \phi^{-1}(x+y)$.

Now. we show that $\phi^{-1}(xy) = \phi^{-1}(x) \cdot \phi^{-1}(y)$.

Let $\phi^{-1}(xy) = p \Rightarrow xy = \phi(p)$, $\phi^{-1}(x) = b \Rightarrow x = \phi(b)$ and $\phi^{-1}(y) = c \Rightarrow y = \phi(c)$.

Then, $xy = \phi(p) \Rightarrow \phi(b)\phi(c) = \phi(p) \Rightarrow \phi(bc) = \phi(p)$ $\hfill (\because \phi$ is homomorphism.$)$

$\Rightarrow bc = p$ $\hfill (\because \phi$ is one-one.$)$

$\Rightarrow \phi^{-1}(x) \cdot \phi^{-1}(y) = \phi^{-1}(xy)$.

Therefore, ϕ^{-1} is homomorphism.

Hence, ϕ^{-1} is an isomorphism. $\hfill \square$

After exploring the fundamental properties of ring homomorphisms, we delve into two key concepts associated with these mappings i.e., the image and the kernel. The image of a ring homomorphism represents the subset of the target ring that is actually covered by the homomorphism, showcasing the range of its influence. Understanding the image provides insights into the essential characteristics of the mapping. On the other hand, the kernel of a ring homomorphism is the set of elements in the source ring that map to the additive identity in the target ring. Understanding the kernel unveils insights into the structure and behavior of the homomorphism.

Definition 6.26 (Image of a Ring Homomorphism) Let $\phi : R \to S$ be a ring homomorphism. Then, the image of ϕ is defined as

$$\text{Im } \phi = \{\phi(x) \mid x \in R\}.$$

Definition 6.27 (Kernel of a Ring Homomorphism) Let $\phi : R \to S$ be a ring homomorphism. Then, the kernel of ϕ is defined as

$$\text{Ker } \phi = \{x \in R \mid \phi(x) = 0\}.$$

The *kernel* of a ring homomorphism is the set of elements that are mapped to 0.

Example 6.28 Refer to the map defined in Example 6.6, the image of the trivial homomorphism is

$\text{Im } \phi = \{\phi(x) \mid x \in R\} = \{0\}$

and its kernel is given by

$\text{Ker } \phi = \{x \in R \mid \phi(x) = 0\} = R$.

Example 6.29 Refer to the map defined in Example 6.7. The image of the identity homomorphism is

$\text{Im } I = \{I(x) \mid x \in R\} = \{x \mid x \in R\} = R$

and its kernel is given by

$\text{Ker } I = \{x \in R \mid I(x) = 0\} = \{x \in R \mid x = 0\} = \{0\}$.

Example 6.30 For mapping $\phi\colon \mathbb{Z} \to \mathbb{Z}_n$ defined by $\phi(k) = k \bmod n$. We have
Ker $\phi = \{x \in \mathbb{Z} \mid \phi(x) = 0\}$.

The additive identity in $\mathbb{Z}_n$ is represented by the equivalence class $[0]_n$, which consists of all integers that are congruent to 0 modulo n. Therefore, the kernel of ϕ is the set of all integers x such that $x \bmod n = 0$, or equivalently, x is a multiple of n. Hence, Ker $\phi = \{x \in \mathbb{Z} \mid x \equiv 0 \bmod n\} = n\mathbb{Z}$.

Example 6.31 For mapping $\phi\colon \mathbb{R}[x] \to \mathbb{R}$ defined as $\phi(f(x)) = f(1)$. The

$$\text{Ker } \phi = \{f(x) \in \mathbb{R}[x] : \phi(f(x)) = 0\}$$
$$= \{f(x) \in \mathbb{R}[x] : f(1) = 0\}$$

If $f(x) = a_0 + a_1 x + a_2 x^2 + ... + a_n x^n$, then $f(1) = a_0 + a_1 + a_2 + ... a_n$ which implies $0 = a_0 + a_1 + a_2 + ... + a_n$. Therefore, Ker ϕ is the set of all those polynomials in $\mathbb{R}[x]$ whose sum of coefficient is zero. Therefore, Ker $\phi = \{a_0 + a_1 x + ... + a_n x^n : a_i \in \mathbb{R} \ \forall \ i$ and $a_0 + a_1 + ... + a_n = 0\}$.

The following proposition shows that the pre-image of an ideal under a ring homomorphism is an ideal, and the kernel of the homomorphism is contained in this pre-image.

Proposition 6.32 *Let ϕ be a ring homomorphism from a ring R to a ring S. Let $B \subseteq S$ be an ideal of S. Then,*

$$\phi^{-1}(B) = \{r \in R \mid \phi(r) \in B\}$$

is an ideal of R and Ker $\phi \subseteq \phi^{-1}(B)$.

Proof Clearly, $\phi^{-1}(B) \subseteq R$. Now, $0 \in R$ as R is a ring. Since $\phi\colon :R \to S$ is a homomorphism and under homomorphism, 0 of domain R is mapped to 0 of S. Therefore, we have $\phi(0) = 0_s \in B$ as B is an ideal of S.
Hence $\phi^{-1}(B) \neq \emptyset$.

Now, let $a, b \in \phi^{-1}(B)$. We need to show that $a - b \in \phi^{-1}(B)$. Since, $a, b \in \phi^{-1}(B)$. So $\phi(a), \phi(b) \in B$.

Then,

$$\phi(a) - \phi(b) \in B$$

$$(\because B \text{ is an ideal of } S\,)$$

Hence,

$$\phi(a - b) \in B.$$

$$(\because \phi \text{ is a homomorphism.})$$

Thus, $a - b \in \phi^{-1}(B)$.
Further, suppose that $a \in \phi^{-1}(B)$ and $r \in R$. We show that $ar \in \phi^{-1}(B)$ and $ra \in \phi^{-1}(B)$. Since, $a \in \phi^{-1}(B)$, so $\phi(a) \in B$. Now, $\phi(ar) = \phi(a)\phi(r) \in B$ because $r \in R$, and $\phi : R \to S$ which implies $\phi(r) \in S$. Also, $\phi(a) \in B$ and B is an ideal.
This implies $ar \in \phi^{-1}(B)$.
Similarly, we have $\phi(ra) \in B$ and hence $ra \in \phi^{-1}(B)$. Thus, $\phi^{-1}(B)$ is an ideal of R.
Further, we need to show that Ker $\phi \subseteq \phi^{-1}(B)$.
If $x \in$ Ker ϕ, then $\phi(x) = 0 \in B$. Thus, $x \in \phi^{-1}(B)$. Hence, Ker $\phi \subseteq \phi^{-1}(B)$. $\qquad\square$

The next proposition suggests that for a map $\phi\colon R \to S$ to be an isomorphism, it must be surjective and have a trivial kernel, ensuring bijection without proving it directly.

Proposition 6.33 *Let ϕ be a ring homomorphism from a ring R to a ring S. Then, ϕ is an isomorphism if and only if ϕ is onto and $\mathrm{Ker}\,\phi = \{0\}$.*

Proof Let ϕ be an onto ring homomorphism such that $\mathrm{Ker}\,\phi = \{0\}$. In order to show that ϕ is an isomorphism, we only need to show that ϕ is one-one.

Let, $\phi(x) = \phi(y)$ for some $x,\, y \in R$ which implies that $\phi(x) - \phi(y) = 0$, that is, $\phi(x - y) = 0$ $\hspace{2cm}$ ($\because \phi$ is homomorphism.)

Thus, we have $x - y \in \mathrm{Ker}\,\phi$. Since $\mathrm{Ker}\,\phi = \{0\}$, therefore $x - y \in \{0\}$ and hence $x = y$. Therefore, ϕ is one-one. Hence, ϕ is an isomorphism.

Conversely, let ϕ be an isomorphism. We have to show that ϕ is onto and $\mathrm{Ker}\,\phi = \{0\}$. Since, ϕ is an isomorphism, then ϕ is one-one, onto and homomorphism. Next, we show that $\mathrm{Ker}\,\phi = \{0\}$.

Let $x \in \mathrm{Ker}\,\phi$ therefore, $\phi(x) = 0 \Rightarrow \phi(x) = \phi(0) \Rightarrow x = 0$ $\hspace{1.5cm}$ (ϕ is one-one.)

Hence, $\mathrm{Ker}\,\phi = \{0\}$. $\hspace{14cm}$ $\square$

The subsequent proposition proves that kernel and image of a ring homomorphism are subrings.

Proposition 6.34 *Let $\phi\colon R \to S$ be a ring homomorphism. Then the kernel of ϕ is a subring of R and the image of ϕ is a subring of S.*

Proof Let $\phi : R \to S$ be a ring homomorphism. Clearly, $0 \in \mathrm{Ker}\,\phi$ as $\phi(0) = 0$. Therefore, $\mathrm{Ker}\,\phi$ is non-empty.

Let $x,\, y \in \mathrm{Ker}\,\phi$. Then, $\phi(x) = 0$, $\phi(y) = 0$.

Consider, $\phi(x - y) = \phi(x) - \phi(y) = 0_S - 0_S = 0_S$. Thus, $x - y \in \mathrm{Ker}\,\phi$.

Also, $\phi(xy) = \phi(x)\phi(y) = 0_S 0_S = 0_S$. Thus, $xy \in \mathrm{Ker}\,\phi$.

Therefore, $\mathrm{Ker}\,\phi$ is a subring of R.

Further, we show that image of ϕ, i.e., $\phi(R)$ is a subring of S.

Let $x,\, y \in \phi(R)$, then there exist $a,\, b \in R$ such that $\phi(a) = x$, $\phi(b) = y$.

Therefore, $x - y = \phi(a) - \phi(b) = \phi(a - b) \in \phi(R)$. Also, $xy = \phi(a)\phi(b) = \phi(ab) \in \phi(R)$. Thus, the $\phi(R)$ is a subring of S. $\hspace{6cm}$ $\square$

Proposition 6.35 *Let $\phi\colon R \to S$ be a surjection. Then for any non-empty subset J of S, $\phi(\phi^{-1}(J)) = J$. In particular, if ϕ is an onto ring homomorphism and J is an ideal of S, then $\phi(\phi^{-1}(J)) = J$.*

Proof Let $x \in \phi(\phi^{-1}(J))$. Then, $x = \phi(y)$, for some $y \in \phi^{-1}(J)$. That is, $\phi(y) \in J$, hence, $x \in J$. Thus, $\phi(\phi^{-1}(J)) \subseteq J$. $\hspace{6cm}$ (6.1)

Now, let $x \in J$. Since ϕ is surjective. Therefore, there exists $y \in R$ such that $\phi(y) = x$. Then $y \in \phi^{-1}(x) \subseteq \phi^{-1}(J)$. Therefore, $x = \phi(y) \in \phi(\phi^{-1}(J))$.

Thus, $J \subseteq \phi(\phi^{-1}(J))$. $\hspace{8cm}$ (6.2)

Hence, from Equations (6.1) and (6.2), $\phi(\phi^{-1}(J)) = J$. $\hspace{4cm}$ $\square$

Now, consider a ring epimorphism $\phi : R \to S$. In the above result, we have shown that if J is an ideal of S, then $J = \phi(\phi^{-1}(J))$.

Also, by property (iii) and (iv) of Theorem 6.25, if I is an ideal of F, then $\phi(I)$ is an ideal of S and $\phi^{-1}(\phi(I))$ is an ideal of R.

However, $I \neq \phi^{-1}(\phi(I))$ in general.

For example, consider $\phi : \mathbb{Z} \to \mathbb{Z}_5$ defined as $\phi(m) = m \bmod 5$. Here ϕ is onto. Let's take $I = 2\mathbb{Z}$. Then $5 \notin I$. But $5 \in \phi^{-1}(\phi(I))$, since $\phi(5) = 0 \in \phi(I)$. Therefore, $I \neq \phi^{-1}(\phi(I))$. But there is a relation between the ideals I and $\phi^{-1}(\phi(I))$ of ring R which has been mentioned in the succeeding theorem.

Theorem 6.36 *Let $\phi \colon R \to S$ be a ring homomorphism, and let I be an ideal of a ring R. Then, $\phi^{-1}(\phi(I)) = I + \operatorname{Ker} \phi$.*

Proof Since, $0 \in \phi(I)$, therefore $\operatorname{Ker} \phi = \phi^{-1}(\{0\}) \subseteq \phi^{-1}(\phi(I))$. Also, for $x \in I$, $\phi(x) \in \phi(I)$. Thus $x \in \phi^{-1}(\phi(I))$ and hence, $I \subseteq \phi^{-1}(\phi(I))$. Therefore,

$$I + \operatorname{Ker} \phi \subseteq \phi^{-1}(\phi(I)). \tag{6.3}$$

Now, we need to show that $\phi^{-1}(\phi(I)) \subseteq I + \operatorname{Ker} \phi$.
For this, let $x \in \phi^{-1}(\phi(I))$. Then $\phi(x) \in \phi(I) \Rightarrow \phi(x) = \phi(y)$, for some $y \in I$.
Thus, $\phi(x - y) = 0 \Rightarrow x - y \in \operatorname{Ker} \phi \Rightarrow x \in y + \operatorname{Ker} \phi \subseteq I + \operatorname{Ker} \phi$.
Therefore, $\phi^{-1}(\phi(I)) \subseteq I + \operatorname{Ker} \phi$. $\tag{6.4}$
From Equations (6.3) and (6.4), we conclude that $\phi^{-1}(\phi(I)) = I + \operatorname{Ker} \phi$. $\qquad \square$

The next corollary directly follows from the above theorem.

Corollary 6.37 If $\phi : R \to S$ is a ring homomorphism, and I is an ideal of R containing $\operatorname{Ker} \phi$, then $\phi^{-1}(\phi(I)) = I$.

Proof Since $\operatorname{Ker} \phi \subseteq I$, $I + \operatorname{Ker} \phi = I$. Therefore, $\phi^{-1}(\phi(I)) = I$. $\qquad \square$

Observe the map $\phi \colon \mathbb{Z} \to \mathbb{Z}_n$ defined by $\phi(k) = k \bmod n$. We have already shown that ϕ is a ring homomorphism with $\operatorname{Ker} \phi = n\mathbb{Z}$. Also we know that $n\mathbb{Z}$ is an ideal of $\mathbb{Z}$ for $n \in \mathbb{N}$. Thus, we observe that in this mapping, kernels are ideals. This gives an intuition to generalize this result to other ring homomorphisms also.

Theorem 6.38 (Kernels Are Ideals) *Let $\phi \colon R \to S$ be a ring homomorphism. Then, $\operatorname{Ker} \phi = \{r \in R \mid \phi(r) = 0\}$ is an ideal of R.*

Proof From Proposition 6.34, $\operatorname{Ker} \phi$ is a subring of ring R. This implies that $x - y \in \operatorname{Ker} \phi$ for all $x, y \in \operatorname{Ker} \phi$.

Further, suppose that $x \in \operatorname{Ker} \phi$ and $r \in R$. Now, we show that $xr \in \operatorname{Ker} \phi$ and $rx \in \operatorname{Ker} \phi$.
Since, $x \in \operatorname{Ker} \phi$, therefore $\phi(x) = 0$.
Consider $\phi(rx) = \phi(r)\phi(x) = \phi(r) \cdot 0 = 0$. Thus, $rx \in \operatorname{Ker} \phi$.
Also, $\phi(xr) = \phi(x) \cdot \phi(r) = 0 \cdot \phi(x) = 0$. Thus, $xr \in \operatorname{Ker} \phi$.
Hence, $\operatorname{Ker} \phi$ is an ideal of R. $\qquad \square$

The above theorem finds a notable application in the context of endomorphisms on $\mathbb{Z}_n$, where n is a positive integer. Understanding the relationship between kernels and ideals is particularly insightful when studying the structure and properties of endomorphisms on $\mathbb{Z}_n$, offering a powerful tool for analyzing the algebraic behavior of these mappings within modular arithmetic. Next we have a result for endomorphisms on $\mathbb{Z}_n$, where $n \in \mathbb{N}$.

Proposition 6.39 *Every ring homomorphism ϕ from $\mathbb{Z}_n$ to itself has the form $\phi(x) = ax$, where $a^2 = a$.*

Proof Let, $\phi : \mathbb{Z}_n \to \mathbb{Z}_n$ be a ring homomorphism. Then, for any $m \in \mathbb{Z}_n$, we have (using the fact that $\phi(x + y) = \phi(x) + \phi(y)$)
$\phi(m) = \phi(m \cdot 1) = m \cdot \phi(1) = m \cdot a$ where $a = \phi(1) \in \mathbb{Z}_n$.
Now, using the fact that $\phi(xy) = \phi(x)\phi(y)$ for any $x, y \in \mathbb{Z}_n$, we see that $a = \phi(1) = \phi(1 \cdot 1) = \phi(1)\phi(1) = a^2$.
Thus, we have shown that $\phi(x) = ax$ for all $x \in \mathbb{Z}_n$, with $a^2 = a$. $\qquad\square$

The preceding result provides insight into the mapping structure of idempotent elements under a homomorphism, as further elaborated in the subsequent proposition.

Proposition 6.40 *A ring homomorphism carries an idempotent to an idempotent.*

Proof Take any ring homomorphism $\phi : R \to S$ and any idempotent $a \in R$. Thus, we have $a^2 = a$, and we need to prove that $\phi(a)$ is idempotent in ring S. Using multiplication preservation of ϕ, and the fact that $a = a^2$ implies $\phi(a) = \phi(a^2) = \phi(a \cdot a) = \phi(a) \cdot \phi(a) = (\phi(a))^2$.
Thus, ring homomorphism carries an idempotent to an idempotent. $\qquad\square$

Observe that for any ring R with unity 1, the element 1 is an idempotent, so for any ring homomorphism $\phi : R \to S$, the element $\phi(1)$ is an idempotent.
Now, let us consider the ring $\mathbb{Z}_2$. Here, we have $0^2 = 0$ and $1^2 = 1$ so the mapping $a \mapsto a^2$ is a ring homomorphism on $\mathbb{Z}_2$. This example gives us an intuition to generalize this result to any ring R with characteristic 2.

Proposition 6.41 *Let R be a commutative ring with characteristic 2. Then the mapping $a \longmapsto a^2$ is a ring homomorphism from R to R.*

Proof Define $\phi : R \to R$ such that $\phi(a) = a^2$ and char $R = 2$. In order to show that ϕ is a ring homomorphism, firstly we need to show that $\phi(a + b) = \phi(a) + \phi(b)$.
Consider, $\phi(a + b) = (a + b)^2 = a^2 + b^2 + 2ab = a^2 + b^2$
$$(\because \text{char } R = 2. \text{ Therefore, } 2x = 0 \ \forall \ x \in R)$$
Thus, we have $\phi(a + b) = \phi(a) + \phi(b)$.
Further, we need to show that $\phi(a \cdot b) = \phi(a) \cdot \phi(b)$.
Consider, $\phi(a \cdot b) = (ab)^2 = a^2 \cdot b^2$ $\qquad\qquad (\because R \text{ is commutative.})$
$\Rightarrow \phi(a \cdot b) = \phi(a) \cdot \phi(b)$.
Hence, ϕ is a ring homomorphism. $\qquad\square$

The above result can be generalized for the commutative rings having characteristic p, where p is a prime which is given in the next theorem.

Theorem 6.42 *Let R be a commutative ring with characteristic p, where p is a prime. Then the mapping $a \longmapsto a^p$ is a ring homomorphism from R to R.*

Proof Let R be a ring with char $R = p$. Let us define a map $\phi : R \to R$ such that $\phi(a) = a^p$. In order to show that ϕ is a ring homomorphism, firstly we need to show that $\phi(a+b) = \phi(a) + \phi(b)$.

Consider, $\phi(a+b) = (a+b)^p = a^p +^p C_1 a^{p-1} b +^p C_2 a^{p-2} b^2 + ... +^p C_{p-1} ab^{p-1} + b^p = a^p + pa^{p-1}b + p(p-1)a^{p-2}b^2 + ... + pab^{p-1} + b^p = a^p + 0 + ... + 0 + b^p$

$$(\because \text{char } R = p \Rightarrow px = 0 \; \forall \; x \in R.)$$

$\Rightarrow \phi(a+b) = \phi(a) + \phi(b)$.

Further, we need to show that $\phi(a \cdot b) = \phi(a) \cdot \phi(b)$.

Consider, $\phi(a \cdot b) = (ab)^p = a^p \cdot b^p$ $\qquad\qquad\qquad (\because R \text{ is commutative.})$

$\Rightarrow \phi(a \cdot b) = \phi(a) \cdot \phi(b)$.

Hence, ϕ is a ring homomorphism. $\qquad\qquad\qquad\qquad\qquad\qquad\qquad\qquad\square$

Let us examine what happens when characteristic of a ring is non-prime.

Consider the ring $\mathbb{Z}_6$ and define the map $\phi : \mathbb{Z}_6 \to \mathbb{Z}_6$ such that $\phi(a) = a^6$ and char $\mathbb{Z}_6 = 6$. Then, we observe that $\phi(2+5) = \phi(7) = \phi(1)$ in $\mathbb{Z}_6$, which implies $\phi(2+5) = 1^6 = 1$. But $\phi(2) + \phi(5) = 2^6 + 5^6 = 5 \pmod 6$. Hence, addition is not preserved. Therefore, ϕ is not a ring homomorphism.

Analogous to fundamental theorem of homomorphism for groups, we have fundamental theorem of homomorphism for rings as well. However, for this, we have to understand the concept of cosets in a ring.

Definition 6.43 (Cosets in Rings) Suppose that I is an ideal of a ring R. Let $a \in R$, then a *coset* of I is a set of the form $a + I = \{a + s \mid s \in I\}$.

The element a mentioned in the above definition is called a *coset representative* of the coset. For elements $a_1, a_2 \in R$, if $a_1 - a_2 \in I$, then the elements a_1 and a_2 represent the same coset of I. In fact, the ideal I is itself a coset, i.e., $0 + I$.

Now, we have a theorem establishing ring homomorphism from ring R to quotient ring R/I, where I is an ideal of R.

Theorem 6.44 *Let I be an ideal of R. The map $\phi : R \to R/I$ defined by $\phi(r) = r + I$ is a ring homomorphism of R onto R/I with kernel I. The map $\phi : R \to R/I$ is often called the natural or canonical homomorphism.*

Proof Let I be an ideal of R. The map $\phi : R \to R/I$ is defined by $\phi(r) = r + I$. Now, we show that ϕ is a ring homomorphism.

Let $r, s \in R$. Then, we have

$\phi(r) + \phi(s) = (r+I) + (s+I) = (r+s) + I = \phi(r+s)$.

Also, we have $\phi(r) \cdot \phi(s) = (r+I) \cdot (s+I) = (r \cdot s) + I = \phi(r \cdot s)$.

Hence, ϕ is a ring homomorphism.

Moreover, $r \in \text{Ker } \phi \Leftrightarrow \phi(r) = I \Leftrightarrow r + I = I \Leftrightarrow r \in I$.

Therefore, Ker $\phi = I$. $\qquad\qquad\qquad\qquad\qquad\qquad\qquad\qquad\qquad\qquad\qquad\square$

In Theorem 6.36, we have shown that every kernel is an ideal in a ring homomorphism. In fact, every ideal is also a kernel of some homomorphism, which is given by the succeeding theorem.

Theorem 6.45 (Ideals Are Kernels) *Every ideal of a ring R is the kernel of a ring homomorphism of R. In particular, an ideal A is the kernel of the mapping $r \mapsto r + A$ from R to R/A.*

Proof Let A be any ideal of ring R. Let $\phi : R \to R/A$ be a mapping defined as $\phi(r) = r + A$, $r \in R$. We show that ϕ is an onto homomorphism. Clearly, ϕ is well-defined. Now, we need to show that ϕ is onto.

Let, $r + A \in R/A$, where $r \in R$. Then, $\phi(r) = r + A$. So, for $r + A \in R/A$, there exists $r \in R$ such that $\phi(r) = r + A$. Therefore, ϕ is onto.

Further, we show that ϕ is a homomorphism. For $r, s \in R$, consider $\phi(r + s) = r + s + A = (r + A) + (s + A) = \phi(r) + \phi(s)$.

Also, $\phi(rs) = rs + A = (r + A) \cdot (s + A) = \phi(r)\phi(s)$. Therefore, ϕ is a homomorphism.

Next we show that A is a kernel, i.e., $A = \mathrm{Ker}\ \phi$. Consider $\mathrm{Ker}\ \phi = \{x \in R : \phi(x) = A\} = \{x \in R : x + A = A\} = \{x \in R : x \in A\} = A$. Hence, ideal is a kernel. $\qquad\square$

Earlier we have shown that an ideal must be a subring, however a subring need not be an ideal. In fact, a subring is an ideal iff it is the kernel of some ring homomorphism.

Next we define the First Isomorphism Theorem for rings. It is a fundamental result that establishes a connection between the structure of a ring and the structure of its quotient ring. It states that given a ring homomorphism between two rings, the kernel of the homomorphism plays a crucial role. The quotient ring of the domain ring by its kernel is shown to be isomorphic to the image of the homomorphism in the codomain ring. This theorem provides valuable insights into understanding the relationship between ring homomorphisms, their kernels, and the corresponding quotient rings, offering a powerful tool for analyzing and classifying ring structures.

Theorem 6.46 (First Isomorphism Theorem for Rings/Fundamental Theorem of Homomorphism) *Let ϕ be a ring homomorphism from ring R to ring S. Then the mapping from $R/\mathrm{Ker}\ \phi$ to $\phi(R)$, given by $r + \mathrm{Ker}\ \phi \mapsto \phi(r)$, is an isomorphism. In symbols, $\dfrac{R}{\mathrm{Ker}\ \phi} \cong \phi(R)$.*

Proof Let $\mathrm{Ker}\ \phi$ be denoted by K. We need to show that $R/K \approx \phi(R)$.

Define $\alpha : R/K \to \phi(R)$ as $\alpha(x + K) = \phi(x)\ \forall\ x \in R$.

Now, $x + K = y + K \Leftrightarrow x - y \in K \Leftrightarrow \phi(x - y) = 0 \Leftrightarrow \phi(x) - \phi(y) = 0$
$$(\because \phi \text{ is ring homomorphism.})$$
$\Leftrightarrow \phi(x) = \phi(y) \Leftrightarrow \alpha(x + K) = \alpha(y + K)$. Thus, the mapping α is well-defined and one-one. Next, we show that α is onto. Let $\phi(r) \in \phi(R)$. Then, there exists $r + K \in R/K$ such that $\alpha(r + K) = \phi(r)$. Therefore, α is onto.

Finally, we show that α is ring homomorphism.

Consider, $\alpha(x + K + y + K) = \alpha(x + y + K) = \phi(x + y) = \phi(x) + \phi(y) = \alpha(x + K) + \alpha(y + K)$.
$$(\because \phi \text{ is ring homomorphism.})$$
Also, we need to show that $\alpha((x + K)(y + K)) = \alpha(x + K)\alpha(y + K)$.

Consider, $\alpha((x + K)(y + K)) = \alpha(xy + K) = \phi(xy) = \phi(x)\phi(y) = \alpha(x + K)\alpha(y + K)$.
$$(\because \phi \text{ is ring homomorphism.})$$
Therefore, α is ring homomorphism.

Since, α is well-defined, one-one, onto and ring homomorphism. Therefore, mapping α is a ring isomorphism. Thus, $R/K \cong \phi(R)$.

Hence, $\dfrac{R}{\text{Ker } \phi} \cong \phi(R)$. $\square$

Using the above theorem, we can find all the possible homomorphic images of R if we know all the quotient rings of R. For example, consider the onto ring homomorphism $\phi : \mathbb{Z} \to \mathbb{Z}_n$ $(n > 1 \in \mathbb{N})$ defined by $\phi(x) = x \bmod n$ for all $x \in \mathbb{Z}$ then Ker $\phi = n\mathbb{Z}$. Thus, by the First Isomophism Theorem of rings, we have $\mathbb{Z}/n\mathbb{Z} \cong \mathbb{Z}_n$.

Next we have a theorem and certain corollaries establishing a ring homomorphism from the ring of integers $\mathbb{Z}$ to an arbitrary ring with unity.

Theorem 6.47 (Homomorphism from $\mathbb{Z}$ to a Ring with Unity) *Let R be a ring with unity 1. The mapping $\phi : \mathbb{Z} \to R$ given by $n \to n \cdot 1$ is a ring homomorphism.*

Proof Let, the mapping $\phi : \mathbb{Z} \to R$ be defined as $\phi(n) = n \cdot 1$. We show that ϕ is a ring homomorphism.

Case 1. When $m \geq 0$, $n \geq 0$, $m, n \in \mathbb{Z}$, we have
$$\phi(m + n) = (m + n) \cdot 1 = \underbrace{1 + 1 + \dots + 1}_{m+n \text{ times}} = \underbrace{(1 + 1 + \dots + 1)}_{m \text{ times}} + \underbrace{(1 + 1 + \dots + 1)}_{n \text{ times}} =$$
$$m \cdot 1 + n \cdot 1 = \phi(m) + \phi(n).$$

Case 2. When $m < 0$, $n < 0$, $m, n \in \mathbb{Z}$, then $-m > 0$, $-n > 0$. Let $m = -m'$ and $n = -n'$, where $-m', -n' > 0$. Thus, we have
$$\phi(m + n) = (m + n) \cdot 1 = ((-m') + (-n')) \cdot 1 = (m' + n') \cdot (-1) =$$
$$\underbrace{-(1 + 1 + \dots + 1)}_{m'+n' \text{ times}} = \underbrace{-(1 + 1 + \dots + 1)}_{m' \text{ times}} + \underbrace{-(1 + 1 + \dots + 1)}_{n' \text{ times}} =$$
$$m' \cdot (-1) + n' \cdot (-1) = (-m') \cdot 1 + (-n') \cdot 1 = m \cdot 1 + n \cdot 1 = \phi(m) + \phi(n).$$

Case 3. When $m \geq 0$, $n < 0$, $m, n \in \mathbb{Z}$, then $-n > 0$. Let, $n = -n'$, where $n' > 0$. Thus, we have
$$\phi(m + n) = (m + n) \cdot 1 = (m + (-n')) \cdot 1 = \underbrace{(1 + 1 + \dots + 1)}_{m+(-n') \text{ times}} =$$
$$\underbrace{(1 + 1 + \dots + 1)}_{m \text{ times}} + \underbrace{-(1 + 1 + \dots + 1)}_{n' \text{ times}} = m \cdot 1 + (n') \cdot (-1) = m \cdot 1 + (-n') \cdot 1 =$$
$$m \cdot 1 + n \cdot 1 = \phi(m) + \phi(n).$$

Case 4. When $m < 0$, $n \geq 0$, $m, n \in \mathbb{Z}$, then $-m > 0$. Let, $m = -m'$, where $m' > 0$. Thus, we have
$$\phi(m + n) = (m + n) \cdot 1 = (-m' + n) \cdot 1 = \underbrace{(1 + 1 + \dots + 1)}_{-m'+n \text{ times}} =$$
$$\underbrace{-(1 + 1 + \dots + 1)}_{m' \text{ times}} + \underbrace{(1 + 1 + \dots + 1)}_{n \text{ times}} = m' \cdot (-1) + (n) \cdot 1 = (-m') \cdot 1 + n \cdot 1 =$$
$$m \cdot 1 + n \cdot 1 = \phi(m) + \phi(n).$$

Thus, from all the four cases, we conclude that ϕ preserves addition.

Now, we show that $\phi(mn) = \phi(m)\phi(n)$.

Consider $\phi(mn) = (mn) \cdot 1 = (m \cdot 1)(n \cdot 1) = \phi(m)\phi(n)$.

Therefore, ϕ preserves multiplication as well.

Hence, ϕ is a ring homomorphism. $\square$

Corollary 6.48 (A Ring with Unity Contains $\mathbb{Z}_n$ or $\mathbb{Z}$) If R is a ring with unity and the characteristic of R is $n > 0$, then R contains a subring isomorphic to $\mathbb{Z}_n$. If the characteristic of R is 0, then R contains a subring isomorphic to $\mathbb{Z}$.

Proof Let, R be a ring with unity 1 and the characteristic of R is $n > 0$. Since, $O(1) = n \Rightarrow n \cdot 1 = 0$. We show that R contains a subring isomorphic to $\mathbb{Z}_n$.

Let, $S = \{n \cdot 1 \mid n \in \mathbb{Z}\}$. We first show that S is a subring of R.

Clearly, $S \subseteq R$. Also, for $n = 1$, $1 \cdot 1 \in S \Rightarrow S \neq \emptyset$.

Let, $n \cdot 1, m \cdot 1 \in S$ for $n, m \in \mathbb{Z}$.

Then, $n \cdot 1 - m \cdot 1 = (n - m) \cdot 1 \in S$. $\qquad\qquad$ ($\because n, m \in \mathbb{Z} \Rightarrow n - m \in \mathbb{Z}$.)

Also, $(n \cdot 1)(m \cdot 1) = (nm) \cdot 1 \in S$. $\qquad\qquad$ ($\because n, m \in \mathbb{Z} \Rightarrow (nm) \in \mathbb{Z}$.)

Therefore, S is a subring of R.

Now, define $\phi : \mathbb{Z} \to S$ such that $\phi(n) = n \cdot 1$, $n \in \mathbb{Z}$.

ϕ is well-defined and onto.

Further, from Theorem 6.47 ϕ is a homomorphism.

So, by First Isomorphism Theorem, $\dfrac{\mathbb{Z}}{\text{Ker } \phi} \cong S$. $\qquad\qquad\qquad\qquad\qquad\qquad$ (6.5)

Now, Ker $\phi = \{k \in \mathbb{Z} \mid \phi(k) = 0\} = \{k \in \mathbb{Z} \mid k.1 = 0\}$.

If Char $R = n$, then n is the least positive integer satisfying $nx = 0 \ \forall \ x \in R$. In particular, for $x = 1$, $n \cdot 1 = 0$. This implies, $n \in Ker \, \phi$, which implies $\langle n \rangle \subseteq Ker \, \phi$.

For $k \in Ker \ \phi$, $k \cdot 1 = 0$, n is the least such positive integer, such that $n.1 = 0 \Rightarrow n | k$, therefore Ker $\phi \subseteq \langle n \rangle$.

Therefore, $Ker \, \phi = \langle \mathrm{n} \rangle$. Hence, $\dfrac{\mathbb{Z}}{\langle n \rangle} \cong S$. Thus, $\mathbb{Z}_n \cong S$. $\qquad$ $\left(\because \dfrac{\mathbb{Z}}{\langle n \rangle} \cong \mathbb{Z}_n. \right)$

Now, if Char $R = 0$, then Ker $\phi = \langle 0 \rangle$. Thus, $\dfrac{\mathbb{Z}}{\{0\}} \cong S \Rightarrow \mathbb{Z} \cong S$. $\qquad\qquad$ $\square$

Corollary 6.49 ($\mathbb{Z}_m$ is a Homomorphic Image of $\mathbb{Z}$) For any positive integer m, the mapping of $\phi : \mathbb{Z} \to \mathbb{Z}_m$ given by $x \mapsto x \bmod m$ is a ring homomorphism.

Proof We know that $\mathbb{Z}_m$ is a ring with unity. It follows directly from Theorem 6.47 that $\phi : \mathbb{Z} \to \mathbb{Z}_m$ defined as $\phi(x) = x \bmod m = x \cdot 1$ is a ring homomorphism. $\qquad$ ($\because$ in $\mathbb{Z}_m$, $x \odot_m 1 = x \cdot 1$.)

$\qquad\qquad\qquad\qquad\qquad\qquad\qquad\qquad\qquad\qquad\qquad\qquad\qquad\qquad\qquad\qquad\qquad$ $\square$

Corollary 6.50 (A Field Contains $\mathbb{Z}_p$ or $\mathbb{Q}$) If F is a field of characteristic p, then F contains a subfield isomorphic to $\mathbb{Z}_p$. If F is a field of characteristic 0, then F contains a subfield isomorphic to the rational numbers.

Proof Let, F be a field then we know that char F is either prime p or 0.

Case 1. Let char $F = p$ (prime). By previous theorem, F contains a subring S isomorphic to $\mathbb{Z}_p$ i.e., $S \cong \mathbb{Z}_p$.

Since, $\mathbb{Z}_p$ is a field, then S is also a field.

$\Rightarrow F$ contains a subfield S isomorphic to $\mathbb{Z}_p$. In other words F contains $\mathbb{Z}_p$.

Case 2. Let char $F = 0$.

Let $T = \{ab^{-1} \mid a, b \in F, b \neq 0\}$. Now, we show that T is a subfield.

Let $x = a_1 b_1^{-1}$ and $y = a_2 b_2^{-1} \in T$ where $a_1, a_2, b_1, b_2 \in F$, $b_1, b_2 \neq 0$.

Consider, $x - y = a_1 b_1^{-1} - a_2 b_2^{-1} = \dfrac{a_1}{b_1} - \dfrac{a_2}{b_2} = \dfrac{a_1 b_2 - a_2 b_1}{b_1 b_2} = (a_1 b_2 - a_2 b_1)$

$(b_1 b_2)^{-1} \in T$

Also, $x \cdot y = a_1 b_1^{-1} \cdot a_2 b_2^{-1} = (a_1 a_2)(b_1 b_2)^{-1} \in T$

Therefore, T is a subfield of F.

Now, we need to show that $T \cong \mathbb{Q}$. Define a map $\phi : T \to \mathbb{Q}$ such that $\phi(ab^{-1}) = \dfrac{a}{b}$. For ab^{-1}, $xy^{-1} \in T$, $\phi(ab^{-1}) = \phi(xy^{-1}) \Rightarrow \dfrac{a}{b} = \dfrac{x}{y} \Rightarrow ay = bx$. Since $y \neq 0$, multiply both sides by y^{-1}. Thus, $ay(y^{-1}) = bx(y^{-1}) \Rightarrow a = bxy^-1$. Now, since b^{-1} exists, therefore multiply both sides by b^{-1}. So, $ab^{-1} = (b^{-1})bxy^{-1} \Rightarrow ab^{-1} = xy^{-1}$. Thus, ϕ is injective. For $\dfrac{a}{b} \in \mathbb{Q}$, there exists $ab^{-1} \in T$ such that $\phi(ab^{-1}) = \dfrac{a}{b}$. Thus, ϕ is onto. Further, for ab^{-1}, $xy^{-1} \in T$,

$$\phi(ab^{-1} + xy^{-1}) = \phi((ay + xb)(by)^{-1}) = \dfrac{ay + xb}{by} = \dfrac{a}{b} + \dfrac{x}{y} = \phi(ab^{-1}) + \phi(xy^{-1})$$

and $\phi((ab^{-1}) \cdot (xy^{-1})) = \phi((ax)(yb)^{-1}) = \dfrac{ax}{by} = \dfrac{a}{b} \cdot \dfrac{x}{y} = \phi(ab^{-1}) \cdot \phi(xy^{-1})$.

Thus, ϕ is homomorphism. Therefore, $T \cong \mathbb{Q}$. Hence, F contains a subfield isomorphic to the rational numbers $\mathbb{Q}$. $\qquad\square$

Analogous to isomorphism theorems in groups, the succeeding theorems give second and third isomorphism theorems for rings.

Theorem 6.51 (Second Isomorphism Theorem) *Let R be a ring, let $S \subset R$ be a subring and let I be an ideal of R. Then,*

   (i) *$S + I = \{s + a : s \in S, a \in I\}$ is a subring of R.*
  (ii) *$S \cap I$ is an ideal of S.*
 (iii) *$(S + I)/I$ is isomorphic to $S/(S \cap I)$.*

Proof   **(i)** S is a subring and I is an ideal so $0 + 0 \in S + I$, i.e., $S + I$ is non-empty. Let $s_1 + a_1$ and $s_2 + a_2$ be elements of $S + I$, where $s_1, s_2 \in S$ and $a_1, a_2 \in I$. Then $(s_1 + a_1) - (s_2 + a_2) = \underbrace{(s_1 - s_2)}_{\in S} + \underbrace{(a_1 - a_2)}_{\in I} \in S + I$.

Also, $(s_1 + a_1)(s_2 + a_2) = \underbrace{s_1 s_2}_{\in S} + \underbrace{s_1 a_2 + a_1 s_2 + a_1 a_2}_{\in I} \in S + I$.

Hence, $S + I$ is a subring of R.

  (ii) The intersection $S \cap I$ is non-empty since 0 is contained in I and S. Let $a_1, a_2 \in S \cap I$ and let $s \in S$. Then, $a_1 + a_2 \in S \cap I$ since S and I are both closed under addition. Furthermore, sa_1 and $a_1 s$ are in $S \cap I$ since I is an ideal, $S \subseteq R$ and S is closed under multiplication. Therefore, $S \cap I$ is an ideal of S.

 (iii) Consider the map $\phi : S \to (S + I)/I$ defined by $\phi(s) = s + I$ for some $s \in S$. This is a ring homomorphism by definition of addition and multiplication in quotient rings.

Let $\alpha + I \in \dfrac{S+I}{I}$ where $\alpha \in S + I$, i.e., $\alpha = s + a$ for some $s \in S$ and $a \in I$. This implies that $\alpha + I = (s+a) + I = (s+I) + (a+I) = s + I = \phi(s)$. Thus, ϕ is surjective.

Now, Ker $\phi = \{s \in S : \phi(s) = I\} = \{s \in S : s + I = I\} = \{s \in S : s \in I\} = S \cap I$. Then, $S \cap I = $ Ker ϕ is an ideal of S and by First Isomorphism Theorem, $\dfrac{S+I}{I} \cong \dfrac{S}{S \cap I}$.

$\square$

Theorem 6.52 (Third Isomorphism Theorem) *Let R be a ring and let $J \subseteq I$ be ideals of R. Then I/J is an ideal of R/J and $\dfrac{R/J}{I/J} \cong R/I$.*

Proof Since I and J are ideals, they are non-empty and so $I/J = \{a + J : a \in I\}$ is also non-empty. Let $a_1, a_2 \in I$ and let $r \in R$. By definition of addition and multiplication of cosets, we have
$(a_1 + J) + (a_2 + J) = (a_1 + a_2) + J$,
$(r + J)(a_1 + J) = ra_1 + J$ and
$(a_1 + J)(r + J) = a_1 r + J$.
Since I is an ideal, $a_1 + a_2$, ra_1 and $a_1 r$ are contained in I so I/J is an ideal of R/J.
Now, consider the map $\phi : R/J \to R/I$ defined as $\phi(r + J) = r + I$. We claim that this is a well-defined surjective homomorphism with kernel equal to I/J.
The map ϕ is well-defined as $r_1 + J = r_2 + J \Rightarrow r_1 - r_2 \in J \subseteq I \Rightarrow r_1 + I = r_2 + I \Rightarrow \phi(r_1 + J) = \phi(r_2 + J)$.
Now, suppose that $r_1 + J, r_2 + J \in R/J$ for $r_1, r_1 \in R$. Then,
$\phi((r_1 + J) + (r_2 + J)) = \phi(r_1 + r_2 + J) = r_1 + r_2 + I = (r_1 + I) + (r_2 + I) = \phi(r_1 + J) + \phi(r_2 + J)$.
Also, $\phi((r_1 + J) \cdot (r_2 + J)) = \phi(r_1 r_2 + J) = r_1 r_2 + I = (r_1 + I) \cdot (r_2 + I) = \phi(r_1 + J) \cdot \phi(r_2 + J)$.
Hence, ϕ is a ring homomorphism.
We claim that ϕ is surjective. Let $\alpha + I \in R/I$ where $\alpha \in R$, then $\alpha + J \in R/J$ such that $\phi(\alpha + J) = \alpha + I$. Thus, ϕ is surjective.
Further, Ker $\phi = \{r + J \in R/J : \phi(r + J) = I\} = \{r + J \in R/J : r + I = I\} = \{r + J \in R/J : r \in I\} = I/J$. Hence, $I/J = $ Ker ϕ. Then, by the First Isomorphism Theorem, we have $\dfrac{R/J}{I/J} \cong R/I$.

$\square$

Theorem 6.53 (Correspondence Theorem) *Let R be a ring and I be an ideal of R.*

Then $S \mapsto S/I$ is a one-to-one correspondence between the set of subrings containing ideal I and the set of subrings of factor ring R/I. Moreover, the ideals of R containing I correspond to ideals of R/I.

Proof Let R be a ring and I be an ideal of R. Suppose that S is a subring of R containing I. Let us define a map $\phi : R \to R/I$ such that $\phi(S) = S/I$. First we show that function ϕ is well defined by showing that S/I is a subring of R/I.
Let $a, b \in S$. Then,
$(a + I) - (b + I) = (a - b) + I \in S/I$. $(\because S$ is closed under subtraction.$)$

Also, $(a + I)(b + I) = (ab) + I \in S/I.$ ($\because S$ is closed under multiplication.)

Thus, S/I is a subring of R/I.

Next we show that ϕ is onto. Suppose that $T \subseteq R/I$ is a subring and the set S is given by $S = \{x \in R : x + I \in T\}$. Let $a, b \in S$ then $(a - b) + I = (a + I) - (b + I)$ and T is closed under subtraction, thus, $a - b \in S$.

Also, since $(ab) + I = (a + I)(b + I)$ and T is closed under multiplication, thus $ab \in S$. Therefore, S is a subring of R.

Moreover, since $\alpha + I = 0 + I \in T$ for every $\alpha \in I$, so S contains I. Thus by construction $S/I = T$.

Further, we show that ϕ is one-one. Let S_1 and S_2 be two subrings of ring R containing I such that $S_1/I = S_2/I$ in R/I.

In order to show that $S_1 = S_2$, we will show that $S_1 \subseteq S_2$ and $S_2 \subseteq S_1$.

Now, we show that $S_1 \subseteq S_2$. Let $x \in S_1$. It has been given that $S_1/I = S_2/I$, then there exist some $y \in S_2$ with $x + I = y + I$. So, there exists an $\alpha \in I$ such that $x = y + \alpha$. Since $I \subseteq S_2$, so $y, \alpha \in S_2$ this implies $x \in S_2$. Therefore, we have $S_1 \subseteq S_2$.

Similarly taking an element say $x \in S_2$, we can easily show that $S_2 \subseteq S_1$.

Hence, $S_1/I = S_2/I \Rightarrow S_1 = S_2$. Therefore, mapping ϕ is one-one.

Now, to show that S/I is an ideal of R/I, suppose that S is an ideal of R.

Suppose that $a + I \in S/I$ and $x + I \in R/I$. Then, we have $xa + I$, $ax + I \in S/I$ as S is an ideal of R.

Finally, suppose that $T \subseteq R/I$ is an ideal. Let, the set $S = \{x \in R : x + I \in T\}$. We have to show that S is an ideal of R, showing that the correspondence restricts to a correspondence on ideals. Now, let $a \in S$ and $x \in R$, then we have $xa + I = (x + I)(a + I) \in S/I$ as S/I is an ideal. Similarly, $ax \in S/I$. Hence, S is an ideal. $\qquad\square$

Theorem 6.54 *A homomorphism from a field onto a ring with more than one element must be an isomorphism.*

Proof Let F be a field and R be a ring. Let $\phi : F \to R$ be an onto homomorphism. So, by First Theorem of Isomorphism, $\dfrac{F}{\mathrm{Ker}\ \phi} \cong R$. Since $\mathrm{Ker}\ \phi$ is an ideal of F and only ideals of F are $\{0\}$ and F itself.

If $\mathrm{Ker}\ \phi = F$, then $\dfrac{F}{F} \cong R$. This implies that R has only one element, which is a contradiction.

Therefore, $\mathrm{Ker}\ \phi = \{0\}$. Thus, ϕ is one-one and hence an isomorphism. $\qquad\square$

Notice that the ring of integers, $\mathbb{Z}$ is an integral domain but not a field; however, it is contained in the field of rational numbers, which is just quotients of integers. This discussion provides an idea to establish a relation between arbitrary integral domain and fields especially the field of quotients of the integral domain. Consequently follows the next theorem.

Theorem 6.55 (Field of Quotients)   *Let D be an integral domain. Then there exists a field F (called the field of quotients of D) that contains a subring isomorphic to D.*

Proof Let D be an integral domain, i.e., a commutative ring with unity. Let A be the set such that $A = \left\{ \dfrac{a}{b} \mid a, b \in D, b \neq 0 \right\}$.

Define a relation $\equiv$ on A as $\dfrac{a}{b} \equiv \dfrac{c}{d}$ iff $ad = bc$. We assert that $\equiv$ is an equivalence relation.

Reflexive: We have $\dfrac{a}{b} \equiv \dfrac{a}{b}$ iff $ab = ba$ which is true. ($\because D$ is an integral domain so commutativity holds in D.)

Thus, the relation $\equiv$ is reflexive.

$\quad$ *Symmetric:* We have $\dfrac{a}{b} \equiv \dfrac{c}{d}$. We need to show that $\dfrac{c}{d} \equiv \dfrac{a}{b}$ i.e., $cb = da$.

So, $\dfrac{a}{b} \equiv \dfrac{c}{d} \Rightarrow ad = bc \Rightarrow bc = ad \Rightarrow cb = da \Rightarrow \dfrac{c}{d} \equiv \dfrac{a}{b}$.

Thus, the relation $\equiv$ is symmetric.

$\quad$ *Transitive:* We have $\dfrac{a}{b} \equiv \dfrac{c}{d}$ and $\dfrac{c}{d} \equiv \dfrac{e}{f}$. To show that $\dfrac{a}{b} \equiv \dfrac{e}{f}$ iff $af = be$.

Since $\dfrac{a}{b} \equiv \dfrac{c}{d}$ iff $ad = bc$ and $\dfrac{c}{d} \equiv \dfrac{e}{f}$ iff $cf = de$.

Consider, $\quad ad = bc \Rightarrow (ad)e = (bc)e \Rightarrow a(de) = b(ce) \Rightarrow a(cf) = b(ec) \Rightarrow (af)c = (be)c \Rightarrow af = be$. (As D is an integral domain, so cancellation law holds in D.)

Therefore, $\dfrac{a}{b} \equiv \dfrac{e}{f}$ iff $af = be$.

Thus, the relation $\equiv$ is transitive.

Hence, the relation $\equiv$ is an equivalence relation on A.

$\quad$ Now, it partitions the set A into mutually disjoint equivalence classes. Let F be the set of equivalence classes of A under the relation $\equiv$ and denote the equivalence class of, $\dfrac{x}{y}$ by $\left[\dfrac{x}{y} \right]$.

$\quad$ Define $\left[\dfrac{a}{b} \right] + \left[\dfrac{c}{d} \right] = \left[\dfrac{ad + bc}{bd} \right]$ and $\left[\dfrac{a}{b} \right] \cdot \left[\dfrac{c}{d} \right] = \left[\dfrac{ac}{bd} \right]$.

Since, D is an integral domain, so $b, d \neq 0 \Rightarrow bd \neq 0$. Therefore, the operations are defined in F.

Now, we show that addition is well defined.

Let $\left[\dfrac{a}{b} \right] = \left[\dfrac{a'}{b'} \right] \Rightarrow ab' = ba'$ and $\left[\dfrac{c}{d} \right] = \left[\dfrac{c'}{d'} \right] \Rightarrow cd' = dc'$.

We have to show that $\left[\dfrac{ad + bc}{bd} \right] = \left[\dfrac{a'd' + b'c'}{b'd'} \right]$, i.e., $(ad + bc)(b'd') = bd(a'd' + b'c')$.

Consider $\quad (ad + bc)(b'd') = ad(b'd') + bc(b'd') = a(db')d' + b(cb')d' = a(b'd)d' + b(b'c)d' = (ab')(dd') + (bb')(cd') = (ba')(dd') + (bb')(c'd) = (dd')(ba') + (c'd)(bb') = bd(a'd') + bd(b'c') = bd(a'd' + b'c')$.

Thus, $\dfrac{ad + bc}{bd} = \dfrac{a'd' + b'c'}{b'd'}$. So, addition is well-defined.

Next, we show that multiplication is well-defined.

Let, $\left[\dfrac{a}{b} \right] = \left[\dfrac{a'}{b'} \right] \Rightarrow ab' = ba'$ and $\left[\dfrac{c}{d} \right] = \left[\dfrac{c'}{d'} \right] \Rightarrow cd' = dc'$.

We have to show that $\left[\dfrac{a}{b} \right] \left[\dfrac{c}{d} \right] = \left[\dfrac{a'}{b'} \right] \left[\dfrac{c'}{d'} \right] \Rightarrow \left[\dfrac{ac}{bd} \right] = \left[\dfrac{a'c'}{b'd'} \right] \Rightarrow (ac)(b'd') = (bd)(a'c')$.

Consider $(ac)(b'd') = a(cb')d' = a(b'c)d' = (ab')(cd') = (ba')(dc') = b(a'd)c' = b(da')c' = (bd)(a'c')$.

Thus, $\dfrac{ac}{bd} = \dfrac{a'c'}{b'd'}$. So, multiplication is well-defined.

Let, $F = \left\{ \left[\dfrac{a}{b}\right] \;\middle|\; a, b \in D, b \neq 0 \right\}$ be the set of equivalence classes of A and $\left[\dfrac{a}{b}\right] + \left[\dfrac{c}{d}\right] = \left[\dfrac{ad + bc}{bd}\right]$, $\left[\dfrac{a}{b}\right] \cdot \left[\dfrac{c}{d}\right] = \left[\dfrac{ac}{bd}\right] \in F$.

Now, we show that F is a field.

Closure: Clearly, closure property with respect to '+' and '·' holds in F.

Commutativity w.r.t. '+': $\left[\dfrac{a}{b}\right] + \left[\dfrac{c}{d}\right] = \left[\dfrac{ad + bc}{bd}\right] = \left[\dfrac{cb + da}{db}\right] = \left[\dfrac{c}{d}\right] + \left[\dfrac{a}{b}\right]$. Thus, commutativity w.r.t. '+' holds in F.

Commutativity w.r.t. '·': $\left[\dfrac{a}{b}\right] \cdot \left[\dfrac{c}{d}\right] = \left[\dfrac{ac}{bd}\right] = \left[\dfrac{ca}{db}\right] = \left[\dfrac{c}{d}\right] \cdot \left[\dfrac{a}{b}\right]$. Thus, commutativity w.r.t. '·' holds in F.

Associativity w.r.t. '+': $\left(\left[\dfrac{a}{b}\right] + \left[\dfrac{c}{d}\right]\right) + \left[\dfrac{e}{f}\right] = \left[\dfrac{ad + bc}{bd}\right] + \left[\dfrac{e}{f}\right] = \left[\dfrac{(ad + bc)f + (bd)e}{(bd)f}\right] = \left[\dfrac{f(ad + bc) + e(bd)}{(bd)f}\right] = \left[\dfrac{fad + b(fc + ed)}{bdf}\right] = \left[\dfrac{fad}{bdf} + \dfrac{b(fc + ed)}{dbf}\right] = \left[\dfrac{a}{b}\right] + \left(\left[\dfrac{c}{d}\right] + \left[\dfrac{e}{f}\right]\right)$. Thus, associativity w.r.t. '+' holds in F.

Associativity w.r.t. '·': $\left[\dfrac{a}{b}\right] \cdot \left(\left[\dfrac{c}{d}\right] \cdot \left[\dfrac{e}{f}\right]\right) = \left[\dfrac{a}{b}\right]\left[\dfrac{ce}{df}\right] = \left[\dfrac{a(ce)}{b(df)}\right] = \left[\dfrac{(ac)e}{(bd)f}\right] = \left(\left[\dfrac{a}{b}\right] \cdot \left[\dfrac{c}{d}\right]\right) \cdot \left[\dfrac{e}{f}\right]$. Thus, associativity w.r.t. '·' holds in F.

Additive inverse: Let, $\left[\dfrac{a}{b}\right] \in F$. Its additive inverse is $\left[\dfrac{-a}{b}\right]$.

So, $\left[\dfrac{a}{b}\right] + \left[\dfrac{-a}{b}\right] = \left[\dfrac{ab - ba}{bb}\right] = \left[\dfrac{0}{b^2}\right] = \left[\dfrac{0}{1}\right]$.

Similarly, $\left[\dfrac{-a}{b}\right] + \left[\dfrac{a}{b}\right] = \left[\dfrac{0}{1}\right]$. Thus, $\left[\dfrac{-a}{b}\right]$ is the additive inverse of $\left[\dfrac{a}{b}\right]$.

Unity: Let, $\left[\dfrac{a}{b}\right] \in F$. Since $\left[\dfrac{a}{b}\right] \cdot \left[\dfrac{1}{1}\right] = \left[\dfrac{a \cdot 1}{b \cdot 1}\right] = \left[\dfrac{a}{b}\right]$ and $\left[\dfrac{1}{1}\right] \cdot \left[\dfrac{a}{b}\right] = \left[\dfrac{a}{b}\right]$, thus $\left[\dfrac{1}{1}\right]$ is the unity.

Multiplicative inverse: Since $\left[\dfrac{a}{b}\right]\left[\dfrac{b}{a}\right] = \left[\dfrac{a \cdot b}{b \cdot a}\right] = \left[\dfrac{1}{1}\right]$ and similarly $\left[\dfrac{b}{a}\right]\left[\dfrac{a}{b}\right] = \left[\dfrac{b \cdot a}{a \cdot b}\right] = \left[\dfrac{1}{1}\right]$. Thus, the multiplicative inverse of $\left[\dfrac{a}{b}\right]$ is $\left[\dfrac{b}{a}\right]$.

Distributivity: Consider $\left[\dfrac{a}{b}\right]\left(\left[\dfrac{c}{d}\right] + \left[\dfrac{e}{f}\right]\right) = \left[\dfrac{a}{b}\right]\left[\dfrac{cf + de}{df}\right] = \left[\dfrac{a(cf + de)}{b(df)}\right] = \left[\dfrac{a(cf) + a(de)}{bdf}\right] = \left[\dfrac{a(cf)}{bdf} + \dfrac{a(de)}{bdf}\right] = \left[\dfrac{ac}{bd} + \dfrac{ae}{bf}\right] = \left[\dfrac{a}{b}\right]\left[\dfrac{c}{d}\right] + \left[\dfrac{a}{b}\right]\left[\dfrac{e}{f}\right]$.

Thus, distributive property holds in F.

Hence, F is a field.

Now, consider the mapping $\phi : D \to F$ defined as $\phi(x) = \left[\dfrac{x}{1}\right]$.

Firstly, we show that the mapping is well-defined. Let, $x = y \Rightarrow x \cdot 1 = 1 \cdot y \Rightarrow \left[\dfrac{x}{1}\right] = \left[\dfrac{y}{1}\right] \Rightarrow \phi(x) = \phi(y)$. Therefore, ϕ is well-defined.

Next we show that ϕ is one-one. Let, $\phi(x) = \phi(y) \Rightarrow \left[\dfrac{x}{1}\right] = \left[\dfrac{y}{1}\right] \Rightarrow x \cdot 1 = 1 \cdot y \Rightarrow x = y$. Therefore, ϕ is one-one.

Further, we show that ϕ is onto. For, $\left[\dfrac{x}{1}\right] \in F$, there exists $x \in D$ such that $\phi(x) = \left[\dfrac{x}{1}\right]$. Therefore, ϕ is onto.

Finally, we show that ϕ is homomorphism. Let, $x, y \in D$.

Consider $\phi(x + y) = \left[\dfrac{x + y}{1}\right] = \left[\dfrac{x \cdot 1 + y \cdot 1}{1}\right] = \left[\dfrac{x}{1}\right] + \left[\dfrac{y}{1}\right] = \phi(x) + \phi(y)$.

Also, we have $\phi(xy) = \left[\dfrac{xy}{1}\right] = \left[\dfrac{x \cdot y}{1 \cdot 1}\right] = \left[\dfrac{x}{1}\right] \cdot \left[\dfrac{y}{1}\right] = \phi(x) \cdot \phi(y)$.

Therefore, ϕ is a homomorphism.

Hence, ϕ is an isomorphism from D to $\phi(D)$ which is a subring of F. $\qquad\square$

Example 6.56 The field of quotients of $\mathbb{Z}[x]$ is

$$\left\{ \left[\dfrac{f(x)}{g(x)}\right] \,\middle|\, g(x) \text{ is not a zero polynomial}, \ f(x), g(x) \in \mathbb{Z}[x] \right\}$$

since $\mathbb{Z}[x]$ is an integral domain.

Example 6.57 The field of quotients of $\mathbb{Z}_p[x]$, where p is prime, is

$$\left\{ \left[\dfrac{f(x)}{g(x)}\right] \,\middle|\, f(x), g(x) \in \mathbb{Z}_p[x], g(x) \neq 0 \right\}, \text{ since } \mathbb{Z}_p \text{ is a field and hence an integral domain,}$$

thus $\mathbb{Z}_p[x]$ is also an integral domain.

Theorem 6.58 *For any integer $n > 1$, $\mathbb{Z}_n[x]/\langle x \rangle$ is isomorphic to $\mathbb{Z}_n$.*

Proof Let n be an integer which is greater than 1. Define a function $\phi : \mathbb{Z}_n[x] \to \mathbb{Z}_n$ by $\phi(a_n x^n + a_{n-1} x^{n-1} + \ldots + a_0) = a_0$.

Now, we will show that ϕ is a ring homomorphism.

Let $f = a_n x^n + a_{n-1} x^{n-1} + \ldots + a_0$ and $g = b_n x^n + b_{n-1} x^{n-1} + \ldots + b_0 \in \mathbb{Z}_n[x]$.

Consider, $\phi(f + g) = \phi((a_n x^n + a_{n-1} x^{n-1} + \ldots + a_0) + (b_n x^n + b_{n-1} x^{n-1} + \ldots + b_0)) = \phi((a_n + b_n)x^n + (a_{n-1} + b_{n-1})x^{n-1} + \ldots + (a_0 + b_0)) = a_0 + b_0 = \phi(a_n x^n + a_{n-1} x^{n-1} + \ldots + a_0) + \phi(b_n x^n + b_{n-1} x^{n-1} + \ldots + b_0) = \phi(f) + \phi(g)$.

Thus, addition is preserved.

Also, $\phi(f \cdot g) = \phi((a_n x^n + a_{n-1} x^{n-1} + \ldots + a_0) \cdot (b_n x^n + b_{n-1} x^{n-1} + \ldots + b_0)) = \phi((a_n b_n) x^{(2n)} + (a_n b_{n-1} + a_{n-1} b_n)x^{2n-1} + \ldots + a_0 b_0) = a_0 b_0 = \phi(a_n x^n + a_{n-1} x^{n-1} + \ldots + a_0) \cdot \phi(b_n x^n + b_{n-1} x^{n-1} + \ldots + b_0) = \phi(f)\phi(g)$.

Thus, multiplication is also preserved.

Hence, ϕ is a ring homomorphism.

Further, $\text{Ker } \phi = \{a_n x^n + \ldots + a_0 \mid \phi(a_n x^n + \ldots + a_0) = 0\}$
$= \{a_n x^n + \ldots + a_0 \mid a_0 = 0\}$
$= \{a_n x^n + \ldots + a_1 x : a_i \in \mathbb{Z}_n\}$
$= \{x(a_n x^{n-1} + \ldots + a_1) : a_n x^{n-1} + \ldots + a_1 \in \mathbb{Z}_n[x]\}$
$= \langle x \rangle.$

Clearly ϕ is surjective since for any $a \in \mathbb{Z}_n$, $a \in \mathbb{Z}_n[x]$ and $\phi(a) = a$. Hence, by the First Isomorphism Theorem,

$$\frac{\mathbb{Z}_n[x]}{\langle x \rangle} = \frac{\mathbb{Z}_n[x]}{\ker \phi} \cong \mathrm{Im}(\phi) = \mathbb{Z}_n. \qquad \square$$

Theorem 6.59 *Any homomorphism of a field is either an isomorphism or takes each element into 0.*

Proof Let F be some field. Let $\phi : F \to F$ be some homomorphism. Let K be kernel of homomorphism ϕ. We know K is an ideal of F. But the only ideals of F are $\{0\}$ or F itself. When $K = \{0\}$, ϕ is one-to-one homomorphism from F to F and hence surjective. Therefore, ϕ is an isomorphism.

But when $K = F$, then $\phi(x) = 0 \ \forall \ x \in F$, or ϕ takes every element of F into 0. Hence any homomorphism of a field is either an isomorphism or takes each element into 0. $\qquad \square$

Illustrative Problems

Problem 1 Show that there does not exist any ring homomorphism from $\mathbb{C}$ to $\mathbb{R}$.

Solution Suppose that $a = \phi(i) \in \mathbb{R}$. Since $i^2 = -1$ in $\mathbb{C}$ and since ϕ is a ring homomorphism, so we have $-1 = \phi(-1) = \phi(i \cdot i) = \phi(i) \cdot \phi(i) = a^2$.

Thus, $a \in \mathbb{R}$ is a real number such that $a^2 = -1$. This is a contradiction since for every $a \in \mathbb{R}$ we have $a^2 \geq 0$. Hence, there does not exist a ring homomorphism from $\mathbb{C}$ to $\mathbb{R}$.

Problem 2 Show that the map $\phi : \mathbb{C} \to M_2(\mathbb{R})$ defined by $\phi(a + ib) = \begin{pmatrix} a & b \\ -b & a \end{pmatrix}$ is a ring homomorphism.

Solution Suppose the map $\phi : \mathbb{C} \to M_2(\mathbb{R})$ is defined by $\phi(a + ib) = \begin{pmatrix} a & b \\ -b & a \end{pmatrix}$. We show that ϕ is a ring homomorphism.

Let $(a + ib), (c + id) \in \mathbb{C}$. Then, consider

$$\phi((a + ib) + (c + id)) = \phi((a + c) + i(b + d)) = \begin{pmatrix} a+c & b+d \\ -b-d & a+c \end{pmatrix}$$

$$= \begin{pmatrix} a & b \\ -b & a \end{pmatrix} + \begin{pmatrix} c & d \\ -d & c \end{pmatrix} = \phi(a + ib) + \phi(c + id).$$

Also, $\phi((a + ib)(c + id)) = \phi((ac - bd) + i(ad + bc)) = \begin{pmatrix} ac - bd & ad + bc \\ -ad - bc & ac - bd \end{pmatrix}$

$$= \begin{pmatrix} a & b \\ -b & a \end{pmatrix} \cdot \begin{pmatrix} c & d \\ -d & c \end{pmatrix} = \phi(a + ib) \cdot \phi(c + id).$$

Hence, ϕ is a ring homomorphism.

Problem 3 Determine all ring homomorphisms from $\mathbb{Z}_{12}$ to $\mathbb{Z}_{30}$.

Solution Any homomorphism is completely determined if the image of generators is known. We know that 1 is a generator of $\mathbb{Z}_{12}$. Let $1 \mapsto a$, where $a \in \mathbb{Z}_{30}$. Since, $1 \in \mathbb{Z}_{12}$. So, by Lagrange's theorem $O(1)$ divides $12 \Rightarrow O(a)$ divides 12. $(\because O(a) \text{ divides } O(1))$
Also, $O(a)$ divides 30 $(\because a \in \mathbb{Z}_{30}.)$
$\Rightarrow O(a)$ divides gcd $(12, 30) = 6$
$\Rightarrow O(a) = 1, 2, 3$ or 6
$\Rightarrow a = 0, 15, 10, 5, 20, 25.$
These values of a give group homomorphism with respect to addition. In order to identify, ring homomorphism, we observe that
$\phi(1 \cdot 1) = \phi(1) \Rightarrow \phi(1)\phi(1) = \phi(1) \Rightarrow a \cdot a = a \Rightarrow a = 0, 10, 15, 25.$
Hence, there are 4 ring homomorphisms from $\mathbb{Z}_{12} \to \mathbb{Z}_{30}$ given by $\alpha_0, \alpha_{10}, \alpha_{15}, \alpha_{25}$ where $\alpha_i : \mathbb{Z}_{12} \to \mathbb{Z}_{30}$ is such that $\alpha_i(1) = i$ for $i = 0, 10, 5, 25.$

Problem 4 Determine all the ring homomorphisms from $\mathbb{Z}$ to itself.

Solution Any homomorphism is completely determined if the image of generators is known. Since 1 is a generator of $\mathbb{Z}$, let $1 \mapsto a$ where $a \in \mathbb{Z}$. So, $a^2 = a$ in $\mathbb{Z}$. This implies that $a(a - 1) = 0 \Rightarrow a = 0$ or $a - 1 = 0 \Rightarrow a = 0$ or $a = 1$. Thus, there exist only two ring homomorphisms from $\mathbb{Z} \to \mathbb{Z}$ and these are identity and zero mapping.

Problem 5 (Test for divisibility by 9) Show that an integer n with decimal representation $n = a_k a_{k-1} \dots a_1 a_0$ is divisible by 9 iff $a_k + a_{k-1} + \dots + a_1 + a_0$ is divisible by 9.

Solution We have $n = a_k a_{k-1} \dots a_1 a_0 = a_k \times 10^k + a_{k-1} \times 10^{k-1} + \dots + a_1 \times 10 + a_0$ (6.6)
Define natural homomorphism from $\mathbb{Z}$ to $\mathbb{Z}_9$ i.e., $\phi : \mathbb{Z} \to \mathbb{Z}_9$ such that $\phi(x) = x \bmod 9$.
n is divisible by 9 iff $\phi(n) = 0$
$\Leftrightarrow \phi(a_k 10^k + a_{k-1} 10^{k-1} + \dots + a_1 10 + a_0) = 0$ (from equation (6.6))
$\Leftrightarrow \phi(a_k)\phi(10^k) + \phi(a_{k-1})\phi(10^{k-1}) + \dots + \phi(a_1)\phi(10) + \phi(a_0) = 0$
$\Leftrightarrow \phi(a_k)[\phi(10)]^k + \dots + \phi(a_1)\phi(10) + \phi(a_0) = 0$
$\Leftrightarrow \phi(a_k) + \dots + \phi(a_1) + \phi(a_0) = 0$ $(\because \phi(10) = 1)$
$\Leftrightarrow \phi(a_k + \dots + a_1 + a_0) = 0$
$\Leftrightarrow a_k + \dots + a_1 + a_0$ is divisible by 9.

Problem 6 Show that none of the elements of the sequence $3, 7, 11, 15, \dots$ can be written as sum of squares of two integers.

Solution A general term of the given sequence is $a_k = 4k + 3$, where k is an integer. Consider the natural homomorphism $\phi : \mathbb{Z} \to \mathbb{Z}_4$ such that $\phi(m) = m \pmod 4$
$\Rightarrow \phi(a_k) = 3.$
Let's assume that there exists an element in the sequence $a_k = 4k + 3$, where k is an integer, which can be expressed as the sum of squares of two integers. So, there exist integers x and y such that $a_k = x^2 + y^2$.
Now, consider the possible remainders when a perfect square is divided by 4. The remainders can only be 0 or 1. Then, both x^2 and y^2 must have a remainder of either 0 or 1 when divided by 4. However, the sum of two such remainders $(0 + 0, 0 + 1, $ or $1 + 1)$ can only result in a remainder of 0, 1, or 2 when divided by 4. It cannot be equal to the remainder 3, which is required for $a_k = 4k + 3$.

This contradiction implies that our assumption was wrong and none of the elements listed in the given sequence can be written as sum of squares of two integers.

Problem 7 Is the mapping from $\mathbb{Z}_5$ to $\mathbb{Z}_{30}$ given by $x \mapsto 6x$ a ring homomorphism?

Solution Consider any two elements a, $b \in \mathbb{Z}_5$. The map $\phi(x) = 6x$ is well-defined as $a = b$ in $\mathbb{Z}_5$ implies that 5 divides $a - b$.
So, 30 divides $6a - 6b$. Moreover, $\phi(a + b) = 6(a + b) = 6a + 6b = \phi(a) + \phi(b)$
and $\phi(ab) = 6ab = 6 \cdot 6ab$ $\hfill (\because 6^2 = 6 \text{ in } \mathbb{Z}_{30}.)$
$= 6a \cdot 6b = \phi(a) \cdot \phi(b)$.
Hence, both multiplication and addition are preserved. Therefore, mapping $\mathbb{Z}_5$ to $\mathbb{Z}_{30}$ given by $x \mapsto 6x$ is a ring homomorphism.

Problem 8 Let $\phi : \mathbb{Z}[x] \to \mathbb{Z}$ be defined as $\phi(f(x)) = f(0)$. Show that ϕ is an onto ring homomorphism, with Ker $\phi = \langle x \rangle$. Further show that $\langle x \rangle$ is a prime ideal of $\mathbb{Z}[x]$ but not a maximal ideal of $\mathbb{Z}[x]$.

Solution Firstly, we show that ϕ is well defined. Let $f(x) = g(x) \Rightarrow f(0) = g(0) \Rightarrow \phi(f(x)) = \phi(g(x))$. Now, we show that ϕ is onto. Let $a \in \mathbb{Z}$, then $f(x) = a \in \mathbb{Z}[x]$ and $\phi(f(x)) = f(0) = a$ which implies that for $a \in \mathbb{Z}$, there exists $f(x) \in \mathbb{Z}[x]$ such that $\phi(f(x)) = a$. Therefore, ϕ is onto.
Next, we show that ϕ is homomorphism. Let $f(x)$, $g(x) \in \mathbb{Z}[x]$. Then, $\phi(f(x) + g(x)) = \phi((f + g)(x)) = (f + g)(0) = f(0) + g(0) = \phi(f(x)) + \phi(g(x))$.
Also, $\phi(f(x) \cdot g(x)) = \phi((f \cdot g)(x)) = (f \cdot g)(0) = f(0) \cdot g(0) = \phi(f(x)) \cdot \phi(g(x))$.
Therefore, ϕ is a homomorphism.

Thus, by the First Isomorphism Theorem, $\dfrac{\mathbb{Z}[x]}{\text{Ker } \phi} \cong \mathbb{Z}.$ $\hfill (6.7)$

Now, $\quad$ Ker $\phi = \{f(x) \in \mathbb{Z}[x] : \phi(f(x)) = 0\} = \{f(x) \in \mathbb{Z}[x] : f(0) = 0\} = $ set $\quad$ of $\quad$ all polynomials whose constant term is zero $= \langle x \rangle$.
$(\because f(x) = a_0 + a_1 x + ... + a_n x^n \Rightarrow f(0) = a_0 \Rightarrow 0 = a_0 \Rightarrow f(x) = a_1 x + a_2 x^2 + ... + a_n x^n$
$= x(a_1 + a_2 x + ... + a_n x^{n-1}) = \langle x \rangle.)$
Further, from Equation (6.7), we have $\dfrac{\mathbb{Z}[x]}{\langle x \rangle} \cong \mathbb{Z}.$

It is clear that $\mathbb{Z}[x]$ is a commutative ring with unity. Since we know that $\mathbb{Z}$ is an integral domain, thus $\dfrac{\mathbb{Z}[x]}{\langle x \rangle}$ is also an integral domain. Therefore, by Theorem 5.51, $\langle x \rangle$ is a prime ideal.

Now as $\mathbb{Z}$ is not a field, this implies that $\dfrac{\mathbb{Z}[x]}{\langle x \rangle}$ is also not a field. Therefore, by Theorem 5.56, $\langle x \rangle$ is not a maximal ideal.

Problem 9 In $\mathbb{Z}$, let $A = \langle 2 \rangle$ and $B = \langle 8 \rangle$. Show that the group A/B is isomorphic to the group $\mathbb{Z}_4$ but that the ring A/B is not ring-isomorphic to the ring $\mathbb{Z}_4$.

Solution In $\mathbb{Z}$, we have
$A = \langle 2 \rangle = \{2k \mid k \in \mathbb{Z}\} \leq \mathbb{Z}$ and $B = \langle 8 \rangle = \{8k \mid k \in \mathbb{Z}\} \leq \mathbb{Z}$.
The factor group $A/B = \{0 + B, 2 + B, 4 + B, 6 + B\}$.
Since, $A/B = \langle 2 + B \rangle$, so A/B is cyclic. Therefore, A/B is a group isomorphic to $\mathbb{Z}_4$.
Now, take $(2 + B) \in A/B$ and $2 \in \mathbb{Z}_4$. Then, $(2 + B) \cdot (2 + B) = (2 \cdot 2) + B = 4 + B$.
However, in $\mathbb{Z}_4$, $2 \cdot 2 = 4 = 0$ in $\mathbb{Z}_4$. Since $(2 + B) \cdot (2 + B) \neq 2 \cdot 2$, thus multiplication is not preserved. Hence, ring A/B is not isomorphic to the ring $\mathbb{Z}_4$.

Problem 10 Show that $(\mathbb{Z} \oplus \mathbb{Z})/(\langle a \rangle \oplus \langle b \rangle)$ is ring-isomorphic to $\mathbb{Z}_a \oplus \mathbb{Z}_b$.

Solution Here, $\langle a \rangle \oplus \langle b \rangle$ can be written as $\{(na, mb) \mid n, m \in \mathbb{Z}\}$.
Define $\phi : \mathbb{Z} \oplus \mathbb{Z} \to \mathbb{Z}_a \oplus \mathbb{Z}_b$ by $\phi(x, y) = (x \bmod a, y \bmod b)$.
We will show that ϕ is an onto homomorphism with $\mathrm{Ker}\ \phi = \langle a \rangle \oplus \langle b \rangle = \{(na, mb) \mid n, m \in \mathbb{Z}\}$.
Let, $(x, y), (z, w) \in \mathbb{Z} \oplus \mathbb{Z}$.
Then, $\phi((x, y) + (z, w)) = \phi(x + z, y + w) = ((x + z) \bmod a, (y + w) \bmod b) = (x \bmod a, y \bmod b) + (z \bmod a, w \bmod b) = \phi(x, y) + \phi(z, w)$.
Also, $\phi((x, y) \cdot (z, w)) = \phi(x \cdot z, y \cdot w) = ((x \cdot z) \bmod a, (y \cdot w) \bmod b) = (x \bmod a, y \bmod b) \cdot (z \bmod a, w \bmod b) = \phi(x, y) \cdot \phi(z, w)$.
   Thus, ϕ is a ring homomorphism.
Let, $(j, k) \in \mathbb{Z}_a \oplus \mathbb{Z}_b$. Then, $j, k \in \mathbb{Z}$ with $0 \leq j < a$ and $0 \leq k < b$. So, $(j, k) \in \mathbb{Z} \oplus \mathbb{Z}$ and $\phi(j, k) = (j, k)$. Thus, ϕ is onto.
Now, $\mathrm{Ker}\ \phi = \{(x, y) \in \mathbb{Z} \oplus \mathbb{Z} : \phi(x, y) = (0, 0)\}$
$= \{(x, y) \in \mathbb{Z} \oplus \mathbb{Z} : x \bmod a = 0, y \bmod b = 0\}$
$= \{(x, y) \in \mathbb{Z} \oplus \mathbb{Z} : x = na, y = mb \text{ for some } m, n \in \mathbb{Z}\}$
$= \{(x, y) \in \mathbb{Z} \oplus \mathbb{Z} : (x, y) \in \langle a \rangle \oplus \langle b \rangle\}$
$= \langle a \rangle \oplus \langle b \rangle$.

   Then, by First Isomorphism Theorem, $(\mathbb{Z} \oplus \mathbb{Z})/(\langle a \rangle \oplus \langle b \rangle)$ is ring-isomorphic to $\mathbb{Z}_a \oplus \mathbb{Z}_b$.

Problem 11 Determine all ring homomorphisms from $\mathbb{Z} \oplus \mathbb{Z}$ to $\mathbb{Z}$.

Solution Suppose that ϕ is a ring homomorphism from $\mathbb{Z} \oplus \mathbb{Z}$ to $\mathbb{Z}$.
Let, $\phi((1, 0)) = a$ and $\phi((0, 1)) = b$. Then,
$\phi((x, y)) = \phi(x(1, 0) + y(0, 1)) = \phi(x(1, 0)) + \phi(y(0, 1)) = x\phi((1, 0)) + y\phi((0, 1)) = ax + by$.
Since, ϕ preserves multiplication, we know
$\phi((x, y)(x', y')) = \phi((xx', yy')) = axx' + byy' = \phi((x, y))\phi((x', y')) = (ax + by)(ax' + by') = a^2 xx' + abxy' + abyx' + b^2 yy'$.
Now, observe that $axx' + byy' = a^2 xx' + abxy' + abx'y + b^2 yy' \ \forall \ x, x', y, y'$ if and only if $a^2 = a$, $b^2 = b$ and $ab = 0$.
This implies that $a = 0$ or 1 and $b = 0$ or 1 but both cannot be 1 simultaneously.

   We claim that the only ring homomorphisms from $\mathbb{Z} \oplus \mathbb{Z}$ to $\mathbb{Z}$ are the functions $\phi_0, \phi_1, \phi_2 : \mathbb{Z} \oplus \mathbb{Z} \to \mathbb{Z}$ defined by

$\phi_0(x, y) = 0$, $\phi_1(x, y) = x$, $\phi_2(x, y) = y$. To see this, suppose $\phi : \mathbb{Z} \oplus \mathbb{Z} \to \mathbb{Z}$ is a ring homomorphism. Note that for $a \in \{(0, 1), (1, 0), (1, 1)\}$ we have that $a^2 = a$, which implies that $(\phi(a))^2 = \phi(a)$ since ϕ is a ring homomorphism. The only $n \in \mathbb{Z}$ such that $n^2 = n$ are 0 and 1. This implies that $\phi(a) \in \{0, 1\}$ for each $a \in \{(0, 1), (1, 0), (1, 1)\}$. We consider the following cases:

Case 1. $\phi(1, 0) = 1$ and $\phi(0, 1) = 0$. Then, for any $(x, y) \in \mathbb{Z} \oplus \mathbb{Z}$ we have $\phi(x, y) = \phi(x \cdot (1, 0) + y \cdot (0, 1)) = x \cdot \phi(1, 0) + y \cdot \phi(0, 1) = x \cdot 1 + y \cdot 0 = x$. Hence in this case, $\phi = \phi_1$.

Case 2. $\phi(1, 0) = 0$ and $\phi(0, 1) = 1$. Then, for any $(x, y) \in \mathbb{Z} \oplus \mathbb{Z}$ we have $\phi(x, y) = \phi(x \cdot (1, 0) + y \cdot (0, 1)) = x \cdot \phi(1, 0) + y \cdot \phi(0, 1) = x \cdot 0 + y \cdot 1 = y$. Hence in this case, $\phi = \phi_2$.

Case 3. $\phi(1, 0) = 0$ and $\phi(0, 1) = 0$. Then, for any $(x, y) \in \mathbb{Z} \oplus \mathbb{Z}$ we have $\phi(x, y) = \phi(x \cdot (1, 0) + y \cdot (0, 1)) = x \cdot \phi(1, 0) + y \cdot \phi(0, 1) = x \cdot 0 + y \cdot 0 = 0$. Hence in this case, $\phi = \phi_0$.

Case 4. $\phi(1, 0) = 1$ and $\phi(0, 1) = 1$. In this case, we have $\phi(1, 1) = \phi((1, 0) + (0, 1)) = \phi(1, 0) + \phi(0, 1) = 1 + 1 = 2$, contradicting our observation above that $\phi(1, 1) \in \{0, 1\}$. Hence, this case cannot happen. Thus, we see that ϕ_0, ϕ_1, ϕ_2 are the only ring homomorphisms from $\mathbb{Z} \oplus \mathbb{Z} \to \mathbb{Z}$.

Problem 12 Let R be a ring with unit element. Using its elements we define a ring $\tilde{R}$ by defining $a \oplus b = a + b + 1$, and $a \cdot b = ab + a + b$, where $a, b \in R$ and where the addition and multiplication on the right-hand side of these relations are those of R. Show that R is isomorphic to $\tilde{R}$.

Solution Define mapping $\phi : R \to \tilde{R}$ such that $\phi(x) = x - 1$. Clearly the mapping is well-defined. We have $\phi(x + y) = (x + y) - 1 = (x - 1) + (y - 1) + 1 = (x - 1) \oplus (y - 1) = \phi(x) \oplus \phi(y)$.
Also, $\phi(xy) = xy - 1 = (x - 1)(y - 1) + (x - 1) + (y - 1) = (x - 1) \cdot (y - 1) = \phi(x) \cdot \phi(y)$.
So the mapping ϕ is a ring homomorphism.
Further, $\phi(x) = \phi(y) \Rightarrow x - 1 = y - 1 \Rightarrow x = y$. So ϕ is one-to-one.
Moreover, for some $y \in \tilde{R}$, $y + 1 \in R$ is its inverse-image. So inverse-image of every element exists. So mapping is onto. Thus, ϕ is an isomorphism from R onto $\tilde{R}$. Hence, $R \approx \tilde{R}$.

Problem 13 Let A be a ring and let $R = \begin{bmatrix} A & A \\ 0 & A \end{bmatrix} = \left\{ \begin{bmatrix} a & b \\ 0 & c \end{bmatrix} \,\middle|\, a, b, c \in A \right\}$ be the upper triangular matrix ring over A. Show that $S = \begin{bmatrix} 0 & A \\ 0 & 0 \end{bmatrix} = \left\{ \begin{bmatrix} 0 & s \\ 0 & 0 \end{bmatrix} \,\middle|\, s \in A \right\}$ is an ideal in A and that $R/S \cong A \times A$.

Solution Consider the map $\phi : R \to A \times A$ defined by $\phi\left(\begin{bmatrix} a & b \\ 0 & c \end{bmatrix} \right) = (a, c)$ for $a, b, c \in A$.

Now, we show that ϕ is a ring homomorphism.

Let, $\begin{bmatrix} a_1 & b_1 \\ 0 & c_1 \end{bmatrix}, \begin{bmatrix} a_2 & b_2 \\ 0 & c_2 \end{bmatrix} \in R.$

Consider, $\phi\left(\begin{bmatrix} a_1 & b_1 \\ 0 & c_1 \end{bmatrix} + \begin{bmatrix} a_2 & b_2 \\ 0 & c_2 \end{bmatrix}\right) = \phi\left(\begin{bmatrix} a_1 + a_2 & b_1 + b_2 \\ 0 & c_1 + c_2 \end{bmatrix}\right)$

$= (a_1 + a_2, c_1 + c_2) = (a_1, c_1) + (a_2, c_2) = \phi\left(\begin{bmatrix} a_1 & b_1 \\ 0 & c_1 \end{bmatrix}\right) + \phi\left(\begin{bmatrix} a_1 & b_1 \\ 0 & c_1 \end{bmatrix}\right).$

Also, $\phi\left(\begin{bmatrix} a_1 & b_1 \\ 0 & c_1 \end{bmatrix} \cdot \begin{bmatrix} a_2 & b_2 \\ 0 & c_2 \end{bmatrix}\right) = \phi\left(\begin{bmatrix} a_1 a_2 & a_1 b_2 + b_1 c_2 \\ 0 & c_1 c_2 \end{bmatrix}\right)$

$= (a_1 a_2, c_1 c_2) = (a_1, c_1) \cdot (a_2, c_2) = \phi\left(\begin{bmatrix} a_1 & b_1 \\ 0 & c_1 \end{bmatrix}\right) \cdot \phi\left(\begin{bmatrix} a_1 & b_1 \\ 0 & c_1 \end{bmatrix}\right).$

Therefore, ϕ is a ring homomorphism.

Further, by definition of ϕ and S we have Ker $\phi = \left\{\begin{bmatrix} a_1 & b_1 \\ 0 & c_1 \end{bmatrix} \;\middle|\; (a_1, c_1) = (0, 0)\right\} = S.$

Note that this already implies that S is an ideal in R.

Finally we observe that ϕ is onto as for arbitrary element s, $s' \in A$, we have $(s, s') \in A \times A$. Then, $\begin{bmatrix} s & 0 \\ 0 & s' \end{bmatrix} \in R$ and $\phi\left(\begin{bmatrix} s & 0 \\ 0 & s' \end{bmatrix}\right) = (s, s')$. Thus $\phi : R \to A \times A$ is an onto ring homomorphism with Ker $\phi = S$. Hence, by the First Isomorphism Theorem for rings, we have $R/S \cong A \times A$.

Exercise

(1) Consider the map $\phi : \mathbb{Z}[x] \to \mathbb{Z}$ defined as $\phi(f(x)) = f(1)$. Show that ϕ is a ring homomorphism. Also find Ker ϕ and Im ϕ.

(2) Show that the mapping $x \mapsto 5x$ from $\mathbb{Z}_4$ to $\mathbb{Z}_{10}$ is a ring homomorphism.

(3) Is the ring $2\mathbb{Z}$ isomorphic to the ring $4\mathbb{Z}$?

(4) Consider the mapping $\phi : \mathbb{Z}_3 \to \mathbb{Z}_6$ defined as $\phi(n(\text{mod } 3)) = 4n(\text{mod } 6)$. Check whether or not ϕ is a ring homomorphism.

(5) Consider the ring $M = \left\{\begin{pmatrix} a & 0 \\ 0 & b \end{pmatrix} \;\middle|\; a, b \in \mathbb{R}\right\}$ under matrix addition and multiplication. Show that the map $\phi : \mathbb{Z} \to M$ defined as $\phi(n) = \begin{pmatrix} a & 0 \\ 0 & b \end{pmatrix}$ is a ring homomorphism. Also, find Ker ϕ. Is ϕ an epimorphism?

(6) Consider the mapping $\phi : \mathbb{Z}[\sqrt{2}] \to \mathbb{Z}[\sqrt{2}]$ defined as $\phi(a + \sqrt{2}b) = a - \sqrt{2}b$. Check whether ϕ is a ring endomorphism.

(7) Is the map $g : \mathbb{Q}[\sqrt{3}] \to \mathbb{Q}[\sqrt{7}]$ defined by $g(a + b\sqrt{3}) = a + b\sqrt{7}$ a ring homomorphism? Give reasons for your answer.

(8) Check whether or not the function $\phi : 3\mathbb{Z} \to 5\mathbb{Z}$ such that $\phi(3n) = 5n$ is a homomorphism.

(9) Determine all ring isomorphisms from $\mathbb{Z}_n$ to itself.

(10) Find all homomorphisms $\phi : \dfrac{\mathbb{Z}}{6\mathbb{Z}} \to \dfrac{\mathbb{Z}}{15\mathbb{Z}}$.

(11) (Test for Divisibility by 3) Let n be an integer with decimal representation $a_k a_{k-1} \ldots a_1 a_0$. Prove that n is divisible by 3 if and only if $a_k + a_{k-1} + \ldots + a + 1 + a_0$ is divisible by 3.

(12) (Test for Divisibility by 4) Let n be an integer with decimal representation $a_k a_{k-1} \ldots a_1 a_0$. Prove that n is divisible by 4 if and only if $a_1 a_0$ is divisible by 4.

(13) (Test for Divisibility by 11) Let n be an integer with decimal representation $a_k a_{k-1} \ldots a_1 a_0$. Prove that n is divisible by 11 if and only if $a_0 - a_1 + a_2 - \ldots (-1)^k a_k$ is divisible by 11.

(14) Let $\mathbb{Z}_3[i] = \{a + bi \mid a, b \in \mathbb{Z}_3\}$. Show that the field $\mathbb{Z}_3[i]$ is ring-isomorphic to the field $\mathbb{Z}_3[x]/\langle x^2 + 1\rangle$.

(15) Let $R = \left\{ \begin{bmatrix} a & b \\ 0 & c \end{bmatrix} \,\middle|\, a, b, c \in \mathbb{Z} \right\}$.

Prove or disprove that the mapping $\begin{bmatrix} a & b \\ 0 & c \end{bmatrix} \mapsto a$ is a ring homomorphism.

(16) Suppose that ϕ is a ring homomorphism from $\mathbb{Z}_m$ to $\mathbb{Z}_n$. Prove that if $\phi(1) = a$, then $a^2 = a$. Give an example to show that the converse is false.

(17) Is the mapping from $\mathbb{Z}_{10}$ to $\mathbb{Z}_{10}$ given by $x \mapsto 2x$ a ring homomorphism?

(18) Determine all ring homomorphisms from $\mathbb{Z}_6$ to $\mathbb{Z}_6$.

(19) Determine all ring homomorphisms from $\mathbb{Z}_{20}$ to $\mathbb{Z}_{30}$.

(20) Determine all ring homomorphisms from $\mathbb{Z} \oplus \mathbb{Z}$ into $\mathbb{Z} \oplus \mathbb{Z}$.

(21) For any integer $n > 1$, prove that $\langle x \rangle$ is a maximal ideal of $\mathbb{Z}_n[x]$ if and only if n is prime.

(22) Suppose that n divides m and that a is an idempotent of $\mathbb{Z}_n$ (i.e., $a^2 = a$). Show that the mapping $x \mapsto ax$ is a ring homomorphism from $\mathbb{Z}_m$ to $\mathbb{Z}_n$. Show that the same correspondence need not yield a ring homomorphism if n does not divide m.

(23) Show that the prime subfield of a field of characteristic p is ring-isomorphic to $\mathbb{Z}_p$ and that the prime subfield of a field of characteristic 0 is ring-isomorphic to $\mathbb{Q}$.

(24) Let $\mathbb{Q}[\sqrt{2}] = \{a + b\sqrt{2} \mid a, b \in \mathbb{Q}\}$ and $\mathbb{Q}[\sqrt{5}] = \{a + b\sqrt{5} \mid a, b \in \mathbb{Q}\}$. Show that these two rings are not ring-isomorphic.

(25) Let p be a prime and let R be the ring of all 2×2 matrices of the form $\begin{bmatrix} a & b \\ pb & c \end{bmatrix}$, where $a, b \in \mathbb{Z}$. Prove that R is isomorphic to $\mathbb{Z}[\sqrt{p}]$.

(26) Let R be the ring of continuous real-valued functions on the interval $(0, 1)$. Let $I = \{f \in R \mid f(1/3) = 0\}$. Prove that I is a maximal ideal in R.

(27) Let R be a ring and $U(R)$ denotes the group of units of R. Show that if m divides n, then the natural ring homomorphism $\mathbb{Z}_n \to \mathbb{Z}_m$ maps $U(\mathbb{Z}_n)$ onto $U(\mathbb{Z}_m)$. Further, with the help of an example show that $U(R)$ does not have to map onto $U(S)$ under a surjective ring homomorphism $R \to S$.

(28) Give an example of a field F and a one-to-one ring homomorphism $\phi : F \to F$ which is not onto.

7 POLYNOMIAL RINGS

In the realm of ring theory, polynomial rings emerge as indispensable algebraic structures, providing a rich and versatile framework for studying a wide array of mathematical concepts. At their core, polynomial rings serve as a natural extension of the familiar concept of polynomials in a single variable, offering a systematic way to explore algebraic expressions involving multiple variables. This chapter delves into the foundational aspects of polynomial rings, elucidating their construction, properties, and significance within the broader landscape of ring theory.

A polynomial ring is constructed by formalizing expressions involving indeterminates and coefficients, embodying a powerful algebraic structure that captures the essence of polynomial manipulation. The algebraic properties of polynomial rings are examined, their role as noncommutative rings is emphasized, and how they form a foundation for understanding diverse mathematical topics is examined. From polynomial factorization to the roots of polynomials, polynomial rings offer insights into the structure and behavior of rings, making them a cornerstone in the exploration of abstract algebra. Furthermore, the chapter will explore connections between polynomial rings and other algebraic structures, shedding light on their significance from the perspective of mathematical theory. Through this exploration, readers will gain a deeper appreciation for the elegance and applicability of polynomial rings in the context of ring theory.

Let us explore the profound implications and applications of polynomial rings, beginning with the notion of a polynomial itself.

Definition 7.1 (Polynomial) A *polynomial f*, in one variable x over a ring R is defined as the formal expression of the form $a_n x^n + a_{n-1} x^{n-1} + \ldots + a_0$, where n is a natural number and the coefficients are elements of the ring R.

Example 7.2 In the ring $\mathbb{Z}_3$, $f(x) = 2x^2 + x + 1$, $g(x) = x^3 + 2x^2 + 2$ are polynomials.

Definition 7.3 (Polynomial Rings) Let R be a commutative ring. The set of all polynomials over R, i.e., $R[x] = \{a_n x^n + a_{n-1} x^{n-1} + \ldots + a_1 x + a_0 \mid a_i \in R, n \text{ is a non-negative integer}\}$ is called the *ring of polynomials* with indeterminate x over R.

Example 7.4 The ring of polynomials with real coefficients $\mathbb{R}[x]$, the ring of polynomials with complex coefficients $\mathbb{C}[x]$, the ring of polynomials with rational coefficients $\mathbb{Q}[x]$ and the ring of polynomials with integer coefficients $\mathbb{Z}[x]$ are polynomial rings.

Definition 7.5 (Degree of Polynomial) Let $f(x) = a_n x^n + a_{n-1} x^{n-1} + \ldots + a_1 x + a_0$ be an element of polynomial ring $R[x]$. The largest $n \in \mathbb{N}$ for which $a_n \neq 0$ is called *degree of the polynomial* and the corresponding coefficient of x^n is called *leading coefficient*.

Example 7.6 In the polynomial ring $\mathbb{Z}_4[x]$, $f(x) = 3x^2 + 2x + 1$ is a polynomial of degree two with leading coefficient 3.

Example 7.7 In the polynomial ring $\mathbb{C}[x]$, $g(x) = \pi x^3 + (1+i)x$ is a polynomial of degree three with leading coefficient π.

Definition 7.8 (Monic Polynomial) A polynomial $f(x)$ of degree n is called *monic polynomial* if it has multiplicative identity as the leading coefficient of $f(x)$.

Example 7.9 The polynomial $f(x) = x^2 - 3x + 2$ of a ring $R[x]$ is a monic polynomial as the coefficient of the highest-degree term, x^2 is multiplicative identity 1.

Remark A polynomial of degree n is a *zero polynomial* if $a_i = 0 \ \forall \ i$. By convention $\deg(0)$ is not defined.

Note: A *constant polynomial* has the form $f(x) = a_0$. A constant polynomial has degree zero if $a_0 \neq 0$.

Example 7.10 In the ring $\mathbb{Z}_3$, define $f : \mathbb{Z}_3 \to \mathbb{Z}_3$ by $f(x) = x^3 + 2x$ and $g : \mathbb{Z}_3 \to \mathbb{Z}_3$ by $g(x) = x^5 + 2x$. We can observe that $f(0) = 0 = g(0), f(1) = 0 = g(1), f(2) = 0 = g(2)$. Thus, $f(a) = g(a) \ \forall \ a \in \mathbb{Z}_3$. This implies $f(x) = g(x)$ as a function but $f(x) \neq g(x)$ in $\mathbb{Z}_3$ as polynomials.

Remark Two elements $a_n x^n + a_{n-1} x^{n-1} + \ldots + a_1 x + a_0$ and $b_m x^m + b_{m-1} x^{m-1} + \ldots + b_1 x + b_0$ of polynomial ring $R[x]$ are considered equal if and only if $a_i = b_i$ for all nonnegative integers i, where $a_i = 0$ when $i > n$ and $b_i = 0$ when $i > m$.

Now, as we embark on the journey of algebra of polynomial rings, we're about to uncover the intricate structures and relationships that emerge when manipulating these mathematical entities. Let us now explain the algebraic properties that govern polynomial rings, offering insights that extend beyond individual polynomials and into the broader aspect of abstract algebra.

Algebra of Polynomial Rings

Let R be a commutative ring. Let $f(x) = a_n x^n + a_{n-1} x^{n-1} + \ldots + a_1 x + a_0$ and $g(x) = b_m x^m + b_{m-1} x^{m-1} + \ldots + b_1 x + b_0 \in R[x]$. Then, we can define addition $(+)$ and multiplication $(\cdot)$ on $R[x]$ as follows:

(i) Addition in $R[x]$

Polynomials are added componentwise as follows:
$$f(x) + g(x) = (a_s + b_s)x^s + (a_{s-1} + b_{s-1})x^{s-1} + \ldots + (a_1 + b_1)x + (a_0 + b_0),$$
where $s = \max\{m, n\}$.

(ii) Multiplication in $R[x]$

Polynomials are multiplied in the usual way as follows:
$$f(x) \cdot g(x) = c_{m+n}x^{m+n} + c_{m+n-1}x^{m+n-1} + \ldots + c_1 x + c_0,$$
where $c_k = a_k b_0 + a_{k-1}b_1 + \ldots + a_1 b_{k-1} + a_0 b_k$ for $k = 0, 1, \ldots, m+n$.

The properties of the ring R extend to the polynomial ring $R[x]$, as demonstrated by the subsequent propositions.

Proposition 7.11 *If a ring R is commutative, then so is the polynomial ring $R[x]$.*

Proof Consider two arbitrary polynomials $f(x), g(x) \in R[x]$. Let, $f(x) = a_n x^n + a_{n-1}x^{n-1} + \ldots + a_0$ and $g(x) = b_m x^m + b_{m-1}x^{m-1} + \ldots + b_0$ where $a_i, b_j \in R$, for every i and j.
We need to show that $R[x]$ is commutative, i.e., $f(x) \cdot g(x) = g(x) \cdot f(x)$.
Consider $f(x) \cdot g(x) = (a_n x^n + a_{n-1}x^{n-1} + \ldots + a_0) \cdot (b_m x^m + b_{m-1}x^{m-1} + \ldots + b_0)$
$= a_n b_m x^{n+m} + a_n b_{m-1}x^{n+m-1} + \ldots + a_0 b_m x^m + \ldots + a_0 b_0$
$= b_m a_n x^{m+n} + b_{m-1}a_n x^{m-1+n} + \ldots + b_m a_0 x^m + \ldots + b_0 a_0$
$= g(x) \cdot f(x).$ $\qquad\qquad (\because R$ is a commutative ring, so $a_i b_j = b_j a_i \ \forall \ a_i, \ b_j \in R.)$
Hence, $R[x]$ is commutative. $\qquad\qquad\qquad\qquad\qquad\qquad\qquad\qquad\qquad\qquad\qquad$ $\square$

Proposition 7.12 *If a ring R has unity, then the polynomial ring $R[x]$ also has unity.*

Proof Let R be a ring with unity and $R[x]$ be the polynomial ring over R. Let $f(x) = a_n x^n + a_{n-1}x^{n-1} + \ldots + a_0$ be an element of $R[x]$. Let $e(x) = 1 \in R[x]$. We need to show that $e(x)$ is the unity of $R[x]$.
Consider, $f(x) \cdot e(x) = (a_n x^n + a_{n-1}x^{n-1} + \ldots + a_0) \cdot 1 = a_n x^n + a_{n-1}x^{n-1} + \ldots + a_0 = f(x)$
and $e(x) \cdot f(x) = 1 \cdot (a_n x^n + a_{n-1}x^{n-1} + \ldots + a_0) = a_n x^n + a_{n-1}x^{n-1} + \ldots + a_0 = f(x).$
Hence, $R[x]$ also has unity. $\qquad\qquad\qquad\qquad\qquad\qquad\qquad\qquad\qquad\qquad\qquad$ $\square$

The notion of integral domains is crucial in understanding the properties of polynomial rings. The concept of an integral domain, stemming from the previously established propositions, is encapsulated in the upcoming theorem.

Theorem 7.13 *If D is an integral domain, then $D[x]$ is an integral domain.*

Proof An integral domain is a commutative ring with unity with no zero divisors. Since D is a commutative ring with unity, then so is $D[x]$, from propositions 7.11 and 7.12. We only need to show that $D[x]$ has no zero divisors.
Let $f(x) = a_n x^n + a_{n-1}x^{n-1} + \ldots + a_1 x + a_0$ and $g(x) = b_m x^m + b_{m-1}x^{m-1} + \ldots + b_1 x + b_0 \in D[x]$. Let $f(x) \cdot g(x) = 0$.
Now, we show that either $f(x) = 0$ or $g(x) = 0$. Let $f(x) \neq 0$ and $\deg(f(x)) = n$. This implies $a_n \neq 0$.
Consider, $f(x) \cdot g(x) = 0$

$$\Rightarrow (a_n x^n + a_{n-1} x^{n-1} + ... + a_1 x + a_0)(b_m x^m + b_{m-1} x^{m-1} + ... + b_1 x + b_0) = 0$$
$$\Rightarrow a_n b_m x^{m+n} + (a_n b_{m-1} + a_{n-1} b_m) x^{m+n-1} + ... + (a_1 b_0 + a_0 b_1) x + a_0 b_0 = 0 x^{m+n} + 0 x^{m+n-1} + ... + 0 x + 0$$

On comparing both the sides, we get $a_n b_m = 0, a_{n-1} b_m + a_n b_{m-1} = 0 ... a_0 b_0 = 0$. Since $a_n \neq 0$ so $b_m = 0$. Further, $b_m = 0, a_n \neq 0$ implies $b_{m-1} = 0$. Similarly, $b_1 = 0, b_0 = 0$. Hence, $g(x) = 0$. Similarly, if $g(x) \neq 0$ then $f(x) = 0$. Therefore, $D[x]$ has no zero divisors. Hence, $D[x]$ is an integral domain. $\qquad\square$

The above theorem underscores the preservation of integrity in the structure of polynomials. Further, we are going to explore the multiplication of polynomials in a polynomial ring, illustrating the connection between polynomial degrees in an integral domain, emphasizing the harmonious interplay between algebraic structures.

Proposition 7.14 (Degree Rule) *Let D be an integral domain and $f(x), g(x) \in D[x]$. Then, $\deg(f(x) \cdot g(x)) = \deg f(x) + \deg g(x)$.*

Proof Let D be an integral domain. Let $f(x), g(x) \in D[x]$, where $f(x) = a_n x^n + a_{n-1} x^{n-1} + ... + a_0$ and $g(x) = b_m x^m + b_{m-1} x^{m-1} + ... + b_0$.
Then, $f(x) \cdot g(x) = a_n b_m x^{m+n} + ... + a_0 b_0$.
Since $a_n \neq 0$, $b_m \neq 0$ and D is an integral domain, this implies that $a_n b_m \neq 0$. Thus, we get $\deg f(x) = n$, $\deg g(x) = m$ and $\deg(f(x) \cdot g(x)) = n + m$. Hence, $\deg (f(x) \cdot g(x)) = \deg f(x) + \deg g(x)$. $\qquad\square$

Remark For a commutative ring R, it is possible that $\deg (f(x) \cdot g(x)) < \deg f(x) + \deg g(x)$, where $f(x)$ and $g(x)$ are nonzero elements in $R[x]$. For instance, in a commutative ring $\mathbb{Z}_6$, consider the polynomials $f(x) = 2x^2 + 4$ and $g(x) = 3x^3 + 2$ of degree 2 and 3 respectively.
Now, $f(x) \cdot g(x) = (2x^2 + 4)(3x^3 + 2) = (6x^5 + 4x^2 + 12x^3 + 8)$
$\Rightarrow f(x) \cdot g(x) = (6x^5 + 4x^2 + 12x^3 + 8) \bmod 6$
$\Rightarrow f(x) \cdot g(x) = 4x^2 + 2 \in \mathbb{Z}_6$
Thus, $\deg(f(x) \cdot g(x)) = 2 < 2 + 3 = \deg f(x) + \deg g(x)$.

Now, let's contemplate the nature of a polynomial ring $F[x]$ as a field. If $F[x]$ were a field, every nonzero polynomial in $F[x]$ should have a multiplicative inverse. However, this clashes with the degree rule, as the product of two polynomials could yield a polynomial of higher degree, violating the requirement for a multiplicative inverse. Therefore, the degree rule serves as a succinct insight into why $F[x]$ can never be a field, highlighting the inherent challenges in extending field properties to polynomial rings.

Proposition 7.15 *Let F be a field, then $F[x]$ is never a field.*

Proof Let F be a field, then since F is commutative, has the unity, then by propositions 7.11 and 7.12, $F[x]$ will be a commutative ring with unity. Indeed, F being an integral domain, $F[x]$ will also be an integral domain by Theorem 7.13. Now, we show that not all nonzero elements of $F[x]$ have multiplicative inverse. Consider the nonzero polynomial, $f(x) = x$.
Suppose, $g(x) = b_0 + b_1 \cdot x + b_2 \cdot x^2 + ...$ is the multiplicative inverse of $f(x)$.

Then, $f(x) \cdot g(x) = c_0 + c_1 \cdot x + c_2 \cdot x^2 + \dots$ should be the unity of $F[x]$ represented by $e(x) = 1 + 0 \cdot x + 0 \cdot x^2 + \dots$.

This implies, $c_0 = 1$, $c_i = 0$ for all $i > 0$, where $c_0 = a_0 \cdot b_0 = 0 \cdot b_0 = 0 \cdot 1 = 0 \neq 1$. Thus, no $g(x)$ can be multiplicative inverse of $f(x) = x$. Hence, $F[x]$ is not a field. $\quad\square$

The subsequent theorem provides an understanding of how ideals extend to polynomial rings, the next theorem showcases the systematic extension of these algebraic structures.

Theorem 7.16 *If I is an ideal of a ring R, then $I[x]$ is an ideal of $R[x]$.*

Proof Let I be an ideal of a ring R. We have to show that $I[x]$ is an ideal of $R[x]$. Since $0 \in I$, then $0 = 0 + 0 \cdot x + 0 \cdot x^2 + \dots \in I[x] \Rightarrow I[x] \neq \phi$.

Let $f(x), g(x) \in I[x]$ such that $f(x) = a_n x^n + a_{n-1} x^{n-1} + \dots + a_1 x + a_0$ and $g(x) = b_m x^m + b_{m-1} x^{m-1} + \dots + b_1 x + b_0$, where $a_i, b_i \in I \; \forall \; i$.

Then, $f(x) - g(x) = (a_s - b_s)x^s + \dots + (a_1 - b_1)x + (a_0 - b_0)$, where $s = \max\{m, n\}$.

Since, I is an ideal, $(a_i - b_i) \in I \; \forall \; i = 1, 2, \dots, n$. Thus, $f(x) - g(x) \in I[x]$.

Consider, $h(x) = c_p x^p + c_{p-1} x^{p-1} + \dots + c_1 x + c_0 \in R[x]$, where $c_i \in R \; \forall \; i$.

Since I is an ideal, $h(x) \cdot f(x) = (c_p a_n)x^{p+n} + \dots + c_0 a_0 \in I[x]$

$$(\because c_i \in R, \; a_j \in I \Rightarrow c_i a_j \in I.)$$

and $f(x) \cdot h(x) = (a_n c_p)x^{n+p} + \dots + a_0 c_0 \in I[x]$. $\qquad (\because c_i \in R, \; a_j \in I \Rightarrow a_j c_i \in I.)$

Hence, $I[x]$ is an ideal of $R[x]$. $\quad\square$

The next proposition underscores the elegance of prime ideals, showcasing their stability and persistence even in the extended domain of polynomial rings. The transcendence of primality from a ring R to a polynomial ring $R[x]$ unveils a nuanced interplay between algebraic structures, emphasizing the enduring nature of prime ideals across different algebraic settings. To establish the upcoming proposition, we require the following lemma.

Lemma 7.17 Let R be a commutative ring with unity. Let A be an ideal of R. Then, $\dfrac{R[x]}{A[x]} \cong \dfrac{R}{A}[x]$.

Proof Let R be a commutative ring with unity. Let A be an ideal of R. Define a mapping $\phi : R[x] \to \dfrac{R}{A}[x]$ such that $\phi(f(x)) = \phi(a_0 + a_1 x + \dots + a_n x^n) = (a_0 + A) + (a_1 + A)x + \dots + (a_n + A)x^n$. Clearly, ϕ is well defined.

Let $f(x) = a_0 + a_1 x + a_2 x^2 + \dots + a_n x^n$, $g(x) = b_0 + b_1 x + b_2 x^2 + \dots + \dots + b_m x^m$, $s = max\{m, n\}$ and $f(x) \cdot g(x) = h(x) = c_0 + c_1 x + c_2 x^2 + \dots c_t x^t$.

Now, we need to show that ϕ is a homomorphism.

Consider, $\phi(f(x) + g(x)) = \phi((a_0 + b_0) + (a_1 + b_1)x + (a_2 + b_2)x^2 + \dots + (a_s + b_s)x^s)$
$= [(a_0 + b_0) + A] + [(a_1 + b_1) + A]x + [(a_2 + b_2) + A]x^2 + \dots + [(a_s + b_s) + A]x^s$
$= (a_0 + A) + (b_0 + A) + (a_1 + A)x + (b_1 + A)x + \dots + (a_s + A)x^s + (b_s + A)x^s$
$= [(a_0 + A) + (a_1 + A)x + \dots + (a_s + A)x^s] + [(b_0 + A) + (b_1 + A)x + \dots + (b_s + A)x^s]$
$\Rightarrow \phi(f(x) + g(x)) = \phi(f(x)) + \phi(g(x))$.

Then, consider $\phi(f(x) \cdot g(x)) = \phi(c_0 + c_1 x + c_2 x^2 + ...)$
$= (c_0 + A) + (c_1 + A)x + (c_2 + A)x^2 + ...$
$= (a_0 b_0 + A) + (a_1 b_0 + a_0 b_1 + A)x + ...$
$= (a_0 + A)(b_0 + A) + [(a_1 b_0 + A) + (a_0 b_1 + A)]x + ...$
$= (a_0 + A)(b_0 + A) + [(a_1 + A)(b_0 + A) + (a_0 + A)(b_1 + A)]x + ... \qquad (7.1)$
Also, $\phi(f(x)) \cdot \phi(g(x)) = [(a_0 + A) + (a_1 + A)x + ...] \cdot [(b_0 + A) + (b_1 + A)x + ...]$
$= (a_0 + A)(b_0 + A) + [(a_1 + A)(b_0 + A) + (a_0 + A)(b_1 + A)]x + ... \qquad (7.2)$
From Equations (7.1) and (7.2), we have $\phi(f(x) \cdot g(x)) = \phi(f(x)) \cdot \phi(g(x))$.
Thus, ϕ is a homomorphism.
Clearly, from the definition of ϕ, it is evident that ϕ is onto. Hence, by Fundamental Theorem
of Homomorphism of Rings, $\dfrac{R[x]}{Ker\phi} \cong \dfrac{R}{A}[x]$.
Now, $f(x) \in \text{Ker } \phi \Leftrightarrow \phi(f(x)) = (0 + A) + (0 + A)x + ...$
$\Leftrightarrow (a_0 + A) + (a_1 + A)x + ... = (0 + A) + (0 + A)x + ...$
$\Leftrightarrow (a_i + A) = A$ for all $i \Leftrightarrow a_i \in A$ for all $i \Leftrightarrow f(x) \in A[x]$. Hence, $\dfrac{R[x]}{A[x]} \cong \dfrac{R}{A}[x]$. $\qquad \square$

Proposition 7.18 *Let R be a commutative ring with unity. If I is a prime ideal of R, then $I[x]$ is a prime ideal of $R[x]$.*

Proof Let I be a prime ideal of R. We know that for a commutative ring R with unity R, R/I is an integral domain iff I is prime.
Since, I be a prime ideal of R, therefore R/I is an integral domain. Thus, by Theorem 7.13, $\dfrac{R}{I}[x]$ is an integral domain. This implies that $\dfrac{R[x]}{I[x]}$ is an integral domain. ($\because \dfrac{R[x]}{I[x]} \cong \dfrac{R}{I}[x]$
for any ideal I of a commutative ring R with unity.)
Hence, $I[x]$ is a prime ideal of $R[x]$. $\qquad \square$

Continuing in the sequence, we encounter a fundamental property of integers known as the Division Algorithm. In embarking upon the proof of the Division Algorithm for $F[x]$, where F is a field, we examine the polynomial rings. Grounded in the notion of fields, this algorithm becomes a powerful tool, offering a systematic method to divide polynomials and express them as a unique quotient and remainder pair.

Theorem 7.19 (Division Algorithm for $F[x]$) *Let F be a field and let $f(x)$, $g(x) \in F[x]$ with $g(x) \neq 0$. Then, there exist unique polynomials $q(x)$ and $r(x)$ in $F[x]$ such that $f(x) = g(x) \cdot q(x) + r(x)$ and either $r(x) = 0$ or $\deg r(x) < \deg g(x)$.*

Proof Let $f(x) = a_0 + a_1 x + a_2 x^2 + ... + a_n x^n$ and $g(x) = b_0 + b_1 x + b_2 x^2 + ... + b_m x^m$. If $n < m$ then take $q(x) = 0$ and $r(x) = f(x)$. Let $n \geq m$. Let $f(x), g(x) \in F[x]$ with $g(x) \neq 0$, we aim to find $q(x)$ and $r(x)$ such that $f(x) = g(x) \cdot q(x) + r(x)$ and either $r(x) = 0$ or $\deg r(x) < \deg g(x)$.

Consider a set of remainder when $f(x)$ is divided by $g(x)$, defined as $S = \{f(x) - g(x) \cdot h(x) \mid h(x) \in F[x]\}$.

Case 1. Let $0 \in S$. For some $q(x) \in F[x]$, $f(x) - g(x) \cdot q(x) = 0 \Rightarrow f(x) = g(x) \cdot q(x) + 0$.

Case 2. Let $r(x) \neq 0 \in S$ is the smallest element in S. Then, $r(x) = f(x) - g(x) \cdot q_1(x)$ for some $q_1(x) \in F[x]$. This implies that $f(x) = g(x) \cdot q_1(x) + r(x)$. Now, we show that $\deg r(x) < \deg g(x)$. Let if possible, $\deg r(x) \geq \deg g(x)$.

Let $r(x) = c_0 + c_1 x + c_2 x^2 + \ldots + c_t x^t; r_t \neq 0, t \geq m$, then $r_1(x) \in F[x]$ such that
$$r(x) = c_t b_m^{-1} x^{t-m} g(x) + r_1(x), \tag{7.3}$$
where $b_m \neq 0, t > m, c_t b_m^{-1} x^{t-m} \in F[x]$.

Now, $r_1(x) = r(x) - c_t b_m^{-1} x^{t-m} g(x)$
$= c_0 + c_1 x + c_2 x^2 + \ldots + c_t x^t - c_t b_m^{-1} x^{t-m} (b_0 + b_1 x + b_2 x^2 + \ldots + b_m x^m)$
$= c_0 + c_1 x + c_2 x^2 + \ldots + c_t x^t - c_t b_m^{-1} b_0 x^{t-m} - c_t b_m^{-1} b_1 x^{t-m+1} - c_t b_m^{-1} b_2 x^{t-m+2} - \ldots - c_t x^t$.

Therefore, $\deg r_1(x) < t = \deg r(x) \Rightarrow \deg r_1(x) < \deg r(x)$.

Now, we prove that $r_1(x) \in S$. From Equation (7.3), $r_1(x) = r(x) - c_t b_m^{-1} x^{t-m} g(x)$. Also, $r(x) = f(x) - g(x)h(x) \in S$ for some $h(x) \in F[x]$.

$\Rightarrow r_1(x) = f(x) - g(x)h(x) - c_t b_m^{-1} x^{t-m} g(x)$
$= f(x) - g(x) \cdot (h(x) - c_t b_m^{-1} x^{t-m})$
$= f(x) - g(x) \cdot h_1(x) \in S \qquad\qquad (\because h_1(x) \in F[x])$
$\Rightarrow r_1(x) \in S$.

Since, $r_1(x) \in S$, $\deg r_1(x) < \deg r(x)$, $r(x) \in S$ which is a contradiction. So, our assumption is wrong. Therefore, $\deg r(x) < \deg g(x)$. Hence, $f(x) = g(x) \cdot q(x) + r(x)$ with $r(x) = 0$ or $\deg r(x) < \deg g(x)$.

Now, we show uniqueness property of this theorem. Suppose the representation is not unique. Then, for nonzero polynomials $f(x), g(x) \in F[x]$,
$$f(x) = g(x) \cdot q_1(x) + r_1(x) \tag{7.4}$$
and $$f(x) = g(x) \cdot q_2(x) + r_2(x) \tag{7.5}$$
for some $q_1(x), q_2(x), r_1(x), r_2(x) \in F[x]$ and $0 \leq \deg r_1(x), \deg r_2(x) < \deg g(x) = m$.

Now, subtracting Equation (7.4) from Equation (7.5), $(q_1(x) - q_2(x)).g(x) + (r_1(x) - r_2(x)) = 0 \Rightarrow g(x)(q_1(x) - q_2(x)) = (r_2(x) - r_1(x))$.

Since, $\deg r_1(x) < m$, and $\deg r_2(x) < m$, this implies that $\deg (r_2(x) - r_1(x)) < m = \deg g(x)$. This is only possible when $r_2(x) - r_1(x) = 0 \Rightarrow r_2(x) = r_1(x)$ and $q_1(x) - q_2(x) = 0 \Rightarrow q_1(x) = q_2(x)$. Thus, the representation is unique. $\qquad\square$

Before discussing the corollaries of this theorem, we need to recall some definitions as follows:

Definition 7.20 (Factor of a Polynomial) In divison algorithm, when $r(x)$ is the zero polynomial, $g(x)$ is called a *divisor* of $f(x)$ and we write $g(x)|f(x)$. In this case, we also call $g(x)$, a *factor* of $f(x)$.

Definition 7.21 (Zero of a Polynomial) If for some $a \in F$, we have $f(a) = 0$, then we call a as a *zero* of the polynomial $f(x)$ in F.

Definition 7.22 (Multiplicity of a Polynomial) Let $a \in F$ be a zero of polynomial $f(x)$. If $(x - a)^k$ is a factor of $f(x)$ but $(x - a)^{k+1}$ is not a factor of $f(x)$, then a is said to be a *zero of multiplicity k*, where $k \geq 1$.

Armed with these definitions, we are poised to unveil several consequential corollaries emanating from the Division Algorithm.

Corollary 7.23 (Remainder Theorem) Let F be a field, $a \in F$, and $f(x) \in F[x]$. Then $f(a)$ is the remainder in the division of $f(x)$ by $x - a$.

Proof Let, F be a field and $f(x) \in F[x]$. Then, by division algorithm over $f(x)$ and $(x - c)$, there exists $q(x)$ and $r(x) \in F[x]$ such that
$$f(x) = (x - a) \cdot q(x) + r(x) \tag{7.6}$$
where $r(x) = 0$ or $\deg r(x) < \deg(x - a) = 1 \Rightarrow \deg r(x) < 1 \Rightarrow \deg r(x) = 0 \Rightarrow r(x) = r$. This implies that $r(x) = r$, which is a constant.
From Equation (7.6), $f(x) = (x - a) \cdot q(x) + r$ $\tag{7.7}$
Take $x = a$ in Equation (7.7), $f(a) = 0 \cdot q(a) + r = r \Rightarrow r = f(a)$.
The remainder is $f(a)$ in the division of $f(x)$ by $x - a$. $\qquad\square$

Corollary 7.24 (Factor Theorem) Let F be a field, $a \in F$, and $f(x) \in F[x]$. Then, a is a zero of $f(x)$ if and only if $(x - a)$ is a factor of $f(x)$.

Proof Let F be a field, $a \in F$, and $f(x) \in F[x]$. Let a be the zero of $f(x)$, i.e., $f(a) = 0$. Using division algorithm over $f(x)$ and $(x - a)$, there exists $q(x), r(x) \in F[x]$ such that $f(x) = (x - a)q(x) + r(x)$, where $r(x) = 0$ or $\deg r(x) < \deg(x - a) = 1 \Rightarrow \deg r(x) < 1 \Rightarrow$ $\deg r(x) = 0 \Rightarrow r(x) = r$ (a constant).
Taking $x = a$ in $f(x)$, we get, $f(a) = 0 \cdot q(x) + r = r \Rightarrow r = f(a) \Rightarrow r = 0$. $\quad (\because f(a) = 0.)$
Thus, $f(x) = (x - a)q(x)$. Therefore, $(x - a)$ is a factor of $f(x)$.
Conversely, let $(x - a)$ is a factor of $f(x)$ which implies that $f(x) = (x - a)h(x)$ for some $h(x) \in F[x]$.
Take $x = a$ in $f(x)$, $f(a) = 0 \cdot h(x) = 0 \Rightarrow f(a) = 0$. Thus, a is a root of $f(x)$. Therefore, a is a zero of $f(x)$. $\qquad\square$

Corollary 7.25 A polynomial of degree n over a field has at most n zeros, counting multiplicity.

Proof The proof is by induction on n. If $n = 0$, then it is a constant polynomial that has no roots.
Let $n = 1$. Then, it is a linear polynomial which has exactly one root. Thus, the result is true for $n = 1$.
Assume that the result holds for all m, such that $m < n$. Now, we have to prove that the result is true for n. Let $f(x) \in F[x]$ such that $\deg f(x) = n$. Let $a \in F$ be the root of $f(x)$ such that the multiplicity is k.
Then, by division algorithm, $f(x) = (x - a)^k q(x)$ $\tag{7.8}$
Suppose that $q(a) = 0$, which implies that a is a root of $q(x)$. So, $(x - a)$ is a factor of $q(x)$. Therefore, $q(x) = (x - a)q_1(x)$.
Putting the value of $q(x)$ in Equation (7.8), $f(x) = (x - a)^{k+1}q_1(x)$.
Therefore, $(x - a)^{k+1}$ is a factor of $f(x)$, which is a contradiction to the assumption that a is a root of $f(x)$ with multiplicity k. Thus, $q(a) \neq 0$.
Now, from Equation (7.8), $f(x) = (x - a)^k q(x) \Rightarrow \deg f(x) = \deg (x - a)^k + \deg q(x) \Rightarrow$ $n = k + \deg q(x)$ $\tag{7.9}$
$\Rightarrow k = n - \deg q(x) \leq n \Rightarrow k \leq n$.

Now, if $f(x)$ has no zeroes other than a, then the maximum number of zeroes of $f(x) = k \le n$.

And if there exists another root, say b of $f(x)$ such that $f(b) = 0, b \ne a$, then, taking $x = b$ in equation (7.9), we get, $f(b) = (b-a)^k q(b) \Rightarrow 0 = (b-a)^k q(b) \Rightarrow q(b) = 0$ $\quad (\because b \ne a.)$

This implies b is a root of $q(x)$. Since, $\deg q(x) = n - k$

So, the induction hypothesis holds for $q(x)$.

This implies that $q(x)$ has at most $(n-k)$ roots which means $f(x)$ has at most $k + (n-k)$ roots, counting multiplicity. Hence, $f(x)$ can have at most n zeroes, counting multiplicity.

$\square$

Remark The above mentioned corollary 7.25 may not hold for arbitrary polynomial rings. For instance, in ring $\mathbb{Z}_6$, a polynomial $x^2 + 3x + 2$ has four zeros $1, 2, 4$, and 5 in $\mathbb{Z}_6$.

Here, $f(0) = 2 \ne 0$,

$f(1) = 1 + 3 + 2 = 6 \mod 6 = 0$,

$f(2) = 4 + 6 + 2 = 12 \mod 6 = 0$

$f(3) = 9 + 9 + 2 = 20 \mod 6 = 2 \ne 0$,

$f(4) = 16 + 12 + 2 = 30 \mod 6 = 0$,

and $f(5) = 25 + 15 + 2 = 42 \mod 6 = 0$.

Therefore, $f(x)$ has four zeroes in $\mathbb{Z}_6$.

Acknowledging the inherent constraints posed by the division algorithm, the following proposition examine the structural complexities of finite fields.

Proposition 7.26 *For any positive integer n, a field F can have at most a finite number of elements of multiplicative order at most n.*

Proof Let F be a field. We have to show that F has at most finite number of elements of order at most n. Suppose, if possible, F has infinitely many elements of order at most n. So, there are infinitely many zeroes in F of $x^n - 1$, which is a contradiction to corollary 7.25 that it has at most n zeroes. Thus, our assumption is wrong. Hence, a field F can have at most a finite number of elements of multiplicative order at most n.

$\square$

Now, we explore the relationship between prime numbers and polynomial expressions in the subsequent proposition.

Proposition 7.27 *For every prime p, we have $x^{p-1} - 1 = (x-1)(x-2)\dots(x-(p-1))$ in $\mathbb{Z}_p[x]$.*

Proof Let $f(x) = x^{p-1} - 1$. Then, by corollary 7.25, $f(x)$ can have at most $(p-1)$ zeroes.

Now, we have to show that $1, 2, \dots, (p-1)$ are zeroes of $f(x)$. Fermat's Little theorem states that if p is a prime number and a is an integer not divisible by p, then $a^{p-1} \equiv 1 (\mod p)$.

We'll prove that $f(x)$ has $p-1$ distinct roots in $\mathbb{Z}_p$, which are $1, 2, \dots, p-1$. For each $i \in \{1, 2, \dots, p-1\}$, consider $x = i$. By Fermat's Little Theorem, we have $i^{p-1} \equiv 1(\mod p)$. This means that $i^{p-1} - 1$ is divisible by p, which implies that i is a root of the polynomial $x^{p-1} - 1$ in $\mathbb{Z}_p$. Since $1, 2, \dots, p-1$ are all distinct residues modulo p, $x^{p-1} - 1$ has $p-1$ distinct roots in $\mathbb{Z}_p$ that can be expressed as the product of linear factors, i.e., $x^{p-1} - 1 = (x - 1)$ $(x - 2) \ldots (x - (p - 1))$. Thus, $x^{p-1} - 1 = (x - 1)(x - 2) \ldots (x - (p - 1))$ in $\mathbb{Z}_p[x]$. $\square$

Continuing through this chapter, we now recall the concept of principal ideal domain. The division algorithm serves as a crucial bridge to understand the profound concept of a Principal Ideal Domain (PID). Along with providing a foundational linkage, it showcases how every ideal in a principal ideal domain can be generated by a single element. For instance, in the ring of integers, the division algorithm illustrates the generation of principal ideals by a unique divisor, paving the way for a broader understanding of this fundamental algebraic structure.

Definition 7.28 (Principal Ideal Domain) *Principal Ideal Domain is an integral domain R in which every ideal has the form $\langle a \rangle = \{ra \mid r \in R\}$ for some a in R.*

Our next theorem asserts the transcendental power of fields. Exploring how polynomials over a field behave, one discovers the profound truth that $F[x]$ is a principal ideal.

Theorem 7.29 *Let F be a field. Then $F[x]$ is a principal ideal domain.*

Proof Let F be a field. Since every field is an integral domain, then by Theorem 7.13, $F[x]$ is an integral domain.

Let I be an ideal of $F[x]$. If $I = 0$, then $I = \langle 0 \rangle$.

If $I \neq 0$, let $g(x) \in I$ such that deg $g(x)$ is minimum in I.

Now, since I is an ideal, $r(x) \cdot g(x) \in I \ \forall \ r(x) \in F[x]$.

This implies that $\langle g(x) \rangle \subseteq I$.   (7.10)

Let $f(x) \in I$, then by division algorithm over $f(x)$ and $g(x)$, there exists $q(x)$, $r(x) \in F[x]$ such that $f(x) = g(x) \cdot q(x) + r(x)$, where $r(x) = 0$ or deg $r(x) <$ deg $g(x)$.

If $r(x) \neq 0$, then $r(x) = f(x) - g(x)q(x) \in I$ (as I is an ideal so $g(x)q(x) \in I$. Also, $f(x) \in I$.) But, deg $r(x) <$ deg $g(x)$ and $g(x)$ is of minimum degree in $F[x]$. Therefore, we arrive at a contradiction. Thus, $r(x) = 0 \Rightarrow f(x) = g(x)q(x) \Rightarrow f(x) \in \langle g(x) \rangle$.

But $f(x)$ is an arbitrary element of I. This implies that $I \subseteq \langle g(x) \rangle$.   (7.11)

From Equations (7.10) and (7.11), $I = \langle g(x) \rangle$. Therefore, I is a principal ideal. But I is an arbitrary ideal of $F[x]$. Hence, $F[x]$ is a principal ideal domain. $\square$

Next, we present a proposition highlighting polynomials without a common divisor. In this scenario, they collectively generate the unit polynomial.

Proposition 7.30 *Let F be a field and let $f(x)$, $g(x) \in F[x]$. If there is no polynomial of positive degree in $F[x]$ that divides both $f(x)$ and $g(x)$ [in this case, $f(x)$ and $g(x)$ are said to be relatively prime]. Then, there exist polynomials $h(x)$ and $k(x)$ in $F[x]$ with the property that $f(x) \cdot h(x) + g(x) \cdot k(x) = 1$.*

Proof Let F be a field. By Theorem 7.29, $F[x]$ is PID. This implies, every ideal is principal ideal. Let $I = \langle f(x), g(x) \rangle = \langle p(x) \rangle$ be an ideal of $F[x]$. Since $f(x), g(x) \in F[x]$ and $F[x]$ is PID, $p(x) \in F[x]$. This implies that $p(x) | f(x)$ and $p(x) | g(x)$. But given that there is no polynomial of positive degree that divides both $f(x)$ and $g(x)$. This implies, $p(x) = p_0, \ p_0 \neq 0, \ p_0 \in F \Rightarrow I = \langle f(x) \cdot g(x) \rangle = \langle p_0 \rangle$.

Then there exists $m(x)$, $n(x) \in F[x]$ such that $f(x) \cdot m(x) + g(x) \cdot n(x) = p_0$

$$\Rightarrow \quad f(x) \cdot \frac{m(x)}{p_0} + g(x) \cdot \frac{n(x)}{p_0} = 1 \quad \Rightarrow \quad f(x) \cdot h(x) + g(x) \cdot k(x) = 1, \quad \text{where} \quad h(x) = \frac{m(x)}{p_0},$$

$$k(x) = \frac{n(x)}{p_0} \in F[x].$$

Hence, if F is a field and $f(x)$, $g(x) \in F[x]$ such that they are relatively prime, then there

exists polynomials $h(x) = \dfrac{m(x)}{p_0}$ and $k(x) = \dfrac{n(x)}{p_0}$ in $F[x]$ with the property that

$f(x) \cdot h(x) + g(x) \cdot k(x) = 1.$ $\hfill\square$

The subsequent theorem is an immediate consequence of Theorem 7.19. To establish the criterion for the ideal inclusion $I \subseteq \langle g(x) \rangle$ in polynomial rings, we leverage the fact that any element in an ideal I must be a multiple of the generator polynomial $g(x)$. For instance, in the polynomial ring $\mathbb{Z}[x]$, the ideal $\langle 2x^2 + 1 \rangle$ implies that any polynomial in the ideal is a multiple of $2x^2 + 1$.

Theorem 7.31 (Criterion for $I = \langle g(x) \rangle$) *Let F be a field, I a nonzero ideal in $F[x]$, and $g(x)$ an element of $F[x]$. Then, $I = \langle g(x) \rangle$ if and only if $g(x)$ is a nonzero polynomial of minimum degree in I.*

Proof Let $g(x)$ be a nonzero polynomial of minimum degree in I. Then as proved is Theorem 7.29, $I = \langle g(x) \rangle$.
Conversely, let, $I = \langle g(x) \rangle$. We have to show that $g(x)$ is a polynomial of minimum degree.
Let $f(x) \in I = \langle g(x) \rangle$. This implies that $f(x) = g(x).h(x)$ for some $h(x) \in F[x]$.
Thus, $\deg f(x) = \deg g(x) + \deg h(x) \Rightarrow \deg g(x) \leq \deg f(x).$ $\hfill (\because \deg h(x) > 0.)$
Since, $f(x) \in I$ is arbitrary, this implies that the polynomial $g(x)$ is of minimum degree.
Hence, $I = \langle g(x) \rangle$ if and only if $g(x)$ is a nonzero polynomial of minimum degree in I. $\hfill\square$

Example 7.32 Define a mapping $\phi : \mathbb{R}[x] \to \mathbb{C}$ by $\phi(f(x)) = f(i)$. Clearly, the map is well defined and one-one as for $f(x)$, $g(x) \in \mathbb{R}[x]$, $f(x) = g(x) \Leftrightarrow f(i) = g(i) \Leftrightarrow \phi(f(x)) = \phi(g(x))$.
Further, the map is onto as for some $f(i) \in \mathbb{C}$, there exists $f(x) \in \mathbb{R}[x]$ such that $\phi(f(x)) = f(i)$.
Also, the map is a ring homomorphism since
$\phi(f(x) + g(x)) = \phi((f + g)(x)) = (f + g)(i) = f(i) + g(i) = \phi(f(x)) + \phi(g(x)),$
and $\phi(f(x) \cdot g(x)) = \phi((f \cdot g)(x)) = (f \cdot g)(i) = f(i) \cdot g(i) = \phi(f(x)) \cdot \phi(g(x)).$ Hence, ϕ is an onto homomorphism. Thus, by Fundamental Theorem of Homomorphism, we have $\dfrac{\mathbb{R}[x]}{\text{Ker } \phi} \cong \mathbb{C}.$

Now, let $f(x) \in \text{Ker } \phi$ and let $f(x) = a_0 + a_1 x + a_2 x^2 + ... + a_n x^n$. If $f(x)$ is a polynomial of degree 1, then $f(x) = ax + b$, $a \neq 0$. Thus, $\phi(f(x)) = f(i) = ai + b \neq 0$ as $a \neq 0$. This implies that a polynomial of degree 1 does not belong to Ker ϕ. Also Ker ϕ does not contain any nonzero, constant polynomial.

Observe that $f(x) = x^2 + 1 \in \text{Ker } \phi$ as $\phi(f(x)) = f(i) = i^2 + 1 = -1 + 1 = 0$. This implies, $x^2 + 1$ is of minimum degree in Ker ϕ. Since, Ker ϕ is an ideal of $\mathbb{R}[x]$, then by Theorem 7.31, Ker $\phi = \langle x^2 + 1 \rangle$. Hence, we get $\dfrac{\mathbb{R}[x]}{\langle x^2 + 1 \rangle} \cong \mathbb{C}.$

The subsequent proposition demonstrates that within the domain of polynomial rings over a field F, when examining polynomials $f(x)$ and $g(x)$ with degrees lower than a given polynomial $p(x)$, if their cosets modulo $p(x)$ coincide under addition, it implies the polynomials themselves are equal. This property showcases the significance of residue classes in understanding polynomial equivalence within the context of ring theory.

Proposition 7.33 *Let F be a field and let $p(x) \in F[x]$. If $f(x)$, $g(x) \in F[x]$ such that deg $f(x) <$ deg $p(x)$ and deg $g(x) <$ deg $p(x)$, then $f(x) + \langle p(x) \rangle = g(x) + \langle p(x) \rangle$ implies $f(x) = g(x)$.*

Proof Let F be a field and let $p(x) \in F[x]$. Let, if possible, $f(x) \neq g(x)$ and we are given that deg $f(x) <$ deg $p(x)$ and deg $g(x) <$ deg $p(x)$
$\Rightarrow$ deg $(f(x) - g(x)) <$ deg $p(x)$.   (7.12)
Also, given that $f(x) + \langle p(x) \rangle = g(x) + \langle p(x) \rangle$
$\Rightarrow$ $f(x) - g(x) \in \langle p(x) \rangle$.   (7.13)
But any nonzero element of $\langle p(x) \rangle$ has degree greater than $\deg p(x)$. So, from Equations (7.12) and (7.13), we get that $f(x) - g(x)$ is a zero polynomial. This implies that $f(x) - g(x) = 0 \Rightarrow f(x) = g(x)$. $\square$

Illustrative Problems

Problem 1 Show that the polynomial $2x + 1$ in $\mathbb{Z}_4[x]$ has a multiplicative inverse in $\mathbb{Z}_4[x]$.

Solution Suppose the polynomial, $f(x) = 2x + 1$ has a multiplicative inverse, say $ax + b$ in $\mathbb{Z}_4[x]$, then $(2x + 1) \cdot (ax + b) = 1 \Rightarrow 2ax^2 + (2b + a)x + b = 1$.
On comparing like terms on both sides, we get $a = 2$ and $b = 1$ in $\mathbb{Z}_4$. Thus, the multiplicative inverse of the polynomial $2x + 1$ is $2x + 1$ in $\mathbb{Z}_4[x]$. Hence, the polynomial $2x + 1$ in $\mathbb{Z}_4[x]$ has a multiplicative inverse in $\mathbb{Z}_4[x]$.

Problem 2 Let F be a field and let $I = \{a_n x^n + a_{n-1} x^{n-1} + a_{n-2} x^{n-2} + \ldots + a_0 \mid a_n, a_{n-1}, a_{n-2}, \ldots, a_0 \in F$ and $a_n + a_{n-1} + a_{n-2} + \ldots + a_0 = 0\}$ be an ideal of $F[x]$. Find a generator for I.

Solution Let $I = \{a_n x^n + a_{n-1} x^{n-1} + a_{n-2} x^{n-2} + \ldots + a_0 \mid a_n, a_{n-1}, a_{n-2}, \ldots, a_0 \in F$ and $a_n + a_{n-1} + a_{n-2} + \ldots + a_0 = 0\}$ be an ideal. Now, since F is a field, then by Theorem 7.29 and $F[x]$ is a principal ideal domain. This implies that $I = \langle h(x) \rangle$ for some $h(x) \in F[x]$. Then by Theorem 7.31, $h(x)$ is a nonzero polynomial of minimum degree. Observe that $h(x)$ cannot be a nonzero constant polynomial. Therefore, $h(x) = x - 1$. Then $h(x) \in I$ is of least degree in I. Hence, $I = \langle x - 1 \rangle$ and $(x - 1)$ is a generator of I.

Problem 3 Prove that $\mathbb{Z}[x]$ is not a principal ideal domain.

Solution Consider the ideal $I = \langle x, 2 \rangle$.
Thus, $I = \{x f(x) + 2 g(x) \mid f(x), g(x) \in \mathbb{Z}[x]\}$ (7.14)
Now, we have to show that $I \neq \langle p(x) \rangle$, for any $p(x) \in \mathbb{Z}[x]$. Let, if possible, $I = \langle x, 2 \rangle = \langle p(x) \rangle$, where $p(x) \in \mathbb{Z}[x]$.

Clearly, $2 \in \langle x, 2 \rangle = I \Rightarrow p(x) = c$ for $c \in \{-2, 2\}$. Therefore, $I = \langle x, 2 \rangle = \langle p(x) \rangle = \langle c \rangle$, $c \in \{-2, 2\}$.

Also, from Equation (7.14), $x \in \langle x, 2 \rangle = I$ $\qquad\qquad$ ($\because$ when $f(x) = 1, g(x) = 0$.)
This implies that there exists $h(x) \in \mathbb{Z}[x]$ such that $x = h(x) \cdot c$, since $c = 2$ or -2, so $c.h(x)$ has even coefficient of x, which contradicts $x = c.h(x)$. So, our assumption is wrong. Therefore $I \neq \langle p(x) \rangle$. Hence, $\mathbb{Z}[x]$ is not a principal ideal domain.

Problem 4 In $\mathbb{Z}_3[x]$, show that the distinct polynomials $x^4 + x$ and $x^2 + x$ determine the same function from $\mathbb{Z}_3$ to $\mathbb{Z}_3$.

Solution Here, the field, F is $\mathbb{Z}_3 = \{0, 1, 2\}$; $f(x) = x^4 + x$ and $g(x) = x^2 + x$. We have to show that $f(a) = g(a) \ \forall \ a \in \mathbb{Z}_3$.
Here, $f(0) = 0, f(1) = 1 + 1 = 2, f(2) = 16 + 2 = 18 \mod 3 = 0$ $\qquad\qquad$ (7.15)
and $g(0) = 0, g(1) = 1 + 1 = 2, g(2) = 4 + 2 = 6 \mod 3 = 0$. $\qquad\qquad$ (7.16)
From Equations (7.15) and (7.16), we get $f(a) = g(a) \ \forall \ a \in \mathbb{Z}_3$. Thus, the distinct polynomials $x^4 + x$ and $x^2 + x$ determine the same function from $\mathbb{Z}_3$ to $\mathbb{Z}_3$.

Problem 5 Prove that the ideal $\langle x \rangle$ in $\mathbb{Q}[x]$ is maximal.

Solution Define $\phi : \mathbb{Q}[x] \to \mathbb{Q}$ such that $\phi(a_0 + a_1 x + ...) = a_0$, where $a_0, a_1, ... \in \mathbb{Q}$.
The map ϕ is well defined as for $a_0 + a_1 x + ...$ and $b_0 + b_1 x + ... \in \mathbb{Q}[x]$, such that $a_0 + a_1 x + ... = b_0 + b_1 x + ...$, on comparing both the sides, we get $a_0 = b_0 \Rightarrow$ $\phi(a_0 + a_1 x + ...) = \phi(b_0 + b_1 x + ...)$.
Further, ϕ is onto as for $y = a_0 \in \mathbb{Q}$ (codomain), there exists $z = a_0 + 0 \cdot x + ... \in \mathbb{Q}[x]$ (domain) such that $\phi(z) = y$.
Also, the map, ϕ is a ring homomorphism because for $a_0 + a_1 x + ...$ and $b_0 + b_1 x + ...$ $\in \mathbb{Q}[x]$, $\phi((a_0 + a_1 x + ...) \cdot (b_0 + b_1 x + ...)) = \phi(a_0 b_0 + (a_1 b_0 + a_0 b_1)x + ...)$
$= a_0 b_0 = \phi(a_0 + a_1 x + ...) \cdot \phi(b_0 + b_1 x + ...)$,
and $\phi((a_0 + a_1 x + ...) + (b_0 + b_1 x + ...)) = \phi((a_0 + b_0) + (a_1 + b_1)x + ...) = a_0 + b_0$
$= \phi(a_0 + a_1 x + ...) + \phi(b_0 + b_1 x + ...)$.

Therefore, by the Fundamental Theorem of Homomorphism, $\dfrac{\mathbb{Q}[x]}{\text{Ker } \phi} \cong \mathbb{Q}$.

Now, Ker $\phi = \{a_0 + a_1 x + ... \in \mathbb{Q}[x] \mid \phi(a_0 + a_1 x + ...) = 0\} = \{a_0 + a_1 x + ... \in \mathbb{Q}[x] \mid a_0 = 0\} = \{a_1 x + a_2 x^2 + ... \in \mathbb{Q}[x]\} = \{x(a_1 + a_2 x + ...) \mid a_1 + a_2 x + ... \in \mathbb{Q}[x]\} = \langle x \rangle$.
Therefore, $\dfrac{\mathbb{Q}[x]}{\langle x \rangle} \cong \mathbb{Q}$. Since, $\mathbb{Q}$ is a field, then $\dfrac{\mathbb{Q}[x]}{\langle x \rangle}$ is also a field. Hence, $\langle x \rangle$ is a maximal ideal of $\mathbb{Q}[x]$.

Problem 6 Give an example of a commutative ring R with unity and a maximal ideal I of R such that $I[x]$ is not a maximal ideal of $R[x]$.

Solution Let R be the ring of integers, $\mathbb{Z}$ which is a commutative ring with unity. Then, $I = \langle 2 \rangle$ is maximal ideal in $\mathbb{Z}$. By using lemma 7.17, $\dfrac{R[x]}{I[x]} \cong \left(\dfrac{R}{I}\right)(x) \cong \left[\dfrac{\mathbb{Z}}{\langle 2 \rangle}\right](x) \cong \mathbb{Z}_2[x]$, which is an integral domain, not a field. Hence, $I[x]$ is not maximal in $Z[x]$.

Problem 7 Let $f \in R[x]$. If $f(a) = 0$, $f'(a) = 0$, show that $(x - a)^2$ divides $f(x)$.

<u>Solution</u> Given that $f(a) = 0$ which implies that a is a zero of $f(x)$. This implies $(x - a)|f(x)$.

So, $f(x)$ can be written as $f(x) = (x - a)g(x)$ for some $g(x) \in R[x]$.   (7.17)

Therefore, $f'(x) = (x - a)g'(x) + g(x)$. Now, $f'(a) = (a - a)g'(a) + g(a) = 0 + g(a) = g(a) \Rightarrow g(a) = 0$. $(\because f'(a) = 0)$.

So, a is a zero of $g(x)$ which implies that $(x - a)|g(x)$. Therefore, $g(x)$ can be written as $g(x) = (x - a)h(x)$ for some $h(x) \in R[x]$.

On substituting $g(x)$ in Equation (7.17), $f(x) = (x - a)^2 h(x)$. Hence, $(x - a)^2$ divides $f(x)$.

Problem 8 Let R be a commutative ring. Show that $R[x]$ has a subring isomorphic to R.

<u>Solution</u> Define a map $\phi : R \to R[x]$ as $\phi(a) = a$. Then, ϕ is a well-defined homomorphism.

By Fundamental Theorem of Homomorphism, $\dfrac{R}{\text{Ker } \phi} \cong \phi(R) \subseteq R[x]$ and Ker $\phi = \{x \in R \mid \phi(x) = 0\} = \{0\} \Rightarrow R \cong \phi(R) \subseteq R[x]$. Since R is a ring therefore $\phi(R)$ is a ring, therefore $R[x]$ contains a subring isomorphic to R.

Problem 9 If R is a commutative ring, show that the characteristic of $R[x]$ is the same as the characteristic of R.

<u>Solution</u> Let $f(x) \in R[x]$. Then characteristic of $R[x]$ is n if $n \cdot f(x) = 0 \ \forall \ f(x) \in R[x]$.

Let Char $R = c$ and $f(x) = a_0 + a_1 x + \ldots + a_n x^n$, $a_i \in R$.

Consider, $c \cdot f(x) = c \cdot (a_0 + a_1 x + \ldots + a_n x^n) = (c \cdot a_0) + (c \cdot a_1)x + \ldots + (c \cdot a_n)x^n = 0 + 0 \cdot x + \ldots + 0 \cdot x^n = 0$. $(\because c = \text{Char } R)$

Therefore, $cf(x) = 0 \ \forall \ f(x) \in R[x] \Rightarrow$ Char $R \leq$ Char $R[x]$.

Since R is isomorphic to the subring of constant polynomial, then Char $R \geq$ Char$R[x]$.

Therefore, Char$R =$ Char$R[x]$.

Problem 10 Let $f(x)$ and $g(x)$ be cubic polynomials with integer coefficients such that $f(a) = g(a)$ for four integer values of a. Prove that $f(x) = g(x)$. Generalize.

<u>Solution</u> Consider two cubic polynomials, $f(x) = a_3 x^3 + a_2 x^2 + a_1 x + a_0$ and $g(x) = b_3 x^3 + b_2 x^2 + b_1 x + b_0$ with integer coefficients a_3, a_2, a_1, a_0, b_3, b_2, b_1, b_0.

Consider $h(x) = f(x) - g(x) \Rightarrow h(x) = (a_3 - b_3)x^3 + (a_2 - b_2)x^2 + (a_1 - b_1)x + (a_0 - b_0)$ which is also a cubic polynomial with integer coefficients $(a_3 - b_3)$, $(a_2 - b_2)$, $(a_1 - b_1)$, $(a_0 - b_0)$. Thus, by corollary 7.25, $h(x)$ has atmost three zeroes (including multiplicity).   (7.18)

Also, $f(a) = g(a)$ for four integer values of a. This implies, $f(a) - g(a) = 0 \Rightarrow h(a) = f(a) - g(a) = 0$ for four integer values of a.   (7.19)

From Equations (7.18) and (7.19), this is only possible when $h(x)$ is zero polynomial with integer coefficients $(a_3 - b_3) = 0 \Rightarrow a_3 = b_3$, $(a_2 - b_2) = 0 \Rightarrow a_2 = b_2$, $(a_1 - b_1) = 0 \Rightarrow a_1 = b_1$, $(a_0 - b_0) = 0 \Rightarrow a_0 = b_0$. Therefore, $a_i = b_i \ \forall \ i = 0, 1, 2, 3$. Hence, $f(x) = g(x)$.

This result can be generalized for polynomials of degree n where $f(a) = g(a)$ for $n + 1$ distinct integer values of a. If $f(x)$ and $g(x)$ are polynomials of degree n with integer coefficients and $f(a) = g(a)$ for $n + 1$ distinct integers a, then $f(x) = g(x)$. This generalization follows from the fact that a polynomial of degree n with $n+1$ distinct roots is the zero polynomial. Therefore, $f(x) - g(x)$ has $n+1$ roots, making it the zero polynomial and implying $f(x) = g(x)$.

Problem 11 Prove that $\mathbb{Q}[x]/\langle x^2 - 2 \rangle$ is ring isomorphic to $\mathbb{Q}[\sqrt{2}]$.

<u>Solution</u> Define a mapping $\phi : \mathbb{Q}[x] \to \mathbb{Q}[\sqrt{2}]$ such that $\phi(f(x)) = f(\sqrt{2})$ for $f(x) \in \mathbb{Q}[x]$. Clearly, ϕ is well defined as for $f(x)$, $g(x) \in \mathbb{Q}[x]$, let $f(x) = g(x) \Rightarrow f(\sqrt{2}) = g(\sqrt{2}) \Rightarrow \phi(f(x)) = \phi(g(x))$.
The map, ϕ is onto as for $f(\sqrt{2}) \in \mathbb{Q}[\sqrt{2}]$, there exists $f(x) \in \mathbb{Q}[x]$ such that $\phi(f(x)) = f(\sqrt{2})$.
Further, ϕ is a ring homomorphism since for $f(x)$, $g(x) \in \mathbb{Q}[x]$,
we have $\phi(f(x) + g(x)) = \phi((f + g)(x)) = (f + g)(\sqrt{2}) = f(\sqrt{2}) + g(\sqrt{2}) = f(\sqrt{2}) + g(\sqrt{2})$
$= \phi(f(x)) + \phi(g(x))$,
and $\phi(f(x) \cdot g(x)) = \phi((f \cdot g)(x)) = (f \cdot g)(\sqrt{2}) = f(\sqrt{2}) \cdot g(\sqrt{2}) = \phi(f(x)) \cdot \phi(g(x))$.
Therefore, ϕ is an onto homomorphism. Thus, by Fundamental Theorem of Homomorphism,
$$\frac{\mathbb{Q}[x]}{\text{Ker } \phi} \cong \mathbb{Q}[\sqrt{2}]. \tag{7.20}$$
Let $f(x) \in \text{Ker } \phi$. Clearly, $f(x)$ cannot be a nonzero constant polynomial, for if $f(x) = a_0$, where $a_0 \in \mathbb{Q}$, then $f(\sqrt{2}) = a_0 \Rightarrow \phi(f(x)) = a_0 \neq 0$.
Also, $f(x)$ cannot be linear, for if $f(x) = a_0 + a_1 x$, where $a_0, a_1 \in \mathbb{Q}$, $a_1 \neq 0$ then $\phi(f(x)) = f(\sqrt{2}) = a_0 + a_1 \sqrt{2} \neq 0$.
Observe that, if $f(x) = x^2 - 2$, then $\phi(f(x)) = \phi(x^2 - 2) = (\sqrt{2})^2 - 2 = 0$. This implies that $(x^2 - 2) \in \text{Ker } \phi$ and is of minimum degree.
Since $\text{Ker } \phi$ is an ideal of $\mathbb{Q}[x]$, then by Theorem 7.31, $\text{Ker } \phi = \langle x^2 - 2 \rangle$. Then, by Equation (7.20), $\dfrac{\mathbb{Q}[x]}{\langle x^2 - 2 \rangle} \cong \mathbb{Q}[\sqrt{2}]$.

Problem 12 List all the polynomials of degree 2 in $\mathbb{Z}_2[x]$. Which of these are equal as functions from $\mathbb{Z}_2$ to $\mathbb{Z}_2$?

<u>Solution</u> The polynomials of degree 2 in $\mathbb{Z}_2[x]$ are of the form $ax^2 + bx + c$, where $a, b, c \in \mathbb{Z}_2$. These are: $x^2 + x + 1$, $x^2 + x$, x^2, $x^2 + 1$.
Let $f(x) = x^2 + x + 1$, $g(x) = x^2 + x$, $h(x) = x^2$, $r(x) = x^2 + 1 \in \mathbb{Z}_2[x]$. Now, the functions are equal if they possess same value for all $x \in \mathbb{Z}_2$.
Here, $f(0) = 1$, $f(1) = 1$, $g(0) = 0$, $g(1) = 0$, $h(0) = 0$, $h(1) = 1$, and $r(0) = 1$, $r(1) = 0$. Hence, none of them are equal as functions from $\mathbb{Z}_2$ to $\mathbb{Z}_2$.

Problem 13 Let $\mathbb{Z}_n[x]$ be the ring of polynomials in x with coefficients from $\mathbb{Z}_n$ and usual addition and multiplication. If n can be written in the form $t^2 m$, show that $tmx + 1$ is a unit in $\mathbb{Z}_n[x]$.

<u>Solution</u> If $tmx + 1$ is a unit in $\mathbb{Z}_n[x]$, then there exists its inverse $(ax + b) \in \mathbb{Z}_n[x]$ such that $(tmx + 1)(ax + b) = 1 \Rightarrow (tma)x^2 + (tmb + a)x + b = 1$.
On comparing coefficients on both the sides, we get $b = 1$, $tmb + a = 0$ $\Rightarrow a = -tmb = -tm$, $tma = 0 \bmod n$ and we are given that $n = t^2 m$.

The equation $tma = 0$ implies that $-t^2m^2 = 0$ in modulo t^2m which is satisfied by $c = -tm$, $b = 1$ Therefore, $ax + b = -tmx + 1$. Thus, the inverse of $(1 + tmx)$ exists, which is $(1 - tmx)$. Hence, $tmx + 1$ is a unit in $\mathbb{Z}_n[x]$.

Problem 14 Divide $3x^4 + 2x^3 + x + 2$ by $x^2 + 4$ in $\mathbb{Z}_5[x]$.

Solution Here, $\mathbb{Z}_5$ is a field. Now, we will divide $3x^4 + 2x^3 + x + 2$ by $x^2 + 4$ in $\mathbb{Z}_5[x]$ by long division method.

$$
\begin{array}{r}
3x^2 + 2x + 3 \\
x^2 + 4\overline{)3x^4 + 2x^3 \quad\quad + x + 2} \\
\underline{3x^4 \quad\quad + 2x^2} \\
2x^3 + 3x^2 + x \\
\underline{2x^3 \quad\quad + 3x} \\
3x^2 + 3x + 2 \\
\underline{3x^2 \quad\quad + 2} \\
3x
\end{array}
$$

Thus, we get $(3x^4 + 2x^3 + x + 2) = (x^2 + 4)(3x^2 + 2x + 3) + 3x$.

Problem 15 Prove that $2x^{51} - 4x^{49} - 2^{51}$ is divisible by $x - 2$ in $\mathbb{Q}[x]$.

Solution Here, we are applying remainder theorem. Let $f(x) = 2x^{51} - 4x^{49} - 2^{51}$, then $f(2) = 2(2)^{51} - 4(2)^{49} - 2^{51} = 2^{52} - 2^{51} - 2^{51} = 2^{52} - 2^{51}(1 + 1) = 2^{52} - 2^{52} = 0$.   Since $f(2) = 0$, so $x - 2$ is a factor of $f(x)$.

Problem 16 (Applying the Remainder Theorem) Suppose $p(x) \in R[x]$ leaves the remainder of 5 when divided by $x - 1$ and -1 when divided by $x + 2$. What is the remainder when $p(x)$ is divided by $(x - 1)(x + 2)$?

Solution By remainder theorem, $p(1) = 5$, $p(-2) = -1$. On dividing $p(x)$ by $(x - 1)(x + 2)$, the remainder $r(x)$ has degree less than $\deg(x - 1)(x + 2) = 2$. So, $r(x) = ax + b$ for some $a, b \in R$. This implies that $p(x) = q(x)(x - 1)(x + 2) + (ax + b)$.
$p(1) = 5$ implies $a + b = 5$ and $p(-2) = -1$ implies $-2a + b = -1$. On solving the equations for a and b, we get $a = 2$, $b = 3$. Thus, the remainder is $2x + 3$.

Problem 17 If $\phi : R \to S$ is a ring homomorphism, define $\overline{\phi} : R[x] \to S[x]$ by $\overline{\phi}(a_n x_n + \dots + a_0) = \phi(a_n)x_n + \dots + \phi(a_0)$. Show that $\overline{\phi}$ is a ring homomorphism.

Solution Let $f(x) = a_n x^n + \dots + a_0$ and $g(x) = b_n x^n + \dots + b_0 \in R[x]$.
Consider, $\overline{\phi}(f(x) + g(x)) = \overline{\phi}((a_n x^n + \dots + a_0) + (b_n x^n + \dots + b_0))$
$= \overline{\phi}((a_n + b_n)x^n + \dots + (a_0 + b_0))$
$= \phi(a_n + b_n)x^n + \dots + \phi(a_0 + b_0)$
$= [\phi(a_n) + \phi(b_n)]x^n + \dots + [\phi(a_0) + \phi(b_0)]$
$= [\phi(a_n)x^n + \dots + \phi(a_0)] + [\phi(b_n)x^n + \dots + \phi(b_0)]$
$= \overline{\phi}(f(x)) + \overline{\phi}(g(x))$.
Further, consider $\overline{\phi}(f(x) \cdot g(x)) = \overline{\phi}((a_n x^n + \dots + a_0) \cdot (b_n x^n + \dots + b_0))$
$= \overline{\phi}((a_n \cdot b_n)x^{2n} + \dots + (a_0 \cdot b_0))$

$$= \phi(a_n \cdot b_n)x^{2n} + ... + \phi(a_0 \cdot b_0)$$
$$= [\phi(a_n) \cdot \phi(b_n)]x^{2n} + ... + [\phi(a_0) \cdot \phi(b_0)]$$
$$= [\phi(a_n)x^n + ... + \phi(a_0)] \cdot [\phi(b_n)x^n + ... + \phi(b_0)]$$
$$= \overline{\phi}(f(x)) \cdot \overline{\phi}(g(x)).$$

Hence, $\overline{\phi}$ is a ring homomorphism.

Problem 18 Find two distinct cubic polynomials over $\mathbb{Z}_2$ that determine the same function from $\mathbb{Z}_2$ to $\mathbb{Z}_2$.

Solution The cubic polynomials over $\mathbb{Z}_2$ are $x^3 + x^2 + x + 1$, $x^3 + x^2 + x$, $x^3 + x^2 + 1$, $x^3 + x + 1$, $x^3 + 1$, x^3.

Let $f(x) = x^3 + x^2 + x + 1$, $g(x) = x^3 + 1$ and given that the field is $\mathbb{Z}_2 = \{0, 1\}$.

Here, $f(0) = 0 + 0 + 0 + 1 = 1$, $f(1) = 1 + 1 + 1 + 1 = 4 \bmod 2 = 0$ and $g(0) = 0 + 1 = 1$, $g(1) = 1 + 1 = 2 \bmod 2 = 0$.

From above, we observe that $f(a) = g(a) \ \forall \ a \in \mathbb{Z}_2$. Hence, $f(x) = x^3 + x^2 + x + 1$ and $g(x) = x^3 + 1$ are two distinct cubic polynomials over $\mathbb{Z}_2$ that determine the same function from $\mathbb{Z}_2$ to $\mathbb{Z}_2$.

Problem 19 Let $f(x)$ be a nonconstant polynomial in $\mathbb{Z}[x]$. Prove that $\langle f(x) \rangle$ is not maximal in $\mathbb{Z}[x]$.

Solution Since $f(x)$ be a nonconstant polynomial in $\mathbb{Z}[x]$, it is possible to find an $a \in \mathbb{Z}$ such that $f(a) \neq \pm 1$ and $f(a) \neq 0$.

Now, every nonzero element of $\langle f(x) \rangle$ has degree at least 1 and $f(a)$ has degree 0, this implies, $\langle f(x) \rangle \subset \langle f(x), f(a) \rangle$.

Moreover, $\langle f(x), f(a) \rangle \neq \mathbb{Z}[x]$ as $1 \notin \langle f(x), f(a) \rangle$. For if so, then there exist $g(x)$ and $h(x)$ $\in \mathbb{Z}[x]$ such that $f(x)g(x) + f(a)h(x) = 1$. On evaluating both sides at $x = a$, we get $f(a)g(a) + f(a)h(a) = 1 \Rightarrow f(a)|1,$

$$(\because f(a)|f(a)g(a), \ f(a)|f(a)h(a) \Rightarrow f(a)|f(a)g(a) + f(a)h(a).)$$

which is a contradiction since $f(a) \neq \pm 1$. Hence, $f(x)$ is not maximal in $\mathbb{Z}[x]$.

Problem 20 Let F be a field. Show that there exist $a, b \in F$ with the property that $x^2 + x + 1$ divides $x^{43} + ax + b$.

Solution Let $f(x) = x^{43}$, $g(x) = x^2 + x + 1$. Now, on applying division algorithm over $f(x)$ and $g(x)$, we get there exists $q(x)$ and $r(x) \in F[x]$ such that $x^{43} = (x^2 + x + 1)q(x) + r(x)$, where either $r(x) = 0$ or $\deg r(x) < 2$. $\hspace{2em} (\because \deg g(x) = 2.)$
So, $r(x) = cx + d$ for some $c, d \in F$. Then the equation becomes $x^{43} = (x^2 + x + 1)q(x) + cx + d \Rightarrow x^{43} - cx - d = (x^2 + x + 1)q(x)$.
Let $a = -c \in F$, $b = -d \in F \Rightarrow x^{43} + ax + b = (x^2 + x + 1)q(x)$.
Hence, there exists $a, b \in F$ such that $(x^2 + x + 1)$ divides $x^{43} + ax + b$.

Problem 21 Are there any nonconstant polynomials in $\mathbb{Z}[x]$ that have multiplicative inverses? Explain your answer.

Solution Let, $f(x)$ be any nonconstant polynomial in $\mathbb{Z}[x]$. This implies that $\deg f(x) > 0$. Let, if possible $g(x) \in \mathbb{Z}[x]$ be the inverse of $f(x)$. So, $f(x)g(x) = 1$. Then, $\deg f(x)g(x) = \deg f(x) + \deg g(x) > 0 \Rightarrow \deg (1) > 0$ which is a contradiction to the fact that $\deg (1) = 0$. Hence, there are no nonconstant polynomials that have multiplicative inverses in $\mathbb{Z}[x]$.

Problem 22 Let $f(x)$ belong to $F[x]$, where F is a field. Let a be a zero of $f(x)$ of multiplicity n, and write $f(x) = (x-a)^n q(x)$. If $b \neq a$ is a zero of $q(x)$, show that b has the same multiplicity as a zero of $q(x)$ as it does for $f(x)$.

Solution Let m be the multiplicity of b in $q(x)$. We have to show that m is multiplicity of b in $f(x)$. Since m be the multiplicity of b in $q(x)$, this implies, $f(x) = (x-a)^n (x-b)^m q'(x)$ for some $q'(x) \in F[x]$ and $q'(b) \neq 0$. $\hspace{2cm}$ (7.21)
Therefore, b is a zero of $f(x)$ of multiplicity at least m.
If b is a zero of $f(x)$ of multiplicity greater than m, then, b is a zero of $g(x) = f(x) / (x-b)^m = (x-a)^n q'(x) \Rightarrow q(b) = 0 = (b-a)^n q'(b)$.
Since $b \neq a$, $q'(b) = 0$, which is a contradiction to Equation (7.21). So, b is a zero of $f(x)$ of multiplicity m. Hence, b has the same multiplicity m as a zero of $q(x)$ as it does for $f(x)$.

Problem 23 Suppose $f(x)$ is a polynomial with odd integer coefficients and even degree. Prove that $f(x)$ has no rational zeros.

Solution Let $f(x) = a_n x^n + a_{n-1} x^{n-1} + \dots + a_1 x + a_0$. Assume that p/q is a zero of $f(x)$ where p and q are integers and n is even. We may assume that p and q are relatively prime.
Now, $\quad f(p/q) = a_n (p/q)^n + a_{n-1}(p/q)^{n-1} + \dots + a_1(p/q) + a_0 \Rightarrow 0 = a_n (p/q)^n + a_{n-1}(p/q)^{n-1} + \dots + a_1(p/q) + a_0$.
Multiplying q^n both sides, we get, $0 = a_n p^n + a_{n-1}p^{n-1}q + \dots + a_1 pq^{n-1} + a_0 q^n \Rightarrow a_n p^n + a_{n-1}p^{n-1}q + \dots + a_1 pq^{n-1} = -a_0 q^n$.

$\quad$ **Case 1.** If p is even, then the $a_n p^n + a_{n-1}p^{n-1}q + \dots + a_1 pq^{n-1}$ is even.
$\quad$ **Case 2.** If p is odd, then each summand on the left side is odd and since there is an even number of summands, the left side is still even.

Because a_0 is odd, we then have that q is even. Therefore, $a_n p^n = -(a_{n-1}p^{n-1}q + \dots + a_1 pq_{n-1} + a_0 q^n)$ is even since the right side is divisible by q. This implies that p is even, which is a contradiction to the assumption that p and q are relatively prime. So, the assumption is wrong and (p/q) is not a zero of $f(x)$. Hence, $f(x)$ has no rational zeros.

Problem 24 For any field F, recall that $F(x)$ denotes the field of quotients of the ring $F[x]$. Prove that there is no element in $F(x)$ whose square is x.

Solution Let $F(x)$ be a field of quotients of ring $F[x]$. We have to show that there does not exist any $a \in F(x)$ such that $a^2 = x$. Let, if possible, there exists $\dfrac{f(x)}{g(x)} \in F(x)$ such that $\left(\dfrac{f(x)}{g(x)}\right)^2 = x.$ $\hspace{2cm}$ (7.22)
Assume that $f(x)$ and $g(x)$ have no common factor (if so, then we can reduce it to the simplest form). From Equation (7.22), $(f(x))^2 = x(g(x))^2 \Rightarrow f(0) = 0$. Thus, $f(x)$ has the form $x \cdot k(x)$. Then, $(x \cdot k(x))^2 = x(g(x))^2 \Rightarrow x^2(k(x))^2 = x(g(x))^2 \Rightarrow x(k(x))^2 = (g(x))^2 \Rightarrow g(0) = 0$.
But then $f(x)$ and $g(x)$ have x as a common factor, which is a contradiction to our assumption that $f(x)$ and $g(x)$ have no common factor. Hence, there does not exist any $a \in F[x]$ such that $a^2 = x$.

Problem 25 Consider $f(x) = 4 \cdot x^3 + 2 \cdot x^2 + x + 3$ and $g(x) = 3 \cdot x^4 + 3 \cdot x^3 + 3 \cdot x^2 + x + 4$, where $f(x), g(x) \in \mathbb{Z}_5[x]$. Compute $f(x) + g(x)$ and $f(x) \cdot g(x)$.

Solution Given that $f(x) = 4 \cdot x^3 + 2 \cdot x^2 + x + 3$ and $g(x) = 3 \cdot x^4 + 3 \cdot x^3 + 3 \cdot x^2 + x + 4$.

Then, $f(x) + g(x) = (0 + 3)x^4 + (4 + 3)x^3 + (2 + 3)x^2 + (1 + 1)x + (3 + 4)$
$= 3x^4 + 7x^3 + 5x^2 + 2x + 7$
$= 3x^4 + 2x^3 + 0x^2 + 2x + 2$
$= 3x^4 + 2x^3 + 2x + 2.$

And, $f(x) \cdot g(x) = (4 \cdot 3)x^7 + (4 \cdot 3 + 2 \cdot 3)x^6 + (4 \cdot 3 + 2 \cdot 3 + 1 \cdot 3)x^5 + (4 \cdot 1 + 2 \cdot 3 + 1 \cdot 3 + 3 \cdot 3)x^4 + (4 \cdot 4 + 2 \cdot 1 + 1 \cdot 3 + 3 \cdot 3)x^3 + (2 \cdot 4 + 1 \cdot 1 + 3 \cdot 3)x^2 + (1 \cdot 4 + 3 \cdot 1)x + (3 \cdot 4)$
$= 12x^7 + 18x^6 + 21x^5 + 22x^4 + 30x^3 + 18x^2 + 7x + 12$
$= 2x^7 + 3x^6 + 1x^5 + 2x^4 + 0x^3 + 3x^2 + 2x + 2$
$= 2x^7 + 3x^6 + x^5 + 2x^4 + 3x^2 + 2x + 2.$

Problem 26 Let $f(x) = 5x^4 + 3x^3 + 1$ and $g(x) = 3x^2 + 2x + 1$ in $\mathbb{Z}_7[x]$. Determine the quotient and remainder upon dividing $f(x)$ by $g(x)$.

Solution We will divide $f(x) = 5x^4 + 3x^3 + 1$ by $g(x) = 3x^2 + 2x + 1$ in $\mathbb{Z}_7[x]$ by long division method.

$$
\begin{array}{r}
4x^2 + 3x + 6 \\
3x^2 + 2x + 1 \overline{\smash{\big)}\ 5x^4 + 3x^3 \qquad\quad + 1} \\
\underline{5x^4 + \ x^3 + 4x^2} \\
2x^3 + 3x^2 \\
\underline{2x^3 + 6x^2 + 3x} \\
4x^2 + 4x + 1 \\
\underline{4x^2 + 5x + 6} \\
6x + 2
\end{array}
$$

Thus, we get $(5x^4 + 3x^3 + 1) = (4x^2 + 3x + 6) \cdot (3x^2 + 2x + 1) + (6x + 2)$. Here, the quotient is $(4x^2 + 3x + 6)$ and the remainder is $(6x + 2)$.

Exercise

(1) Let R be a ring and x be an indeterminate. Prove that the rings $R[x]$ and $R[x^2]$ are ring-isomorphic.

(2) Show that $\mathbb{Q}[x]/(x^4 + 2 \cdot x^2 + 1)$ is a local ring (a local ring is a ring which has a unique maximal ideal), and that the non-units are precisely the images of those polynomials $f(x)$ which vanish at $\pm i$ (imaginary unit).

(3) If the rings R and S are isomorphic, show that $R[x]$ and $S[x]$ are isomorphic.

(4) Let $f(x)$ be a nonconstant element of $\mathbb{Z}[x]$. Prove that $f(x)$ takes on infinitely many values in $\mathbb{Z}$.

(5) Find the remainder when x^{51} is divided by $x + 4$ in $\mathbb{Z}_7[x]$.

(6) Let F be a field and let $I = \{f(x) \in F[x] \mid f(a) = 0 \; \forall \; a \in F\}$. Prove that I is an ideal in $F[x]$. Prove that I is infinite when F is finite and $I = \{0\}$ when F is infinite. When F is finite, find a monic polynomial $g(x)$ such that $I = \langle g(x) \rangle$.

(7) Let F be a field and let $f(x) = a_n \cdot x^n + a_{n-1} \cdot x^{n-1} + \ldots + a_0 \in F[x]$. Prove that $x - 1$ is a factor of $f(x)$ if and only if $a_n + a_{n-1} + \ldots + a_0 = 0$.

(8) Let m be a fixed positive integer. For any integer a, let $\overline{a}$ denote $a \bmod m$. Show that the mapping of $\phi : \mathbb{Z}[x] \to \mathbb{Z}_m[x]$ given by $\phi(a_n x^n + a_{n-1} x^{n-1} + \ldots + a0) = \overline{a_n} x^n + \overline{a_{n-1}} x^{n-1} + \ldots + \overline{a_0}$ is a ring homomorphism.

(9) Show that there are infinitely many polynomials $f(x)$ in $\mathbb{Z}_3[x]$ such that $f(a) = 0$ for all a in $\mathbb{Z}_3$.

(10) Let $f(x)$ belong to $\mathbb{Z}_p[x]$. Prove that if $f(b) = 0$, then $f(b^p) = 0$.

(11) Let $f(x) = a_m x^m + a_{m-1} x^{m-1} + \ldots + a_0$ and $g(x) = b_n x^n + b_{n-1} x^{n-1} + \ldots + b_0$ belong to $\mathbb{Q}[x]$ and suppose that $f(x) \cdot g(x)$ belongs to $\mathbb{Z}[x]$. Prove that $a_i \cdot b_j$ is an integer for every i and j.

(12) Let $f(x)$ belong to $\mathbb{Z}[x]$. If $a \bmod m = b \bmod m$, prove that $f(a) \bmod m = f(b) \bmod m$. Prove that if both $f(0)$ and $f(1)$ are odd, then f has no zero in $\mathbb{Z}$.

(13) Find sum and product of the polynomials $f(x) = 4x - 5$ and $g(x) = 2x^2 - 4x + 2$ in $\mathbb{Z}_8[x]$.

(14) Let $\mathbb{Z}$ denote the ring of integers and consider the three commutative rings
$R_1 = \mathbb{Z} \times \mathbb{Z}$,
$R_2 = \mathbb{Z}[\sqrt{5}]$,
$R_3 = \mathbb{Z}[x]/x^2\mathbb{Z}[x]$.

For $1 \le i < j \le 3$ either prove that R_i and R_j are isomorphic rings, or prove that they are not.

(15) **(a)** Prove that the three additive groups $\mathbb{Z} \times \mathbb{Z}$, $\mathbb{Z}[i]$, and $\mathbb{Z}[x]/\langle x^2 \rangle$ are all isomorphic to each other.

(b) Prove that no two of the rings $\mathbb{Z} \times \mathbb{Z}$, $\mathbb{Z}[i]$, and $\mathbb{Z}[x]/\langle x^2 \rangle$ are isomorphic to each other.

(16) Find a nonzero polynomial in $\mathbb{Z}_3[x]$ that induces the zero function on $\mathbb{Z}_3$.
Note that each polynomial $a_0 + a_1 \cdot x + a_2 \cdot x^2 + \ldots + a_n \cdot x^n$ in $R[x]$, R an arbitrary commutative ring is associated with the mapping $f : R \to R$ defined by $f(r) = a_0 + a_1 \cdot r + a_2 \cdot r^2 + \ldots + a_n \cdot r^n$ for all $r \in R$.

(17) Find all the complex zeroes of $x^n - 1$.

(18) In the polynomial ring $\mathbb{Z}_7[x]$, consider the polynomial $f(x) = x^3 + 2x^2 + 2x + 1$. Determine all roots of $f(x)$ in $\mathbb{Z}_7[x]$.

8 FACTORIZATION OF POLYNOMIALS

In ring theory, factorization of polynomials constitutes a fundamental aspect of algebraic study, revealing a subtle linkage between the arithmetic properties of polynomials and their structural characteristics. At the core of this exploration lies the distinction between reducible and irreducible polynomials. Reducible polynomials can be expressed as products of two or more nonconstant polynomials, while irreducible polynomials resist such factorization, existing as prime elements within the polynomial ring. Understanding the nature of factorization sheds light on diverse algebraic phenomena and serves as a fundamental building block for various applications across mathematics and its applications.

In this chapter, we explore polynomial factorization in ring theory. We shall discuss reducible and irreducible polynomials, explore the significance of primitive polynomials, and explain the tests and algorithms used to identify the factorization properties of polynomials. Thoroughly examining the polynomial factorization, we aim to provide readers with the tools and insights necessary to understand algebraic structures and properties of polynomial rings.

Firstly, we discuss the factorization of polynomials, which involves an analogous concept to prime integers, termed as follows:

Definition 8.1 (Irreducible Polynomial) Let D be an integral domain. A polynomial $f(x)$ from $D[x]$ is said to be an *irreducible polynomial* over D if whenever $f(x) = g(x) \cdot h(x)$ for $g(x), h(x) \in D[x]$, then either $g(x)$ or $h(x)$ is a unit in $D[x]$.

Definition 8.2 (Reducible Polynomial) A nonzero, non-unit element of $D[x]$ that is not irreducible over D is called reducible over D.

Example 8.3 The polynomial $f(x) = 3x^2 + 9$ is irreducible over $\mathbb{Q}$ but reducible over $\mathbb{Z}$, since $3x^2 + 9 = 3(x^2 + 3)$ and neither 3 nor $x^2 + 3$ is a unit in $\mathbb{Z}[x]$.

Example 8.4 The polynomial $f(x) = x^2 + 3$ is irreducible over $\mathbb{R}$ but reducible over $\mathbb{C}$, since $x^2 + 3 = (x - \sqrt{3}i)(x + \sqrt{3}i)$ and neither $\sqrt{3}i$ nor $-\sqrt{3}i$ belongs to R.

Example 8.5 The polynomial $x^2 - 3$ is irreducible over $\mathbb{Q}$ but reducible over $\mathbb{R}$, since $x^2 - 3 = (x - \sqrt{3})(x + \sqrt{3})$ and $\pm\sqrt{3}$ do not belong to $\mathbb{Q}$.

Example 8.6 The polynomial $x^2 + 2x + 1 = (x - 4)(x - 4)$ in $\mathbb{Z}_5[x]$. However, $x^2 + 2x + 1$ is irreducible over $\mathbb{Z}_3$.

Irreducible polynomials play a fundamental role in understanding factorization properties. In the next proposition, a relationship between the irreducibility of a polynomial and its scalar multiple is established.

Proposition 8.7 *Let F be a field and let a be a nonzero element of F. If $af(x)$ is irreducible over F, then $f(x)$ is irreducible over F.*

Proof Suppose $f(x)$ is not irreducible over F. So, $f(x)$ is reducible over F. Then there exist non-unit polynomials $g(x)$ and $h(x)$ such that, $f(x) = g(x) \cdot h(x)$ where $0 < \deg g(x) < \deg f(x)$ and $0 < \deg h(x) < \deg f(x)$. Observe that if $\deg (g(x)) = 0$, then $g(x)$ is a unit. Therefore, $\deg g(x)$ and $\deg h(x)$ are nonzero.

This implies that $af(x) = ag(x)h(x)$, where $a \neq 0 \in F$. $\hspace{2cm}$ (8.1)

Since $0 < \deg g(x) < \deg f(x)$, it follows that

$0 < a \deg g(x) < a \deg f(x)$ $\hspace{2cm}$ $(\because a \neq 0 \in F)$ (8.2)

Also, we have $0 < \deg h(x) < \deg f(x)$

$\Rightarrow 0 < \deg h(x) < a \deg f(x)$ $\hspace{2cm}$ $(\because a \neq 0 \in F)$ (8.3)

So, by Equations (8.1), (8.2), and (8.3), $af(x)$ is reducible over F, which is a contradiction to the given fact that $af(x)$ is irreducible over F. Thus, our assumption is wrong. Hence, $f(x)$ is irreducible over F. $\hspace{1cm}$ $\square$

The subsequent proposition is based on understanding the behavior of polynomial roots under scalar multiplication.

Proposition 8.8 *If $f(ax)$ is irreducible over a field F where a be a nonzero element of F, then $f(x)$ is irreducible over F.*

Proof Suppose $f(x)$ is not irreducible over F. Then there exist non-unit polynomials $g(x)$ and $h(x)$ such that, $f(x) = g(x) \cdot h(x)$ where $0 < \deg g(x) < \deg f(x)$ and $0 < \deg h(x) < \deg f(x) \ \forall \ x \in F$.

In particular, since, $a, x \in F$, then $ax \in F$

$\Rightarrow f(ax) = g(ax)h(ax)$, where $a \neq 0 \in F$. $\hspace{2cm}$ (8.4)

Since $0 < \deg g(x) < \deg f(x)$, it follows that

$0 < \deg g(ax) < \deg f(ax)$. $\hspace{2cm}$ $(\because a \neq 0 \in F)$ (8.5)

Also, we have $0 < \deg h(x) < \deg f(x)$

$\Rightarrow 0 < \deg h(ax) < \deg f(ax)$. $\hspace{2cm}$ $(\because a \neq 0 \in F)$ (8.6)

So, by Equations (8.4), (8.5), and (8.6), $f(ax)$ is reducible over F, which is a contradiction to the fact that $f(ax)$ is irreducible over F. Thus, our assumption is wrong. Hence, $f(x)$ is irreducible over F. $\hspace{1cm}$ $\square$

The next proposition essentially conveys us that shifting the variable of an irreducible polynomial by a constant amount doesn't affect its irreducibility over the field.

Proposition 8.9 *Suppose $f(x + a)$ is irreducible over a field F. Then, $f(x)$ is irreducible over F. (Here, a is a nonzero element of F.)*

Proof Suppose $f(x)$ is not irreducible over F. Then, $f(x)$ is reducible over F. Thus there exist non-unit polynomials $g(x)$ and $h(x)$ such that, $f(x) = g(x) \cdot h(x)$, where $0 < \deg g(x) < \deg f(x)$ and $0 < \deg h(x) < \deg f(x) \; \forall \; x \in F$.

Now, if $a, x \in F$, then $x + a \in F$ (since F is a field and closure holds).

In particular, for $x + a \in F$, $f(x + a) = g(x + a)h(x + a)$, where $a \neq 0 \in F$. $\hspace{1cm}$ (8.7)

Since $0 < \deg g(x) < \deg f(x)$, it follows that

$0 < \deg g(x + a) < \deg f(x + a)$ $\hspace{3cm}$ $(\because x + a \in F)$ (8.8)

Also, we have $0 < \deg h(x) < \deg f(x)$

$\Rightarrow 0 < \deg h(x + a) < \deg f(x + a)$ $\hspace{2.5cm}$ $(\because x + a \in F)$ (8.9)

So, by Equations (8.7), (8.8), and (8.9), we get, $f(x + a)$ is reducible over F, which is a contradiction to the fact that $f(x + a)$ is irreducible over F. Thus, our assumption is wrong. Hence, $f(x)$ is irreducible over F. $\hspace{1cm}$ $\square$

After establishing properties of irreducibility for the function $f(x)$, reducibility tests in ring theory typically begin by exploring factorization possibilities of $f(x)$ into simpler components. This approach aims to distinguish whether $f(x)$ can be factored into non-trivial factors, thereby providing insights into the algebraic structure of the ring. The theorem underlying this process seeks to determine whether $f(x)$ possesses prime factors or remains irreducible within the ring under consideration. Although, verifying the reducibility of a polynomial from an infinite set can be challenging, certain special cases exist where the task is simplified.

Theorem 8.10 (Reducibility Test for Degrees 2 and 3) *Let F be a field. If $f(x) \in F[x]$ and $\deg f(x)$ is 2 or 3, then $f(x)$ is reducible over F if and only if $f(x)$ has a zero in F.*

Proof Let $f(x)$ be reducible over F. This implies, $f(x) = g(x) \cdot h(x)$ where $g(x), h(x) \in F[x]$. Also, $\deg g(x) < \deg f(x)$, $\deg h(x) < \deg f(x)$. Since $\deg f(x) = 2$ or 3, so one of $g(x)$ or $h(x)$ is definitely of degree 1. Without loss of generality, let $h(x)$ be of degree 1.

Let $h(x) = ax + b$. Then, zero of $h(x)$ is $\dfrac{-b}{a} \in F$.

Now, we have $f(x) = g(x) \cdot h(x) \Rightarrow f\left(\dfrac{-b}{a}\right) = g\left(\dfrac{-b}{a}\right) \cdot h\left(\dfrac{-b}{a}\right) = g\left(\dfrac{-b}{a}\right) \cdot 0 = 0$. So,

$x = \dfrac{-b}{a}$ is a zero of $f(x)$. Therefore, $f(x)$ has a zero in F.

Conversely, suppose $f(x)$ has a zero in F. Let a be its zero in $F \Rightarrow f(a) = 0$. This implies that $(x - a)$ is a factor of $f(x)$ in $F[x]$. Therefore, $f(x) = (x - a) \cdot g(x)$. Hence, $f(x)$ is reducible over $F[x]$. $\hspace{1cm}$ $\square$

Example 8.11 Consider $f(x) = x^3 + 3x + 2$ over the field $\mathbb{Z}_5$. Here, $\mathbb{Z}_5$ is a field and $f(x) = x^3 + 3x + 2$ is a polynomial of degree 3. So, by Theorem 8.10, in order to check reducibility of $f(x)$, we check if $f(x)$ has a zero over the field $\mathbb{Z}_5$. Given that the field is $\mathbb{Z}_5 = \{0, 1, 2, 3, 4\}$. Then,

$f(0) = 0 + 0 + 2 = 2,$

$f(1) = 1 + 3 + 2 = 6 \mod 5 = 1,$

$f(2) = 8 + 6 + 2 = 16 \mod 5 = 1,$

$f(3) = 27 + 9 + 2 = 38 \mod 5 = 3,$
$f(4) = 64 + 12 + 2 = 78 \mod 5 = 3.$
Since $f(x)$ has no zeroes over the field $\mathbb{Z}_5$, $f(x) = x^3 + 3x + 2$ is not reducible over the field $\mathbb{Z}_5$. Hence, $f(x) = x^3 + 3x + 2$ is irreducible over the field $\mathbb{Z}_5$.

Remark In general $\mathbb{Z}_p$ is a field where p is prime. Then, by Theorem 8.10, we can simply check whether a polynomial of degree 2 or 3 is reducible or not over $\mathbb{Z}_p$ by verifying if it has a zero in $\{0, 1, 2, \dots, p-1\}$.

A key concept in polynomial factorization is that of primitive polynomials. These polynomials are characterized by their coefficients, which are relatively prime integers. Primitive polynomials hold particular significance, as they generate distinct irreducible factors when considered over fields of finite characteristic. Introducing the properties and behavior of primitive polynomials reveals a significant understanding of the structure of polynomial rings and their factorization patterns, facilitating deeper comprehension of algebraic structures and their applications. Additionally, the concept of the content of a polynomial has been introduced, offering valuable insights into the common factors present within the polynomial.

Definition 8.12 (Content of a Polynomial) Let $f(x) = a_n x^n + a_{n-1} x^{n-1} + \dots + a_0$ where $a_i \in \mathbb{Z}$ be a nonzero polynomial in $\mathbb{Z}[x]$. Then the content of the polynomial $f(x)$ is the greatest common divisor of the integers $a_n, a_{n-1}, \dots, a_0$, i.e., content of $f(x) = \gcd(a_n, a_{n-1}, \dots, a_0)$.

Definition 8.13 (Primitive Polynomial) A nonzero polynomial with content 1 is called primitive polynomial.

Example 8.14 A polynomial $f(x) = 5x^2 + 1$ is a primitive polynomial as content of $f(x) = \gcd(5, 1) = 1$.

Example 8.15 A polynomial $g(x) = 4x^2 + 8x + 6$ is not a primitive polynomial as content of $g(x) = \gcd(4, 8, 6) = 2 \neq 1$.

Observe that for $f(x) = 2x^2 + 1$ and $g(x) = 3x^3 + 6x^2 + 4$, are primitive polynomials. Here, $f(x)g(x) = 6x^5 + 12x^4 + 3x^3 + 14x^2 + 4$ and content of $f(x) \cdot g(x)$ is 1. Thus, the product of two primitive polynomials is primitive which has been proved in the following lemma.

Lemma 8.16 (Gauss's Lemma) The product of two primitive polynomials is primitive.

Proof Let $f(x)$ and $g(x)$ be primitive polynomials. Let $h(x) = f(x) \cdot g(x)$. Now, we have to prove that $h(x)$ is primitive. Suppose $h(x)$ is not primitive. Then there exists a prime divisor p of the content of $h(x)$. Let $\overline{f(x)g(x)}$, $\overline{f(x)}$ and $\overline{g(x)}$ be polynomials obtained from polynomials $f(x)g(x)$, $f(x)$ and $g(x)$ respectively by reducing the coefficients by modulo p. Here, $\overline{f(x)}, \overline{g(x)} \in \mathbb{Z}_p[x]$. Now, $\overline{f(x).g(x)} = \overline{f(x)} \cdot \overline{g(x)} = 0$ element of $\mathbb{Z}_p[x]$ which implies that $\overline{f(x)} \cdot \overline{g(x)} = 0$ element of $\mathbb{Z}_p[x]$.
So, either all the coefficients of $f(x)$ or $g(x)$ are divisible by p. Therefore, either $f(x)$ or $g(x)$ is not primitive, which is a contradiction. Thus, our assumption is wrong that $f(x) \cdot g(x)$ is not primitive polynomial. Therefore, $f(x) \cdot g(x)$ is a primitive polynomial. Hence, the product of two primitive polynomials is primitive. $\qquad\square$

The subsequent theorem asserts that if a polynomial with integer coefficients is factorable into polynomials with rational coefficients, then it can also be factored into polynomials with integer coefficients. This theorem is crucial for understanding how polynomial factorization over rational numbers relates to factorization over integers.

Theorem 8.17 *Let $f(x) \in \mathbb{Z}[x]$. If $f(x)$ is reducible over $\mathbb{Q}$, then it is reducible over $\mathbb{Z}$.*

Proof Let $f(x) = g(x) \cdot h(x)$ where $g(x), h(x) \in \mathbb{Q}[x]$. Without loss of generality, assume that $f(x)$ is primitive. Let a and b be the lcm of the denominators of the coefficients of $g(x)$ and $h(x)$ respectively.

Consider $a \cdot b \cdot f(x) = a \cdot b \cdot g(x) \cdot h(x) = ag(x) \cdot bh(x)$. Now, $ag(x), bh(x) \in \mathbb{Z}[x]$. Let c_1 be the content of $ag(x)$ and c_2 be the content of $bh(x)$. This implies $ag(x) = c_1 g_1(x)$ and $bh(x) = c_2 h_1(x)$, where $g_1(x), h_1(x)$ are primitive.

Now, $abf(x) = c_1 g_1(x) \cdot c_2 h_1(x) = c_1 c_2 g_1(x) h_1(x)$. Since $f(x)$ is primitive, content of $abf(x) = ab$.

Thus, by Gauss' lemma, content of $c_1 c_2 g_1(x) h_1(x)$ is $c_1 c_2$. Therefore, $ab = c_1 c_2$ and $f(x) = g_1(x) \cdot h_1(x)$, where $g_1(x), h_1(x) \in \mathbb{Z}[x]$ such that deg $(g_1(x)) = $ deg $(g(x))$ and deg $(h_1(x)) = $ deg $(h(x))$. Therefore, $f(x) = g_1(x) \cdot h_1(x)$. Hence, $f(x)$ is reducible over $\mathbb{Z}$. $\square$

Up to this point, various tests for reducibility have been discussed. Moving forward, we'll look into different methods for determining irreducibility. Exploring the reducibility and irreducibility of polynomials necessitates the development of systematic tests to ascertain their factorization properties. Various techniques and algorithms have been devised to identify the reducibility status of polynomials, ranging from elementary methods such as inspection and substitution to advanced methodologies like mod p irreducibility test and Eisenstein's criterion.

In Theorem 8.10, we addressed the reducibility criteria for polynomials of degree 2 and 3. However, when dealing with irreducible polynomials or those of higher degrees, the following test called mod p irreducibility test is essential. The theorem aims to provide a criterion for determining whether a polynomial over a field of prime characteristic is irreducible. By examining the polynomial's congruence modulo a prime p, the theorem offers a practical approach to assess its irreducibility without directly factoring it.

Theorem 8.18 (Mod p Irreducibility Test) *Let p be a prime and suppose that $f(x) \in \mathbb{Z}[x]$ with deg $f(x) \geq 1$. Let $\overline{f(x)}$ be the polynomial in $\mathbb{Z}_p[x]$ obtained from $f(x)$ by reducing all the coefficients of $f(x)$ modulo p. If $\overline{f(x)}$ is irreducible over $\mathbb{Z}_p$ and deg $\overline{f(x)} = $ deg $f(x)$, then $f(x)$ is irreducible over $\mathbb{Q}$.*

Proof Let $f(x)$ be reducible over $\mathbb{Q}[x]$. Then, by Theorem 8.17, $f(x)$ is reducible over $\mathbb{Z}[x]$, which implies that $f(x) = g(x) \cdot h(x)$, where $g(x), h(x) \in \mathbb{Z}[x]$ such that deg $g(x) < $ deg $f(x)$ and deg $h(x) < $ deg $f(x)$. $\hspace{2cm}$ (8.10)

Let $\overline{f(x)}, \overline{g(x)}, \overline{h(x)}$ be polynomials obtained by reducing $f(x), g(x)$ and $h(x)$ modulo p. Given that deg $f(x) = $ deg $\overline{f(x)}$. $\hspace{2cm}$ (8.11)

Also, we have, deg $\overline{g(x)} \leq $ deg $g(x)$. $\hspace{2cm}$ (8.12)

Using Equations (8.10), (8.11), and (8.12), deg $\overline{g(x)} \leq $ deg $g(x) < $ deg $f(x) = $ deg $\overline{f(x)} \Rightarrow$ deg $\overline{g(x)} < $ deg $\overline{f(x)}$. $\hspace{2cm}$ (8.13)

Similarly, $\deg \overline{h(x)} \leq \deg h(x) < \deg f(x) = \deg \overline{f(x)} \Rightarrow \deg \overline{h(x)} < \deg \overline{f(x)}$.   (8.14)
Now, $f(x) = g(x) \cdot h(x) \Rightarrow \overline{f(x)} = \overline{g(x)} \cdot \overline{h(x)}$, where $\deg \overline{g(x)} < \deg \overline{f(x)}$ and $\deg \overline{h(x)} <$
$\deg \overline{f(x)}$.   (By (8.13) and (8.14).)
This implies that $\overline{f(x)}$ is reducible over $\mathbb{Z}_p$, which is a contradiction to the fact that $\overline{f(x)}$ is
irreducible over $\mathbb{Z}_p$. Thus, our assumption is wrong. Hence, $f(x)$ is irreducible over $\mathbb{Q}$. $\square$

The above test is used to check the irreducibility of polynomials of degree greater than 3
and polynomials with integral coefficients. The following example demonstrates the usage
of the above test.

Example 8.19 A polynomial $f(x) = x^3 + 7x^2 + 13x - 4$ is irreducible over $\mathbb{Q}$. In order to
verify this, let's take $p = 2$, so $\overline{f(x)} = x^3 + x^2 + x$, then
$\overline{f(0)} = 0 + 0 + 0 = 0$ and $\overline{f(1)} = 1 + 1 + 1 = 3 \mod 2 = 1$. Thus, $\overline{f(x)}$ has a zero in $\mathbb{Z}_2[x]$
which means that $\overline{f(x)}$ is reducible over $\mathbb{Z}_2$.
Now, take $p = 3$, so $\overline{f(x)} = x^3 + x^2 + x + 2$, then $\overline{f(0)} = 2$, $\overline{f(1)} = 2$ and $\overline{f(2)} = 8 + 4 + 2 +$
$2 = 16 \mod 3 = 1$. Since there are no zeroes, $\overline{f(x)}$ is irreducible over $\mathbb{Z}_3$. Also, $\deg f(x) =$
$\deg \overline{f(x)}$. Hence, by using mod 3 irreduciblity test, $f(x)$ is irreducible over $\mathbb{Q}$.

Remark To conclude that a particular $f(x)$ in $\mathbb{Z}[x]$ is irreducible over $\mathbb{Q}$, all we need to
do is find $\deg f(x) = \deg \overline{f(x)}$ and a single p for which the corresponding polynomial
$\overline{f(x)}$ is irreducible over $\mathbb{Z}_p$. However, there exist polynomials that are irreducible over $\mathbb{Q}$
but reducible over $\mathbb{Z}_p$ for every prime. For example $f(x) = x^4 + 1$ is irreducible over $\mathbb{Q}$ but
reducible over $\mathbb{Z}_p$ for every prime p.

Another significant test for irreducibilty is Eisenstein's criterion. The following test gives
us a sufficient condition for a polynomial with integer coefficients to be irreducible over the
rational numbers.

Theorem 8.20 (Eisenstein's Criterion)   *Let $f(x) = a_n x^n + a_{n-1} x^{n-1} + \dots + a_1 x + a_0$*

$\in \mathbb{Z}[x]$. If there is a prime p such that $p \nmid a_n$, $p \mid a_{n-1}, \dots , p \mid a_0$ and $p^2 \nmid a_0$, then $f(x)$ is
irreducible over $\mathbb{Q}$.

Proof Let $f(x) = a_n x^n + a_{n-1} x^{n-1} + \dots + a_1 x + a_0$: where $a_i \in \mathbb{Z}$. Let p be a prime such
that $p \nmid a_n$ but $p \mid a_{n-1}, \dots , p \mid a_0$ and $p^2 \nmid a_0$.
On contrary, if possible let $f(x)$ be reducible over $\mathbb{Q}$. Then, by Theorem 8.17, $f(x)$ is reducible
over $\mathbb{Z}$. This implies, there exist $g(x)$, $h(x) \in \mathbb{Z}[x]$ such that $f(x) = g(x) \cdot h(x)$ (8.15)
with $1 \leq \deg g(x)$, $\deg h(x) < \mathrm{def}\, f(x) = n$.
Let $g(x) = b_r x^r + b_{r-1} x^{r-1} + \dots + b_0$ and $h(x) = c_s x^s + c_{s-1} x^{s-1} + \dots + c_0$ in $\mathbb{Z}[x]$, where
$1 \leq r, s < n$.
Now, $g(x) \cdot h(x) = b_r c_s x^{r+s} + \dots + (b_1 c_0 + c_1 b_0)x + b_0 c_0$.   (8.16)
By Equations (8.15) and (8.16), we get, $a_0 = b_0 \cdot c_0$.
Also given that $p \mid a_0$ and $p^2 \nmid a_0$, which implies that $p \mid b_0 \cdot c_0$ and $p^2 \nmid b_0 \cdot c_0$. So, either
$p \mid b_0$ or $p \mid c_0$ but not both.
Without loss of generality, let, $p \mid b_0$ but $p \nmid c_0$.   (8.17)
Also, given that $p \nmid a_n$, which, on comparing Equations (8.15) and (8.16), implies that
$p \nmid b_r c_s \Rightarrow p \nmid b_r$ and $p \nmid c_s$.
Let $j = \min \{i \mid p \nmid b_i\}$. So, $p \nmid b_j$ but $p \mid b_0, b_1, \dots , b_{j-1}$.

Now, $a_j = c_j b_0 + c_{j-1} b_1 + ... + c_1 b_{j-1} + c_0 b_j$. Since $p \mid a_j$, $p \mid b_0$, $p \mid b_{j-1}$, so $p \mid c_0 b_j \Rightarrow p \mid c_0$ or $p \mid b_j$ which is a contradiction. Thus, our assumption was wrong and hence $f(x)$ is irreducible over $\mathbb{Q}$. $\square$

Example 8.21 The polynomial $6x^5 + 5x^4 - 25x^3 + 15x + 10$ is irreducible over $\mathbb{Q}$ using the prime $p = 5$ while following Eisenstein's criteria since there exists a prime $p = 5$ such that $5 \nmid 6$, $5 \mid 5$, $5 \mid (-25)$, $5 \mid 15$, $5 \mid 10$, $5^2 (= 25) \nmid 10$.

Using Eisenstein's criterion is crucial as it's a robust method for verifying the irreducibility of polynomials over the rationals. By carefully choosing a prime number and checking if the polynomial meets its irreducibility over $\mathbb{Q}$, this approach uses the structure of cyclotomic polynomials and the distinctive properties of prime numbers, as explained in the subsequent corollary.

Corollary 8.22 For any prime p, the cyclotomic polynomial $\phi_p(x) = \dfrac{x^p - 1}{x - 1} = x^{p-1} + x^{p-2} + ... + x + 1$ is irreducible over $\mathbb{Q}$.

Proof Let $f(x) = \phi_p(x + 1) = \dfrac{(x+1)^p - 1}{(x+1) - 1} = \dfrac{(x+1)^p - 1}{x}$

$$\Rightarrow f(x) = \frac{1}{x}[x^p + {}^pC_1 x^{p-1} + {}^pC_2 x^{p-2} + ... + {}^pC_{p-1} x + 1 - 1]$$

$$= \frac{1}{x}[x^p + px^{p-1} + \frac{p(p-1)}{2} x^{p-2} + ... + px]$$

$$= 1 \cdot x^{p-1} + p \cdot x^{p-2} + \frac{p(p-1)}{2} \cdot x^{p-3} + ... + p.$$

Now, since $p \nmid 1 (= a_n)$, $p \mid p (= a_{n-1})$, $p \mid \dfrac{p(p-1)}{2} (= a_{n-2}), ..., p \mid p (= a_0)$ and $p^2 \nmid p (= a_0)$, using Eisenstein's criteria over $f(x)$, we get that $f(x)$ is irreducible over $\mathbb{Q}$. Now, if $\phi_p(x)$ is reducible, then $\phi_p(x) = g(x)h(x)$, which implies $f(x) = \phi_p(x+1) = g(x+1)h(x+1)$. So, $f(x)$ is reducible, which is a contradiction. Therefore, $\phi_p(x)$ is irreducible over $\mathbb{Q}$. Hence, $\phi_p(x) = x^{p-1} + x^{p-2} + ... + x + 1$ is irreducible over $\mathbb{Q}$. $\square$

In the forthcoming theorem, we establish the connection between maximal ideals and irreducible polynomials over F. Specifically, we can explore how the maximal ideal generated by polynomial, $p(x)$ relates to the factorization properties of $p(x)$ within $F[x]$, the set of all polynomials over field F and its implications for the structure of the ring. By examining how irreducibility impacts the properties of maximal ideals, we can establish the equivalence between $p(x)$ being irreducible over F and $\langle p(x) \rangle$ being a maximal ideal in $F[x]$.

Theorem 8.23 *Let F be a field and $p(x) \in F[x]$. Then, $\langle p(x) \rangle$ is a maximal ideal in $F[x]$ if and only if $p(x)$ is irreducible over F.*

Proof Suppose $\langle p(x) \rangle$ is a maximal ideal in $F[x]$. This implies, $p(x)$ is neither the zero polynomial nor a unit in $F[x]$ since neither $\{0\}$ nor $F[x]$ is a maximal ideal in $F[x]$. Let, if possible, $p(x)$ be reducible over F.

So, there exist non-unit polynomials $g(x)$ and $h(x)$ such that as $p(x) = g(x) \cdot h(x)$, (8.18)
where $\deg g(x) < \deg p(x)$, $\deg h(x) < \deg p(x)$ (8.19)
This implies that $\langle p(x) \rangle \subseteq \langle g(x) \rangle \subseteq F[x]$. Since $\langle p(x) \rangle$ is maximal in $F[x]$, we have either $\langle p(x) \rangle = \langle g(x) \rangle$ or $\langle g(x) \rangle = F[x]$.

Case 1. If $\langle p(x) \rangle = \langle g(x) \rangle$, then deg $p(x) =$ deg $g(x)$ therefore, $h(x)$ is a unit, which is a contradiction to relation (8.18).

Case 2. If $F[x] = \langle g(x) \rangle$, then $g(x)$ is a unit, which leads to a contradiction.

Thus, our assumption is wrong that $p(x)$ is reducible over F. Hence, $p(x)$ is irreducible in $F[x]$.

Conversely, suppose $p(x)$ is irreducible over F.

Let I be any ideal of $F[x]$ such that $\langle p(x) \rangle \subseteq I \subseteq F[x]$ $\qquad$ (8.20)

Since $F[x]$ is principal ideal domain and I is any ideal in $F[x]$, $I = \langle g(x) \rangle$ for some $g(x) \in F[x]$ $\qquad$ (8.21)

$\Rightarrow p(x) \in \langle g(x) \rangle$.

Therefore, $p(x) = g(x) \cdot h(x)$, where $h(x) \in F[x]$. $\qquad$ (8.22)

Also, given that $p(x)$ is irreducible over F. So, either $g(x)$ or $h(x)$ is a unit.

Case 1. Suppose $g(x)$ is unit. Then, $I = F[x]$ $\qquad$ (by Equation (8.22))

Case 2. Suppose $h(x)$ is unit. Then $(h(x))^{-1}p(x) = g(x)$. Therefore, $g(x) \in \langle p(x) \rangle$. This implies $\langle g(x) \rangle \subseteq \langle p(x) \rangle$. Thus we get $\langle p(x) \rangle = \langle g(x) \rangle = I$. Therefore, $I = \langle p(x) \rangle$.

Thus, either $I = F[x]$ or $\langle p(x) \rangle = I$ when $\langle p(x) \rangle \subseteq I \subseteq F[x]$. Hence, $\langle p(x) \rangle$ is maximal in $F[x]$. $\qquad\square$

Remark The above result can be used to construct fields of the desired order.

The following corollary directly stems from the preceding theorem, as the quotient ring inherits field properties from the maximal ideal arising from the irreducible polynomial. This emphasizes a close relationship between a polynomial's irreducibility and the maximal property of the ideal it produces.

Corollary 8.24 Let F be a field and $p(x)$ be an irreducible polynomial over F. Then $\dfrac{F[x]}{\langle p(x) \rangle}$ is a field.

Proof Let F be a field and $p(x)$ be an irreducible polynomial over F. By Theorem 8.23, $p(x) \in F[x]$ is irreducible over F if and only if $\langle p(x) \rangle$ is maximal. Hence, by Theorem 5.56, $F[x]/\langle p(x) \rangle$ is a field. $\qquad\square$

To prove the subsequent proposition, we use the notion of quotient rings. Consider the irreducible polynomial $f(x) \in \mathbb{Z}_p[x]$ of degree n over the prime field $\mathbb{Z}_p$. By studying the field extension $\mathbb{Z}_p[x]/\langle f(x) \rangle$, we reveal a finite field with precisely p^n elements, demonstrating how different algebraic structures work together.

Proposition 8.25 *Suppose that $f(x) \in \mathbb{Z}_p[x]$ be irreducible over $\mathbb{Z}_p$, where p is a prime. If* $\deg f(x) = n$, then $\dfrac{\mathbb{Z}_p[x]}{\langle f(x) \rangle}$ *is a field with p^n elements.*

Proof Given that $f(x)$ is irreducible over $\mathbb{Z}_p$. Then, by Theorem 8.23 and corollary 8.24, $\dfrac{\mathbb{Z}_p}{\langle f(x) \rangle}$ is a field. Let $g(x) + \langle f(x) \rangle \in \dfrac{\mathbb{Z}_p}{\langle f(x) \rangle}$.

By division algorithm, there exists $q(x)$, $r(x) \in \mathbb{Z}_p[x]$ such that $g(x) = q(x) \cdot f(x) + r(x)$, where either $r(x) = 0$ or deg $r(x) <$ deg $f(x) = n$.

Therefore, $g(x) + \langle f(x)\rangle = q(x).f(x) + r(x) + \langle f(x)\rangle = r(x) + \langle f(x)\rangle = a_{n-1}x^{n-1} + \ldots + a_1 x + a_0 + \langle f(x)\rangle$. Since, $a_{n-1}, \ldots, a_1, a_0 \in \mathbb{Z}_p$, therefore the field has p^n elements. $\square$

Proposition 8.26 *For every prime p, there exists a field of order p^2.*

Proof We know that for each prime p, there is an irreducible polynomial $p(x) = x^2 + 1$ of degree 2 over $\mathbb{Z}_p$. Therefore, by Theorem 8.23, $\langle p(x)\rangle$ is maximal ideal in $\mathbb{Z}_p[x]$. Then, by corollary 8.24, $\mathbb{Z}_p[x]/\langle p(x)\rangle$ is a field.

Consider, $f(x) + \langle p(x)\rangle \in \dfrac{\mathbb{Z}_p[x]}{\langle p(x)\rangle}$. By division algorithm, there exists $q(x)$, $r(x) \in \mathbb{Z}_p[x]$ such that $f(x) = q(x) \cdot p(x) + r(x)$ with either $r(x) = 0$ or $\deg r(x) < \deg p(x) = 2$. Let $r(x) = ax + b$, then $f(x) + \langle p(x)\rangle = q(x) \cdot p(x) + r(x) + \langle p(x)\rangle = r(x) + \langle p(x)\rangle$.

Therefore, $\dfrac{\mathbb{Z}_p[x]}{\langle p(x)\rangle} = \{ax + b + \langle p(x)\rangle \mid a, b \in \mathbb{Z}_p\}$ is a field with p^2 elements. Hence, for every prime p, there exists a field of order p^2. $\square$

Example 8.27 Consider the quotient ring $Q[x]/\langle x^2 - 2\rangle$. To check whether $\mathbb{Q}[x]/\langle x^2 - 2\rangle$ is a field, all we have to do is check whether the polynomial $x^2 - 2$ is irreducible in the field $\mathbb{Q}[x]$ or not.
Now, $x^2 - 2 = (x + \sqrt{2})(x - \sqrt{2})$. But $\sqrt{2} \notin \mathbb{Q}$.
Therefore, $x^2 - 2$ is irreducible over $\mathbb{Q}$.
Hence, $Q[x]/\langle x^2 - 2\rangle$ is a field.   (by corollary 8.24)

Example 8.28 Consider the quotient ring $R[x]/\langle x^2 - 2\rangle$. Here $p(x) = x^2 - 2$. But $x^2 - 2 = (x + \sqrt{2})(x - \sqrt{2})$ and also $\sqrt{2}, -\sqrt{2} \in R$.
So, $p(x) = x^2 - 2$ is reducible in $R[x]$. This implies that $p(x) = x^2 - 2$ is not irreducible in $R[x]$. Hence, using corollary 8.24, $R[x]/\langle x^2 - 2\rangle$ is not a field.

Corollary 8.29 Let F be a field and let $p(x), a(x), b(x) \in F[x]$. If $p(x)$ is irreducible over F and $p(x) \mid a(x) \cdot b(x)$, then $p(x) \mid a(x)$ or $p(x) \mid b(x)$.

Proof Let F be a field and let $p(x), a(x), b(x) \in F[x]$. Given that $p(x)$ is irreducible over F. Then, by corollary 8.24, $F[x]/\langle p(x)\rangle$ is a field. This implies that $F[x]/\langle p(x)\rangle$ is an integral domain. Then, $\langle p(x)\rangle$ is a prime ideal.
Since $p(x) \mid a(x) \cdot b(x)$, we have, $a(x) \cdot b(x) \in \langle p(x)\rangle$. Thus, $a(x) \in \langle p(x)\rangle$ or $b(x) \in \langle p(x)\rangle$. So, either $p(x) \mid a(x)$ or $p(x) \mid b(x)$. Hence, if $p(x)$ is irreducible over F and $p(x) \mid a(x) \cdot b(x)$, then either $p(x) \mid a(x)$ or $p(x) \mid b(x)$. $\square$

The next theorem is an immediate consequence of all the ideas discussed in this chapter. We all are familiar with the polynomial rings over intergers, $\mathbb{Z}[x]$. This polynomial ring has an important factorization property that is discussed in the following theorem.

Theorem 8.30 (Unique Factorization in $\mathbb{Z}[x]$) *Every polynomial in $\mathbb{Z}[x]$ that is not the zero polynomial or a unit in $\mathbb{Z}[x]$ can be written in the form $b_1 b_2 \ldots b_s p_1(x)p_2(x) \ldots p_m(x)$, where the b_i's are irreducible polynomials of degree 0 and the $p_i(x)$'s are irreducible polynomials of positive degree. Furthermore, if $b_1 b_2 \ldots b_s p_1(x)p_2(x) \ldots p_m(x) = c_1 c_2 \ldots c_t q_1(x)q_2(x) \ldots q_n(x)$, where the b_i's and c_i's are irreducible polynomials of degree 0 and the $p_i(x)$'s and $q_i(x)$'s are irreducible polynomials of positive degree, then $s = t$, $m = n$,* *and after renumbering the c's and $q(x)$'s, we have $b_i = \pm c_i$ for $i = 1, \ldots, s$ and $p_i(x) = \pm q_i(x)$ for $i = 1, \ldots, m$.*

Proof Let $f(x)$ be a nonzero, nonunit polynomial from $\mathbb{Z}[x]$.

If $\deg f(x) = 0$, then $f(x)$ is constant.

If $\deg f(x) > 0$, let b denote the content of $f(x)$, and let $b_1 b_2 \ldots b_s$ be the factorization of b as a product of primes. Then, $f(x) = b_1 b_2 \ldots b_s f_1(x)$, where $f_1(x) \in \mathbb{Z}[x]$, is primitive and $\deg f_1(x) = \deg f(x)$.

Thus, to prove the existence part of the theorem, it suffices to show that a primitive polynomial $f(x)$ of positive degree can be written as a product of irreducible polynomials of positive degree. So, we use induction on $\deg f(x)$.

If $\deg f(x) = 1$, then $f(x)$ is already irreducible and we are done.

Now suppose that every primitive polynomial of degree less than $\deg f(x)$ can be written as a product of irreducibles of positive degree.

Case 1. If $f(x)$ is irreducible, there is nothing to prove.

Case 2. If $f(x)$ is reducible, this implies that $f(x) = g(x) \cdot h(x)$, where both $g(x)$ and $h(x)$ are primitive and have degree less than that of $f(x)$. Thus, by induction, both $g(x)$ and $h(x)$ can be written as a product of irreducibles of positive degree. Clearly, then, $f(x)$ is also such a product.

Now, we show that the representation is unique. Suppose that the representation of $f(x)$ is not unique.

Let $f(x) = b_1 b_2 \ldots b_s p_1(x) p_2(x) \ldots p_m(x)$ and $f(x) = c_1 c_2 \ldots c_t q_1(x) q_2(x) \ldots q_n(x)$, where b_i's and c_i's are irreducible polynomials of degree 0 and the $p_i(x)$'s and $q_i(x)$'s are irreducible polynomials of positive degree. Let $b = b_1 b_2 \ldots b_s$ and $c = c_1 c_2 \ldots c_t$. Since $p(x)$'s and $q(x)$'s are primitive, then by Gauss's Lemma, $p_1(x) p_2(x) \ldots p_m(x)$ and $q_1(x) q_2(x) \ldots q_n(x)$ are primitive. Hence, both b and c must equal plus or minus the content of $f(x)$ and, therefore, are equal in absolute value. This implies that $s = t$. After renumbering, $b_i = \pm c_i$ for $i = 1, 2, \ldots, s$. Thus, by cancelling the constant terms in the two factorizations for $f(x)$, we have $p_1(x) p_2(x) \ldots p_m(x) = \pm q_1(x) q_2(x) \ldots q_n(x)$.

Now, viewing the $p(x)$'s and $q(x)$'s as elements of $\mathbb{Q}[x]$ and noting that $p_1(x)$ divides $q_1(x) \ldots q_n(x)$, it follows from corollary 8.29 and induction that $p_1(x) \mid q_i(x)$ for some i. By renumbering, we may assume $i = 1$. Then, since $q_1(x)$ is irreducible, we have $q_1(x) = (r/s) \cdot p_1(x)$, where $r, s \in \mathbb{Z}$.

However, since both $q_1(x)$ and $p_1(x)$ are primitive, we must have $r/s = \pm 1$. So, $q_1(x) = \pm p_1(x)$. After cancellation, we have $p_2(x) \ldots p_m(x) = \pm q_2(x) \ldots q_n(x)$.

Now, repeat the argument above with $p_2(x)$ in place of $p_1(x)$.

Case 1. If $m < n$, then, after m such steps, we would have 1 on the left and a nonconstant polynomial on the right. Clearly, this is impossible.

Case 2. If $m > n$, then, after n steps, we would have ± 1 on the right and a nonconstant polynomial on the left. This is also impossible.

Hence, $m = n$ and $p_i(x) = \pm q_i(x)$ after suitable renumbering of the $q(x)$'s. $\qquad \square$

Illustrative Problems

Problem 1 Show that a polynomial $f(x) = 2x^2 + 4$ is irreducible over $\mathbb{R}$ but reducible over $\mathbb{C}$.

__Solution__ Here, $f(x) = 2x^2 + 4$ can be factorized as $f(x) = 2x^2 + 4 = (\sqrt{2}x + 2i)(\sqrt{2}x - 2i)$. Let, $g(x) = (\sqrt{2}x + 2i) \in \mathbb{C}[x]$ and $h(x) = (\sqrt{2}x - 2i) \in \mathbb{C}[x]$. This implies, $f(x) = g(x) \cdot h(x)$ where $g(x), h(x) \in \mathbb{C}[x]$ are non-unit elements. Hence, f is reducible over $\mathbb{C}$.

Now, $g(x), h(x) \notin \mathbb{R}$. Also, $f(x) = 2x^2 + 4 = 2(x^2 + 2)$, where 2 is a unit in $\mathbb{R}$.

$$(\because 2^{-1} = 1/2 \in \mathbb{R}.)$$

Therefore, $f(x)$ is irreducible over $\mathbb{R}$.

Problem 2 Show that a nonconstant polynomial from $\mathbb{Z}[x]$ that is irreducible over $\mathbb{Z}$ is primitive.

__Solution__ Let $f(x) = a_n x^n + a_{n-1} x^{n-1} + ... + a_1 x + a_0 \in \mathbb{Z}[x]$ be a nonconstant polynomial that is irreducible over $\mathbb{Z}[x]$. We have to show that $f(x)$ is primitive.

Let, if possible, $f(x)$ be irreducible over $\mathbb{Z}$ but not primitive. This means that the gcd of the coefficients of $f(x)$ is some integer, say, $d > 1$. Thus, each coefficient of $f(x)$ can be written as $a_i = db_i$, where $b_i \in \mathbb{Z}$. Then, we can factor $f(x)$ as $f(x) = d \cdot g(x)$, where $g(x) = b_n x^n + b_{n-1} x^{n-1} + ... + b_1 x + b_0 \in \mathbb{Z}[x]$.

Now, since $d > 1$ and $f(x)$ is irreducible over $\mathbb{Z}$, $g(x)$ must be a constant polynomial. If $g(x)$ is nonconstant, then $f(x)$ can be factored into nonconstant polynomials d and $g(x)$, contradicting the irreducibility of $f(x)$.

Thus, $g(x)$ must be constant. But this implies that $f(x) = d \cdot g(x)$ is a constant polynomial, which contradicts our assumption that $f(x)$ is a nonconstant polynomial. Thus, our assumption is wrong and hence, $f(x)$ is primitive.

Problem 3 Suppose that there is a real number 'r' with the property that $r + \dfrac{1}{r}$ is an odd integer. Show that r is irrational.

__Solution__ It is given that $r + \dfrac{1}{r}$ is an odd integer. This implies that $r + \dfrac{1}{r} = 2k + 1$, $k \in \mathbb{Z} \Rightarrow r^2 + 1 = 2kr + r \Rightarrow r^2 - 2kr - r + 1 = 0 \Rightarrow r^2 - (2k+1)r + 1 = 0$.

Let $f(r) = r^2 - (2k+1)r + 1$. Then, over $\mathbb{Z}_2$, $\overline{f(r)} = r^2 - r + 1 = r^2 + r + 1$, the poly-nomial obtained by reducing $f(r)$ by modulo 2. Here, $\mathbb{Z}_2$ is a field with elements 0 and 1. Now, $\overline{f(0)} = 1$ and $\overline{f(1)} = 1$. Thus, $\overline{f(r)}$ is irreducible over $\mathbb{Z}_2$ and deg $\overline{f(r)} = $ deg $f(r)$. Then, by mod p irreduciblity test, $f(r)$ is irreducible over $\mathbb{Q}$. Therefore, $f(r)$ has no zeroes in $\mathbb{Q} \Rightarrow r \notin \mathbb{Q}$. Also, given that $r \in R$. This implies that $r \in Q^c$. Hence, r is irrational.

Problem 4 Construct a field of order 25.

__Solution__ Consider the field $\mathbb{Z}_5 = \{0, 1, 2, 3, 4\}$. Let, $f(x) = x^2 + x + 1 \in \mathbb{Z}_5[x]$.
Now, $f(0) = 1$, $f(1) = 3$,
$f(2) = 4 + 2 + 1 = 7 \mod 5 = 2$,
$f(3) = 9 + 3 + 1 = 13 \mod 5 = 3$,
$f(4) = 16 + 4 + 1 = 21 \mod 5 = 1$.

So, $f(x)$ has no zeroes in $\mathbb{Z}_5$. Therefore, $f(x)$ is an irreducible polynomial over $\mathbb{Z}_5$. Thus, by Theorem 8.23, we get that $\langle x^2 + x + 1 \rangle$ is a maximal ideal in $\mathbb{Z}_5[x]$.

Therefore, by corollary 8.24, $\dfrac{\mathbb{Z}_5[x]}{\langle x^2 + x + 1 \rangle}$ is a field . $\hspace{4cm}$ (8.23)

Consider, $f(x) + \langle x^2 + x + 1 \rangle \in \dfrac{\mathbb{Z}_5[x]}{\langle x^2 + x + 1 \rangle}$. By division algorithm, there exists $q(x)$, $r(x) \in \mathbb{Z}_5[x]$ such that
$f(x) = q(x) \cdot (x^2 + x + 1) + r(x)$ with either $r(x) = 0$ or $\deg r(x) < 2$.
Let $r(x) = ax + b$, then
$f(x) + \langle x^2 + x + 1 \rangle = q(x) \cdot (x^2 + x + 1) + r(x) + \langle x^2 + x + 1 \rangle = r(x) + \langle x^2 + x + 1 \rangle.$

Therefore, $\dfrac{\mathbb{Z}_5[x]}{\langle x^2 + x + 1 \rangle} = \{ax + b + \langle x^2 + x + 1 \rangle \mid a, b \in \mathbb{Z}_5\}$ is a field with 25 elements.

Problem 5 Show that $x^3 + x^2 + x + 1$ is reducible over $\mathbb{Q}$. Does this fact contradict corollary 8.22?

Solution Let $f(x) = x^3 + x^2 + x + 1$. Then $\deg f(x) = 3$. We know that, $f(x)$ is reducible over $\mathbb{Q}$ iff $f(x)$ has a zero in $\mathbb{Q}$. Here, $f(-1) = -1 + 1 - 1 + 1 = 0$. So, $-1 \in \mathbb{Q}$ is a zero of $f(x)$. This implies that $f(x)$ is reducible over $\mathbb{Q}$. Now, $f(x) = x^3 + x^2 + x + 1 = \dfrac{x^4 - 1}{x - 1}$. Here, $p = 4$ is not prime. So, it does not contradict corollary 8.22.

Problem 6 Show that $8x^3 - 6x + 1$ is irreducible over $\mathbb{Q}$.

Solution Let $f(x) = 8x^3 - 6x + 1 \in \mathbb{Z}[x]$. Take $a = 1 \in \mathbb{Q}$.
Then, $f(x + a) = f(x + 1) = 8(x + 1)^3 - 6(x + 1) + 1$
$= 8(x^3 + 1 + 3x^2 + 3x) - 6x - 6 + 1$
$= 8x^3 + 8 + 24x^2 + 24x - 6x - 5$
$= 8x^3 + 24x^2 + 18x + 3 \in \mathbb{Z}[x]$.
Also, there exists a prime, $p = 3$ such that $p \nmid 8(= a_3)$, $p \mid 24(= a_2)$ $p \mid 18(= a_1)$ $p \mid 3(= a_0)$ and $p^2 = 9 \nmid 3(= a_0)$. By using Eisenstein's criterion, $f(x + 1)$ is irreducible over $\mathbb{Q}$. Since $f(x + a)$ is irreducible over $\mathbb{Q}$, it follows from proposition 8.9 that $f(x)$ is irreducible over $\mathbb{Q}$. Hence, $8x^3 - 6x + 1$ is irreducible over $\mathbb{Q}$.

Problem 7 Determine which of the polynomials below is/are irreducible over $\mathbb{Q}$.

   (a) $x^5 + 9x^4 + 12x^2 + 6$.
   (b) $x^4 + x + 1$

Solution   **(a)** Here, $f(x) = x^5 + 9x^4 + 12x^2 + 6 \in \mathbb{Z}[x]$. Also, there exists a prime, $p = 3$ such that $p \nmid 1(= a_5)$, $p \mid 9(= a_4)$, $p \mid 12(= a_2)$, $p \mid 6(= a_0)$, $p^2 = 9 \nmid 6(= a_0)$. So, by Eisenstein's Criterian, $f(x)$ is irreducible over $\mathbb{Q}$.

   (b) Let $f(x) = x^4 + x + 1 \in \mathbb{Z}[x]$. We use mod 2 irreduciblity test. Consider $\overline{f(x)} = x^4 + x + 1$, a polynomial obtained by reducing $f(x)$ by modulo 2. Then, $\overline{f}(0) = 1$, $\overline{f}(1) = 1$. This implies that $\overline{f(x)}$ has no zeroes in $\mathbb{Z}_2$ and hence no linear factor in $\mathbb{Z}_2[x]$. Now, we have to show that $\overline{f(x)}$ has no quadratic factors in $\mathbb{Z}_2[x]$. Suppose $f(x)$ has a quadratic factor of form $x^2 + ax + b$. Since a, $b \in \mathbb{Z}_2$, we have the following four possibilities: x^2, $x^2 + x$, $x^2 + 1$, $x^2 + x + 1$. Now, x^2, $x^2 + x$, $x^2 + 1$ have zeroes in Z_2, whereas $\overline{f(x)}$ does not have. Therefore, they can be ruled out. Also, $x^2 + x + 1$ is not a factor as it can be ruled out by long division. Therefore, $\overline{f(x)}$ has no quadratic factors in $\mathbb{Z}_2[x]$.

Since, there are no linear or quadratic factors, so there are no cubic factors also. This implies that $\overline{f(x)}$ is irreducible over $\mathbb{Z}_2$. Hence, $f(x)$ is irreducible over $\mathbb{Q}$.

Problem 8 Prove that, for every integer n, there are infinitely many polynomials of degree n in $\mathbb{Z}[x]$ that are irreducible over $\mathbb{Q}$.

Solution Let, $f(x) = x^n + p$, where p is a prime. Now, there exists a prime, p such that $p \nmid 1 (= a_n)$, $p \mid p (= a_0)$, $p^2 \nmid p (= a_0)$. So, by Eisenstein's criterion, $x^n + p$ is irreducible over $\mathbb{Q}$. Hence, there are infinitely many polynomials of degree n in $\mathbb{Z}[x]$ that are irreducible over $\mathbb{Q}$.

Problem 9 Prove that the ideal $\langle x^2 + 1 \rangle$ is prime in $\mathbb{Z}[x]$ but not maximal in $\mathbb{Z}[x]$.

Solution Let $f(x) \cdot g(x) \in \langle x^2 + 1 \rangle$. This implies that $f(x) \cdot g(x) = (x^2 + 1) \cdot h(x)$, where $h(x) \in \mathbb{Z}[x]$. Thus, $(x^2 + 1) \mid f(x) \cdot g(x)$. Since, $(x^2 + 1)$ is irreducible over $\mathbb{Z}$ and $(x^2 + 1) \mid f(x) \cdot g(x)$, corollary 8.29 implies that $(x^2 + 1) \mid f(x)$ or $(x^2 + 1) \mid g(x)$.
Thus we get, $f(x) = (x^2 + 1) \cdot a(x)$ or $g(x) = (x^2 + 1) \cdot b(x)$ for some $a(x), b(x) \in \mathbb{Z}[x]$ which means that either $f(x) \in \langle x^2 + 1 \rangle$ or $g(x) \in \langle x^2 + 1 \rangle$. Therefore, $\langle x^2 + 1 \rangle$ is a prime ideal in $\mathbb{Z}[x]$.
Now, let $A = \{ f(x) \in \mathbb{Z}[x] \mid f(1) \text{ is even} \}$. Then, A is a proper ideal in $\mathbb{Z}[x]$ ($\because 1 \notin A$) and $\langle x^2 + 1 \rangle \subset A \subset \mathbb{Z}[x]$
($\because f(x) \in \langle x^2 + 1 \rangle \Rightarrow f(x) = (x^2 + 1) \cdot g(x) \quad \Rightarrow f(1) = 2g(x)$ which is even. Thus, $f(x) \in A \Rightarrow \langle x^2 + 1 \rangle \subseteq A$).
Also, $(x + 1) \in A$ but $(x + 1) \notin \langle x^2 + 1 \rangle$. This implies, $\langle x^2 + 1 \rangle \subset A$. Therefore, $\langle x^2 + 1 \rangle$ is not a maximal ideal in $\mathbb{Z}[x]$.

Problem 10 If p is a prime, prove that $x^{p-1} - x^{p-2} + x^{p-3} - \ldots - x + 1$ is irreducible over $\mathbb{Q}$.

Solution Let $f(x) = x^{p-1} - x^{p-2} + x^{p-3} - \ldots - x + 1$, where p is a prime. By corollary 8.22, for any prime p, $g(x) = \phi_p(x) = \dfrac{x^p - 1}{x - 1} = x^{p-1} + x^{p-2} + x^{p-3} + \ldots + x + 1$ is irreducible over $\mathbb{Q}$.

By proposition 8.8, we get, $g(ax) = \phi_p(ax) = \dfrac{(ax)^p - 1}{ax - 1} = (ax)^{p-1} + (ax)^{p-2} + (ax)^{p-3}$ $+ \ldots + ax + 1$ is irreducible over $\mathbb{Q}$. On substituting $a = -1$, we obtain $g(-x) = \phi_p(-x)$ $= x^{p-1} - x^{p-2} + x^{p-3} + \ldots - x + 1 = f(x)$ is irreducible over $\mathbb{Q}$.
For $p = 2$, $g(-x) = 1 - x$ is irreducible over $\mathbb{Q}$.

Problem 11 Show that $x^4 + 1$ is irreducible over $\mathbb{Q}$.

Solution Let, $f(x) = x^4 + 1$. Now, we have to show that $f(x)$ is irreducible over $\mathbb{Q}$. So, using mod 3 irreducibilty test, over $\mathbb{Z}_3$, $\overline{f(x)} = x^4 + 1$ with $\deg \overline{f(x)} = \deg f(x)$, then, $\overline{f(0)} = 1$, $\overline{f(1)} = 2$, $\overline{f(2)} = 17 \mod 3 = 2$.
Since, $\overline{f(x)}$ has no zeroes in $\mathbb{Z}_3[x]$, $\overline{f(x)}$ is irreducible over $\mathbb{Z}_3[x]$.
So, there exists a prime $p = 3$ such that $\overline{f(x)}$ is irreducible over $\mathbb{Z}_p$ with $\deg \overline{f(x)} = \deg f(x)$. By mod 3 irreducibility test, $f(x) = x^4 + 1$ is irreducible over $\mathbb{Q}$.

Problem 12 Show that the number of reducible polynomials over $\mathbb{Z}_p$ of the form $x^2 + ax + b$ is $p(p+1)/2$, where p be a prime.

Solution Let $f(x) = x^2 + ax + b$. Given that $f(x)$ is reducible over $\mathbb{Z}_p$. This implies that $f(x) = g(x) \cdot q(x)$ where $0 < \deg g(x) < \deg f(x) = 2$ and $0 < \deg q(x) < \deg f(x) = 2$ $\Rightarrow \deg g(x) = 1$, $\deg q(x) = 1$. Therefore, $f(x)$ has the form: $f(x) = (x - a)(x - b)$.

Case 1. If $a \neq b$, this implies that $g(x) \neq q(x)$.
Now, the total number of choices for a is p. So, number of choices $(x - a)$ have over $\mathbb{Z}_p$ is p and number of choices $(x - b)$ have over $\mathbb{Z}_p$ is $p - 1$. Therefore, the number of choices to choose $g(x) \cdot q(x) = (x - a) \cdot (x - b)$ have over $\mathbb{Z}_p$ is $p(p - 1)$.
However, since $(x - a)(x - b) = (x - b)(x - a)$, there are only $p(p-1)/2$ distinct such expressions. $\hspace{2cm} (8.24)$

Case 2. If $a = b$, this implies that $g(x) = q(x)$. So, here total number of choices over $\mathbb{Z}_p$ is p.

Thus, the total number of choices are $p(p - 1)/2 + p = \dfrac{p(p + 1)}{2}$.

Hence, the number of reducible polynomials over $\mathbb{Z}_p$ of the form $x^2 + ax + b$ is $p(p + 1)/2$, where p is a prime.

Problem 13 Determine the number of reducible quadratic polynomials over $\mathbb{Z}_p$.

Solution Consider a general quadratic polynomial, $f(x) = ax^2 + bx + c \in \mathbb{Z}_p$, where $a, b, c \in \mathbb{Z}_p$ and $a \neq 0$. Further, $f(x)$ is reducible if it can be factored into the product of two linear polynomials in $\mathbb{Z}_p$. Specifically, it is reducible if it has a root in $\mathbb{Z}_p$, i.e., if there exists some $r \in \mathbb{Z}_p$ such that $f(r) = 0$.
For $f(x)$ to be reducible, the discriminant $d = b^2 - 4ac$ of $f(x)$ must be nonnegative in $\mathbb{Z}_p$.

Case 1. When discriminant, d is nonnegative (including 0), then there are two roots (which could be identical) in $\mathbb{Z}_p$. So, there are $\dfrac{p+1}{2}$ possibilities, including 0, in $\mathbb{Z}_p$ for $b^2 - 4ac$ to be nonnegative when considering b^2 and $4ac$.

Case 2. When discriminant, d is negative, then $f(x)$ does not have roots in $\mathbb{Z}_p$, so it is irreducible. Then there are $\dfrac{p - 1}{2}$ possibilities for d.

So, there are p choices each for a, b, and c, but since $a \neq 0$, there are $p \cdot p \cdot (p - 1) = p^2(p - 1)$ quadratic polynomials in $\mathbb{Z}_p[x]$. However, reducible quadratic polynomial has $\dfrac{p(p + 1)}{2}$ choices, considering this along with $(p - 1)$ choices of a, the total number of reducible quadratic polynomials over $\mathbb{Z}_p$ is $\dfrac{p(p + 1)(p - 1)}{2} = \dfrac{p(p^2 - 1)}{2}$.

Problem 14 Construct a field of order 27.

Solution Consider the field $\mathbb{Z}_3$. Let $f(x) = x^3 + 2x^2 + 1 \in \mathbb{Z}_3[x]$.
Now, $f(0) = 1$, $f(1) = 1 + 2 + 1 = 4$ mod $3 = 1$ and $f(2) = 8 + 8 + 1 = 17$ mod $3 = 2$. So, $f(x)$ has no zeroes in $\mathbb{Z}_3$ which implies that $f(x)$ is an irreducible polynomial over $\mathbb{Z}_3$.

Therefore, by Theorem 8.23, $\langle x^3 + 2x^2 + 1 \rangle$ is a maximal ideal in $\mathbb{Z}_3[x]$. Hence, by corollary 8.24, $\dfrac{\mathbb{Z}_3[x]}{\langle x^3 + 2x^2 + 1 \rangle}$ is a field. $\hspace{2cm}$ (8.25)

Now, we know that if $f(x) \in \mathbb{Z}_p[x]$ with deg $f(x) = n$ and is irreducible over $\mathbb{Z}_p$ where p is a prime. Then, by proposition 8.25, $\dfrac{\mathbb{Z}_p[x]}{\langle f(x) \rangle}$ is a field with p^n elements. So, here $f(x) = x^3 + 2x^2 + 1 \in \mathbb{Z}_3[x]$ with deg $f(x) = 3$ and is irreducible over $\mathbb{Z}_3$, where $p = 3$ is a prime. Therefore, $\dfrac{\mathbb{Z}_3[x]}{\langle x^3 + 2x^2 + 1 \rangle}$ is a field with $p^n = 3^3 = 27$ elements.

Problem 15 Let $f(x) = \dfrac{5}{2}x^5 + \dfrac{9}{2}x^4 + 15x^3 + \dfrac{3}{7}x^2 + 6x + \dfrac{3}{14}$. Show that $f(x)$ is irreducible over $\mathbb{Q}$.

<u>**Solution**</u> Here, $f(x) = \dfrac{5}{2}x^5 + \dfrac{9}{2}x^4 + 15x^3 + \dfrac{3}{7}x^2 + 6x + \dfrac{3}{14}$. Now, we have to show that $f(x)$ is irreducible over $\mathbb{Q}$.

Consider, $h(x) = 14f(x) \in \mathbb{Z}[x]$. This implies that $h(x) = 14f(x) = 35x^5 + 63x^4 + 210x^3 + 6x^2 + 84x + 3 \in \mathbb{Z}[x]$.

Also, there exists a prime, $p = 3$ such that $p \nmid 35(= a_5)$, $p \mid 63(= a_4)$, $p \mid 210(= a_3)$, $p \mid 6(= a_2)$, $p \mid 84(= a_1)$, $p \mid 3(= a_0)$, $p^2 = 9 \nmid 3$. So, by using Eisenstein's criterion, $h(x) = 14f(x)$ is irreducible over $\mathbb{Q}$. Since $af(x)$ is irreducible over $\mathbb{Q}$, therefore, $f(x)$ is irreducible over $\mathbb{Q}$.

Problem 16 Let $f(x) \in \mathbb{Z}_p[x]$. Prove that if $f(x)$ has no factor of the form $x^2 + ax + b$, then it has no quadratic factor over $\mathbb{Z}_p$.

<u>**Solution**</u> Given that $f(x) \in \mathbb{Z}_p[x]$ and $f(x)$ has no factor of the form $x^2 + ax + b$. This implies that $(x^2 + ax + b) \nmid f(x)$ $\hspace{2cm}$ (8.26)

Now, we have to show that $f(x)$ has no quadratic factor over $\mathbb{Z}_p$. Let, if possible, $f(x)$ has quadratic factor over $\mathbb{Z}_p$, say $a_2x^2 + a_1x + a_0$, $a_2 \neq 0$.

Thus, $(a_2x^2 + a_1x + a_0) \mid f(x) \Rightarrow a_2^{-1}(a_2x^2 + a_1x + a_0) \mid a_2^{-1}f(x)$

$\hspace{2cm}$ ($\because a_2 \neq 0$ and $\mathbb{Z}_p$ is a field, so a_2 has a multiplicative inverse in $\mathbb{Z}_p$).

Continuing in a similar manner, we get $(x^2 + ax + b) \mid a_2^{-1}f(x)$, where $a = (a_2)^{-1} \cdot a_1$, $b = (a_2)^{-1} \cdot a_0$. This implies $(x^2 + ax + b) \mid f(x)$, which is a contradiction to Equation (8.26). So, our assumption is wrong that $f(x)$ has quadratic factor over $\mathbb{Z}_p$, say $a_2x^2 + a_1x + a_0$, $a_2 \neq 0$. Hence, $f(x)$ has no quadratic factor over $\mathbb{Z}_p$.

Problem 17 Prove that $\mathbb{R}[x]/\langle (x^2 + 1) \rangle$ is a field isomorphic to the field of complex numbers.

<u>**Solution**</u> Let $p(x) = x^2 + 1$. Note that $p(x)$ is an irreducible polynomial over $\mathbb{R}[x]$ since it has no real roots, i.e., the equation $x^2 + 1 = 0 \Rightarrow x^2 = -1$, which has no solution in $\mathbb{R}$. So, $p(x)$ cannot be factored into polynomials of lower degree with real coefficients.

Now, since $p(x)$ is irreducible, the ideal generated by $x^2 + 1$ is a maximal ideal in $\mathbb{R}[x]$.

Since $\mathbb{R}[x]$ is a commutative ring with unit element, $\dfrac{\mathbb{R}[x]}{\langle x^2 + 1 \rangle}$ is a field.

Moreover, every element in $\dfrac{\mathbb{R}[x]}{\langle x^2 + 1 \rangle}$ can be written in the form: $a_0 + a_1 x + \langle x^2 + 1 \rangle$ for some $a_0, a_1 \in \mathbb{R}$.

(Using the division algorithm, we can write $f(x) = g(x)(x^2 + 1) + h(x)$, $g(x), h(x) \in \mathbb{R}[x]$ and either $h(x) = 0$ or $\deg h(x) \le 1$.

We obtain $f(x) + \langle x^2 + 1 \rangle = g(x)(x^2 + 1) + h(x) + \langle x^2 + 1 \rangle = h(x) + \langle x^2 + 1 \rangle$, where $\deg(h(x)) \le 1$. Hence, every element of $\dfrac{\mathbb{R}[x]}{\langle x^2 + 1 \rangle}$ is of the form $a_0 + a_1 x + \langle x^2 + 1 \rangle$. Then,

$$\frac{\mathbb{R}[x]}{\langle x^2 + 1 \rangle} = \{ a_0 + a_1 x + < x^2 + 1 > \mid a_0, a_1 \in \mathbb{R} \}.)$$

Let us define a map $\phi : \mathbb{R}[x]/< x^2 + 1 > \rightarrow \mathbb{C}$ such that $\phi(a + bx + 1) = a + ib$. Then ϕ is an isomorphism.

Problem 18 Show that the polynomial $x^5 + 5x^2 + 1$ is irreducible over $\mathbb{Q}$.

Solution Let $f(x) = x^5 + 5x^2 + 1$. Now, we have to show that $f(x)$ is irreducible over $\mathbb{Q}$. Consider the field $\mathbb{Z}_2$. Then, $\overline{f(x)} = x^5 + x^2 + 1$, the polynomial obtained by reducing it by modulo 2.

Now, $\overline{f(0)} = 0 + 0 + 1 = 1$,

$\overline{f(1)} = 1 + 1 + 1 = 3 \bmod 2 = 1$,

So, $\overline{f(x)}$ has no zeroes in $\mathbb{Z}_2[x]$. Therefore, it has no linear factors in $\mathbb{Z}_2[x]$.

Now, we will show that $\overline{f(x)}$ has no quadratic factors in $\mathbb{Z}_2[x]$.

Suppose, $\overline{f(x)}$ has a quadratic factor in $\mathbb{Z}_2[x]$ of the form $x^2 + ax + b$. Since $a, b \in \mathbb{Z}_2$, there are four possibilities: $x^2 + x + 1$, $x^2 + x = x(x + 1)$, $x^2 + 1$ and x^2.

Now, x^2, $x^2 + x = x(x + 1)$, $x^2 + 1$ have zeroes in $\mathbb{Z}_2$, whereas $\overline{f(x)}$ does not have. Therefore, they can be ruled out.

Also, $x^2 + x + 1$ is not a factor since $x^5 + x^2 + 1 = (x^2 + x + 1) \cdot (x^3 + x^2) + 1$, which can be shown by long divison method. Therefore, $\overline{f(x)}$ has no quadratic factors in $\mathbb{Z}_2[x]$.

Since there are no linear or quadratic factors, therefore there are no cubic or fourth degree factors also. So, $\overline{f(x)}$ is irreducible over $\mathbb{Z}_2$ with $\deg \overline{f(x)} = \deg f(x)$. Hence, using mod 2 irreducibility test, $f(x)$ is irreducible over $\mathbb{Q}$.

Problem 19 Factor $x^6 + x^4 + 2x^2 + 2 \in \mathbb{Z}_3[x]$ into a product of irreducibles. Say why each of the factors you have are irreducible.

Solution Let, $f(x) = x^6 + x^4 + 2x^2 + 2 \in \mathbb{Z}_3[x]$. Then, $f(1) = 6 \bmod 3 = 0$ and $f(2) = 90 \bmod 3 = 0$. So, by the factor theorem, $(x - 1)$, $(x - 2)$ are factors, i.e., $(x + 2)$, $(x + 1)$ are factors of $f(x)$.

After doing long division, we get that $f(x) = (x + 1)(x + 2)(x^4 + 2x^2 + 1)$. But $x^4 + 2x^2 + 1 = (x^2 + 1)(x^2 + 1)$. So we get $f(x) = (x + 1)(x + 2)(x^2 + 1)^2$.

Now, each linear factor is irreducible since it is of degree 1, and $h(x) = x^2 + 1$ is irreducible since it has no zeroes over $\mathbb{Z}_3$ $\qquad (\because h(0) = 1, h(1) = 2, h(2) = 5 \bmod 3 = 1)$.

Problem 20 Show that there are infinitely many integers n such that $f(x) = x^9 + 12x^5 - 21x + n$ is irreducible in $\mathbb{Q}[x]$.

Solution Given that $f(x) = x^9 + 12x^5 - 21x + n$. Let, $n = 3 \cdot 2^k$, $k \in \mathbb{N}$. Then, $f(x) = x^9 + 12x^5 - 21x + 3.2^k \in \mathbb{Z}[x]$, where $k \in \mathbb{N}$.

Also, there exists a prime $p = 3$ such that $p \nmid 1 (= a_9)$, $p \mid 0 (= a_8, a_7, a_6, a_4, a_3, a_2)$, $p \mid 12 (= a_5)$, $p \mid (-21) (= a_1)$, $p \mid 3 \cdot 2^k (= a_0)$ and $p^2 = 9 \nmid 3 \cdot 2^k (= a_0)$. Therefore, by using Eisenstein's criterion, $f(x) = x^9 + 12x^5 - 21x + 3 \cdot 2^k \in \mathbb{Z}[x]$, where $k \in \mathbb{N}$, is irreducible over $\mathbb{Q}$. Hence, there are infinitely many integers n such that $f(x) = x^9 + 12x^5 - 21x + n$ is irreducible in $\mathbb{Q}[x]$.

Problem 21 Determine whether $f(x) = x^4 + x^2 + x + 1$ is irreducible over $\mathbb{Z}_5$.

Solution Let $f(x) = x^4 + x^2 + x + 1$. Then, $f(0) = 1$, $f(1) = 4$, $f(2) = 3$, $f(3) = 4$ and $f(4) = 2$. Thus, $f(x)$ has no linear factors.

Now, assume that $f(x) = (x^2 + ax + b)(x^2 + cx + d)$. Then we have $x^4 + x^2 + x + 1 = x^4 + (a + c)x^3 + (ac + b + d)x^2 + (ad + bc)x + bd$. On comparing the constant terms, we get $(b, d) = (1, 1)$, $(2, 3)$, $(3, 2)$ or $(4, 4)$. From $a + c = 0$, we get $a = -c$. Then, $(ad + bc) = 1$ gives $a(d - b) = 1 \Rightarrow (b, d, a) = (2, 3, 1)$ or $(3, 2, 4)$. But $ac + b + d = 4 \neq 1$. Thus, $f(x)$ is irreducible over $\mathbb{Z}_5$.

Problem 22 Factor $f(x) = x^4 + 10x^3 + 13x^2 - 32x + 12 \in \mathbb{Z}[x]$ into a product of irreducible polynomials over $\mathbb{Q}$.

Solution Let $f(x) = x^4 + 10x^3 + 13x^2 - 32x + 12 \in \mathbb{Z}[x]$. First, we check if it has a linear factor. We know that if $f(x) = x^n + a_{n-1}x_{n-1} + \ldots + a_0 \in \mathbb{Z}[x]$, $a_0 \neq 0$ and if $f(x)$ has a zero in $\mathbb{Q}$, then the zero is in $\mathbb{Z}$ and satisfies $m \mid a_0$.

It follows that such a factor is of the form $x - m$ for some integer $m \mid 12$. By a direct computation, we find ± 1, ± 2, ± 3, ± 4, ± 6, ; ± 12 are not zeros of $f(x)$. Thus, $f(x)$ has no linear factors. Assume that $f(x) = (x^2 + ax + b)(x^2 + cx + d)$. Then by Gauss's lemma, $(b, d) = \pm(1, 12)$, $\pm(2, 6)$, or $\pm(3, 4)$.

First consider $(b, d) = (1, 12)$. We have, $x^4 + 10x^3 + 13x^2 - 32x + 12 = x^4 + (a + c)x^3 + (ac + 13)x^2 + (12a + c)x + 12$. On comparing the coefficients, we get $a + c = 10$, $ac = 0$, and $12a + c = -32$. However, no integers can satisfy these equations. Similarly, the choices $(b, d) = (-1, -12)$ and $(2, 6)$ do not work either.

Now, when $(b, d) = (-2, -6)$, we find $a + c = 10$, $ac - 8 = 13$, $-6a - 2c = -32 \Rightarrow (a, c) = (3, 7)$. Thus, $f(x) = (x^2 + 3x - 2)(x^2 + 7x - 6)$.

Problem 23 Prove that $x^3 + 3x - 2$ is irreducible over $\mathbb{Z}$.

Solution Let, if possible, $f(x) = x^3 + 3x - 2$ is reducible over $\mathbb{Z}$. Then, $(x^3 - 3x + 2) = g(x) \cdot h(x)$, where $g(x)$, $h(x) \in \mathbb{Z}[x]$ with $\deg g(x) < 3$ and $\deg h(x) < 3$.

Without loss of generality, let $\deg g(x) = 1$, $\deg h(x) = 2$. Then, $f(x)$ can be of the form $x^3 + 3x - 2 = (x + a)(x^2 + bx + c)$, where a, b, $c \in \mathbb{Z}$. It follows that $ac = -2$, so that a divides 2. In this case, either $a = \pm 1$ or $a = \pm 2$. This implies, ± 1 or ± 2 must be a zero of $f(x) = x^3 + 3x - 2$.

Now, $f(1) = 1 + 3 - 2 = 2$,
$f(-1) = -1 - 3 - 2 = -6$,
$f(2) = 8 + 6 - 2 = 12$,
$f(-2) = -8 - 6 - 2 = -16$.

Thus, neither ± 1 nor ± 2 is a zero of $f(x)$, which is a contradiction. Therefore, $f(x)$ is irreducible over $\mathbb{Z}$.

Problem 24 Prove that $x^5 - ax - 1 \in \mathbb{Z}[x]$ is irreducible unless $a = 0$, 2 or -1.

Solution Let $f(x) = x^5 - ax - 1$. If f is reducible, there are two possible cases: either it has a linear factor or it factors as the product of an irreducible quadratic with an irreducible cubic.

Case 1. If $f(x)$ has a linear factor. It follows that f has a root $r \in \mathbb{Z}$. By the rational root theorem, we know that r divides the constant term. So, $r = \pm 1$. Now, $f(1) = 0 \Rightarrow a = 0$ and $f(-1) = 0 \Rightarrow a = 2$.

Case 2. If $f(x)$ factors as the product of an irreducible quadratic with an irreducible cubic.

Assume that $f(x) = (Ax^2 + bx + c)(Bx^3 + dx^2 + ex + g)$. Since f is monic, we must have $A = B = 1$ or $A = B = -1$. Without loss of generality, assume that $A = B = 1$. Then, $f(x) = x^5 + (b + d)x^4 + (c + e + bd)x^3 + (g + cd + be)x^2 + (bg + ce)x + cg$. Therefore, $d = -b$, $c + e = b^2$, $b(c - e) = g$, $bg + ce = -a$, $cg = -1$.
If $c = -1$, then $g = 1$ and thus $-b(e + 1) = 1$. This implies that $e = 0$ or $e = -2$. In either case, $b^2 = c + e < 0$, which is a contradiction.
If $c = 1$, then $g = -1$ and thus $b(e - 1) = 1 \Rightarrow e = 2$ or $e = 0$. If $e = 2$ then $b^2 = 3$, which is a contradiction. So $e = 0$ and hence $b = -1$ and $a = -1$, giving the factorization, $f(x) = (x^2 - x + 1)(x^3 + x^2 - 1)$.

Problem 25 Find $D[x]$ such that the polynomial $x^3 - 312312x + 123123$ is irreducible in $D[x]$.

Solution Let $f(x) = x^3 - 312312x + 123123 \in \mathbb{Z}[x]$. Also, there exists a prime $p = 13$ such that $p \nmid 1(= a_3)$, $p \mid 0(= a_2)$ $p \mid -312312(= a_1)$, $p \mid 123123(= a_0)$ and $p^2 = 169 \nmid 123123(= a_0)$. So, by Eisenstein's criterion, $f(x) = x^3 - 312312x + 123123$ is irreducible over $\mathbb{Q}$.
Hence, $D[x]$ is $\mathbb{Q}[x]$ such that the polynomial $x^3 - 312312x + 123123$ is irreducible in $D[x]$.

Exercise

(1) Suppose that D is an integral domain and F is a field containing D. If $f(x) \in D[x]$ and $f(x)$ is irreducible over F but reducible over D, what can you say about the factorization of $f(x)$ over D?

(2) Let F be a field and $f(x) \in F[x]$. Show that, as far as deciding upon the irreducibility of $f(x)$ over F is concerned, we may assume that $f(x)$ is monic.

(3) Let p be a prime. Determine the number of irreducible polynomials over $\mathbb{Z}_p$ of the form $x^2 + ax + b$.

(4) Determine the number of irreducible quadratic polynomials over $\mathbb{Z}_p$.

(5) Let F be a field and let $f(x)$ be a polynomial in $F[x]$ that is reducible over F. Prove that $\langle f(x) \rangle$ is not a prime ideal in $F[x]$.

(6) Find all the zeros and their multiplicities of $x^5 + 4x^4 + 4x^3 - x^2 - 4x + 1$ over $\mathbb{Z}_5$.

(7) Prove that $x^5 - 6x + 3$ is irreducible over $\mathbb{Q}$.

(8) Prove that $x^7 + 48x - 24$ is irreducible in $\mathbb{Q}[x]$.

(9) Not only does Eisenstein's criterion (with Gauss' lemma) fail to prove that $x^4 + 4$ is irreducible in $\mathbb{Q}[x]$, but, also, this polynomial does factor into two irreducible quadratics in $\mathbb{Q}[x]$. Find them.

(10) Prove that $x^3 + y^3 + z^3$ is irreducible in $k[x, y, z]$ when k is a field not of characteristic 3.

(11) Prove that $x^2 + y^3 + z^5$ is irreducible in $k[x, y, z]$ the underlying field k is of characteristic 2, 3 or 5.

(12) Prove that $x^3 + y + y^5$ is irreducible in $\mathbb{C}[x, y]$.

(13) Prove that $x^n + y^n + 1$ is irreducible in $k[x, y]$ when the characteristic of k does not divide n.

(14) Let k be a field with characteristic not dividing n. Show that any polynomial $x^n - P(y)$, where $P(y)$ has no repeated factors, is irreducible in $k[x, y]$.

(15) Show that $x^2 + y^2 = 2003$ has no solutions in the integers.

(16) Show that the polynomial $x^5 + ax^4 - 3x^3 + bx^2 + cx + 7$ is irreducible over $\mathbb{Q}$ for any choice of integers a, b, c all divisible by 7.

(17) Show that $f(x) = x^4 - 10x^2 + 1$ is irreducible in $\mathbb{Q}[x]$, but reducible in $\mathbb{Z}_p[x]$ for every prime p.

(18) Do the following polynomials satisfy Eisenstein's criterion for irreducibility? Explain.
    (a) $x^7 + 4x^6 - 18x^5 + 42x^2 - 6$
    (b) $2x^5 + 18x^3 + 3x$
    (c) $x^4 + 9$
    (d) $5x^4 + 7x + 35$.

(19) Find an irreducible factor of $x^{12} - 1$ over $\mathbb{Q}$.

(20) Prove that $x^n + 1$ is not irreducible over $\mathbb{Q}[x]$ if n is odd.

(21) Give example of a polynomial, which is
    (i) primitive and irreducible.
    (ii) primitive and reducible.
  (iii) not primitive but irreducible.
    (iv) not primitive but reducible.

(22) Show that $4x^2 + 6x + 2$ is not a primitive polynomial in $\mathbb{Z}[x]$. Is it primitive over $\mathbb{Q}$? Justify.

(23) Show that $x^2 + 1$ and $x^2 + x + 4$ are irreducible over F, the field of integers modulo 11. Prove also that $\dfrac{F[x]}{\langle x^2 + 1 \rangle}$ and $\dfrac{F[x]}{\langle x^2 + x + 4 \rangle}$ are isomorphic fields, each having 121 elements.

9 DIVISIBILITY IN INTEGRAL DOMAINS

In an integral domain, the notion of divisibility plays a crucial role due to its fundamental properties and implications. Divisibility provides insights into the structure of elements within the integral domain. It facilitates factorization of elements into irreducible components, which further allows to understand the structure of the domain. The concept of divisibility underpins many aspects of algebraic structures and computations. It provides a framework for understanding factorization, greatest common divisors, algebraic manipulations, prime elements and congruences, making it indispensable in various applications.

In this chapter, divisibility in an integral domain is examined in a more generalized form. The reader will be introduced to concepts like associates, prime and irreducible elements. Subsequently, the concept of Euclidean Domain is discussed. Euclidean domains emerge as captivating entities to explore the intricacies of division and factorization. Further, the chapter describes Unique Factorization Domains and Noetherian Domains. At the core of unique factorization domain, lies the concept of prime factorization, which reveals the underlying structure of numbers and polynomials alike. Within a unique factorization domain, every nonzero, non-unit element can be uniquely factored into irreducible elements in a manner that transcends the boundaries of abstract mathematics to find application in a myriad of practical domains, from cryptography and coding theory to computational algebra and beyond.

In an integral domain, associates encapsulate the idea of multiplicative equivalence - a relationship where two elements, a and b, are deemed associates if there exists a unit u within the domain such that $a = ub$. For example, in $\mathbb{Z}$, the elements 2 and -2 are associates as $2 = (-1)(-2)$, where -1 is a unit in $\mathbb{Z}$. In $\mathbb{Z}_5$, observe that every nonzero element is associate of each nonzero element.

Definition 9.1 (Associates) Elements a and b of an integral domain D are called *associates* if $a = ub$, where u is a unit in D.

Example 9.2 In the integral domain $\mathbb{Z}$, 1 and -1 are the only unit elements. So, for any integer $k \in \mathbb{Z}$, k and $-k$ are associates.

Example 9.3 In the integral domain $\mathbb{Q}$, all nonzero elements are associates as every nonzero element is a unit in $\mathbb{Q}$.

Example 9.4 In the integral domain $F[x]$, every nonzero constant polynomial is a unit. So, the associates of a nonconstant polynomial $f(x)$ is $uf(x)$ where $u \in F$ is a unit.

Observe that, in the integral domain $\mathbb{Z}$, the elements 2 and -2 are associates and the ideals generated by them, $\langle 2 \rangle$ and $\langle -2 \rangle$ are equal. In fact, 2 and 3 are not associates and $\langle 2 \rangle \neq \langle 3 \rangle$. Thus, by determining the relationship between two elements, we can infer whether the ideals generated by them are equal or differ by a unit. This helps in understanding the structure of the integral domain and its ideal lattice. The relationship between elements and their corresponding principal ideals is given in the following proposition.

Proposition 9.5 *In an integral domain D, two nonzero elements a and b are associates if and only if $\langle a \rangle = \langle b \rangle$.*

Proof Let a and b be associates in D. Then, $a = b \cdot u$, for some unit $u \in D$.
We know that $\langle a \rangle = \{a \cdot x \mid x \in D\}$
and $\langle b \rangle = \{b \cdot y \mid y \in D\}$
For any $a \cdot x \in \langle a \rangle$, $a \cdot x = b \cdot u \cdot x = b \cdot y$, where $y = u \cdot x \in D$. This implies, $a \cdot x \in \langle b \rangle$.
Therefore, $\langle a \rangle \subseteq \langle b \rangle$. $\hspace{4cm}$ (9.1)
Now for any $b \cdot y \in \langle b \rangle$, $b \cdot y = a \cdot u^{-1} \cdot y = a \cdot x$ where $x = u^{-1} \cdot y \in D$. This implies,
$b \cdot y \in \langle a \rangle$. Therefore, $\langle b \rangle \subseteq \langle a \rangle$. $\hspace{4cm}$ (9.2)
From (9.1) and (9.2), we get $\langle a \rangle = \langle b \rangle$.
Conversely, if $\langle a \rangle = \langle b \rangle$, then we need to show that a and b are associates in D.
Clearly, $a \in \langle a \rangle = \langle b \rangle$. This implies $a \in \langle b \rangle$. So, $a = b \cdot y$ for some $y \in D$. $\hspace{1cm}$ (9.3)
Also, $b \in \langle b \rangle = \langle a \rangle$. This implies $b \in \langle a \rangle$. So, $b = a \cdot x$ for some $x \in D$. $\hspace{1cm}$ (9.4)
From (9.3) and (9.4), we get $a = a \cdot x \cdot y$. Then, $a - a \cdot x \cdot y = 0 \Rightarrow a(1 - x \cdot y) = 0 \Rightarrow$
$1 - x \cdot y = 0$ as D is an integral domain and $a \neq 0$. So, $x \cdot y = 1$. Therefore, x and y are
units. Hence, a and b are associates. $\hspace{6cm}$ □

The elements of an integral domain can be characterized as - reducible and irreducible; prime and non-prime. This characterization enables to understand properties of individual elements within the integral domain, which further helps in analyzing the behavior of the ring under multiplication.

Definition 9.6 (Irreducibles) A nonzero element a of an integral domain D is called *irreducible* if a is not a unit and, whenever $b, c \in D$ with $a = bc$, then either b or c is a unit.

Definition 9.7 (Primes) A nonzero element a of an integral domain D is called *prime* if a is not a unit and $a \mid bc$ implies $a \mid b$ or $a \mid c$.

In the integral domain $\mathbb{Z}$, the notion of prime and irreducible are equivalent. However, this may not always be true when the integral domain is other than $\mathbb{Z}$. To determine whether an element is prime or irreducible in $\mathbb{Z}[\sqrt{d}]$, where $d(\neq 1)$ and d is not divisible by square of a prime, we define a norm function, $N : \mathbb{Z}[\sqrt{d}] \to \mathbb{Z}$ given by $N(a + b\sqrt{d}) = |a^2 - db^2|$. This function has some properties, which are given in the next proposition.

Proposition 9.8 (Norm Function) *Define a function N, called the norm, from $\mathbb{Z}[\sqrt{d}]$ into the nonnegative integers by $N(a + b\sqrt{d}) = |a^2 - db^2|$, where $d(\neq 1)$ and d is not divisible by square of prime. It satisfies the following four properties:*

   (i) $N(x) = 0$ *if and only if* $x = 0$.
  (ii) $N(x \cdot y) = N(x) \cdot N(y)$.
 (iii) x *is a unit if and only if* $N(x) = 1$.
 (iv) *If* $N(x)$ *is prime, then* x *is irreducible in* $\mathbb{Z}[\sqrt{d}]$.

Proof   **(i)** Let $N(x) = 0$, where $x \in \mathbb{Z}[\sqrt{d}]$. So, for $x = a + b\sqrt{d}$, we have $N(a + b\sqrt{d}) = 0 \Leftrightarrow |a^2 - db^2| = 0 \Leftrightarrow a^2 - db^2 = 0 \Leftrightarrow a^2 = db^2 \Leftrightarrow a = 0,\ b = 0 \Leftrightarrow x = 0$. ($\because$ if $a \neq 0$ and $b \neq 0$ then either $d = 1$ when $a = b$ or $d = \dfrac{a^2}{b^2}$ if $a \neq b$. The case $d = 1$ contradicts the given fact that $d \neq 1$. Similarly, the case $d = \dfrac{a^2}{b^2}$ also contradicts the given fact that d is not divisible by square of prime).

      Hence, $N(x) = 0$ if and only if $x = 0$.

(ii) Let $x = a + b\sqrt{d}$, $y = c + e\sqrt{d}$. Then, $x \cdot y = (a + b\sqrt{d}) \cdot (c + e\sqrt{d}) = (ac + bed) + (ae + bc)\sqrt{d}$.

Consider, $N(x \cdot y) = N((ac + bed) + (ae + bc)\sqrt{d})$
$\Rightarrow N(x \cdot y) = |(ac + bed)^2 - d(ae + bc)^2|$
$\Rightarrow N(x \cdot y) = |a^2c^2 + e^2b^2d^2 + 2abcde - d(a^2e^2 + b^2c^2 + 2abce)|$
$\Rightarrow N(x \cdot y) = |a^2c^2 + e^2b^2d^2 + 2abcde - a^2de^2 - b^2dc^2 - 2abcde|$
$\Rightarrow N(x \cdot y) = |a^2(c^2 - de^2) - db^2(c^2 - de^2)|$
$\Rightarrow N(x \cdot y) = |(a^2 - db^2) \cdot (c^2 - de^2)| = |a^2 - db^2| \cdot |c^2 - de^2|$
$\Rightarrow N(x \cdot y) = N(x) \cdot N(y)$

(iii) Let x be a unit, then there exists $y \in \mathbb{Z}[\sqrt{d}]$ such that $x \cdot y = 1 \Rightarrow N(xy) = N(1) \Rightarrow N(x) \cdot N(y) = N(1) = 1$ (by part 2)
$\Rightarrow N(x) \cdot N(y) = 1 \Rightarrow N(x) = 1$.   ($\because N(x),\ N(y) \in \mathbb{Z}$ and $N(x),\ N(y) \geq 0$.)
Conversely, if $N(x) = 1$, then for $x = a + b\sqrt{d} \in \mathbb{Z}[\sqrt{d}]$, it follows that $N(a + b\sqrt{d}) = 1 \Rightarrow |a^2 - db^2| = 1 \Rightarrow a^2 - db^2 = \pm 1 \Rightarrow a^2 - db^2 = 1$ ($\because N(x) \in \mathbb{Z}^+ \ \forall \ x \in \mathbb{Z}[\sqrt{d}]$.)
$\Rightarrow (a + b\sqrt{d}) \cdot (a - b\sqrt{d}) = 1$. This implies that $(a + b\sqrt{d}) = x$ is a unit. Therefore, x is a unit if and only if $N(x) = 1$.

(iv) Let, $x = y \cdot z$. Then, in order to prove x is irreducible, we have to show that either y or z is a unit.
Here, we have $x = y \cdot z$ which implies, $N(x) = N(y \cdot z) = N(y) \cdot N(z)$ by part (ii). So, $N(x) = N(y) \cdot N(z)$.
Since, $N(x) \mid N(x)$, then $N(x) \mid N(y) \cdot N(z)$ and also we have given that $N(x)$ is a prime, then either $N(x) \mid N(y)$ or $N(x) \mid N(z)$.

Case 1. If $N(x) \mid N(y)$, then $N(y)$ can be written as $N(y) = N(x) \cdot N(a)$ for some $N(a) \in \mathbb{Z}^+,\ a \in \mathbb{Z}[\sqrt{d}] \Rightarrow N(y) = N(x) \cdot N(a) \Rightarrow N(y) = N(y) \cdot N(z) \cdot N(a)$ ($\because N(x) = N(y) \cdot N(z)$.)
      So, by left cancellation law, we have $1 = N(z) \cdot N(a) \Rightarrow N(z) \cdot N(a) = 1 = 1 \cdot 1 \Rightarrow N(z) = 1$. Thus, by (iii), z is a unit.

Case 2. If $N(x) \mid N(z)$, then $N(z)$ can be written as $N(z) = N(b) \cdot N(x)$ for some $N(b) \in \mathbb{Z}^+$, $b \in \mathbb{Z}[\sqrt{d}] \Rightarrow N(z) = N(b) \cdot N(x) \Rightarrow N(z) = N(b) \cdot N(y) \cdot N(z)$

$$(\because N(x) = N(y) \cdot N(z).)$$

So, by right cancellation law, $1 = N(b) \cdot N(y) \Rightarrow N(b) \cdot N(y) = 1 = 1 \cdot 1 \Rightarrow N(y) = 1$. Thus, by (iii), y is a unit.

In either cases, it follows that if $x = y \cdot z$, then either x or y is a unit and hence, x is irreducible in $\mathbb{Z}[\sqrt{d}]$. $\qquad\square$

Example 9.9 There exhibits an irreducible in $\mathbb{Z}[\sqrt{-5}]$ that is not prime. Here, $N(a + b\sqrt{-5}) = a^2 + 5b^2$. Consider $2 \in \mathbb{Z}[\sqrt{-5}]$. Suppose that we can factor this as xy, where neither x nor y is a unit. Then, $N(x \cdot y) = N(x) \cdot N(y) = N(2) = 4 = 2 \cdot 2$, and it follows that $N(x) = 2$. But there are no integers a and b that satisfy $a^2 + 5b^2 = 2$. Thus, x or y is a unit and 2 is an irreducible. To verify that it is not prime, we observe that $(1 + \sqrt{-5}) \cdot (1 - \sqrt{-5}) = 1 + 5 \cdot 1 = 6$. Now, since $2 \mid 6$, therefore, 2 divides $(1 + \sqrt{-5}) \cdot (1 - \sqrt{-5})$. But $2 \nmid (1 + \sqrt{-5})$ and $2 \nmid (1 - \sqrt{-5})$. ($\because$ For integers a and b to exist so that $(1 \pm \sqrt{-5}) = 2 \cdot (a + b\sqrt{-5})$, we must have $2a = 1$ and $2b = \pm 1$, which is impossible.)

In the succeeding theorem, the relationship between prime and irreducible elements in an integral domain is given.

Theorem 9.10 *In an integral domain, every prime is irreducible.*

Proof Let a be a nonzero, non-unit element of an integral domain D. Let a be a prime. Let $a = bc$, where $b, c \in D$. We need to show that either b or c is a unit.
Now, $a \mid a$ implies $a \mid bc$ and since a is a prime, it follows that either $a \mid b$ or $a \mid c$.
With no loss of generality, assume $a \mid b$. Then $b = a \cdot x$ for some $x \in D$
$\Rightarrow b = b \cdot c \cdot x$ $\hfill (\because a = bc)$
$\Rightarrow 1 = c \cdot x$. $\hfill \text{(by cancellation law)}$
This implies, c is a unit and hence a is irreducible. $\qquad\square$

The above theorem illustrates that in an integral domain, every prime element is irreducible. Is every irreducible element prime in an integral domain? No, every irreducible element need not be prime in an integral domain as seen in Example 9.9. However, every irreducible element is a prime element in a principal ideal domain, which is proved in the following theorem.

Theorem 9.11 *In a principal ideal domain, an element is irreducible if and only if it is prime.*

Proof Let D be a principal ideal domain. Since a principal ideal domain is an integral domain so D is an integral domain. By Theorem 9.10, every prime is irreducible in an integral domain. So, the result holds for D as well.
Conversely, let $a \in D$ be an irreducible element. We have to show that a is prime.
Let $a \mid bc$, where $b, c \in D$. Define $I = \{ax + by \mid x, y \in D\}$. Clearly, I is an ideal of D. Since, D is a principal ideal domain, so there exists $d \in D$ such that $I = \langle d \rangle$.
Now, $a = a.1 + b.0 \in I = \langle d \rangle$. This implies, $a = d \cdot t$ for some $t \in D$. But a is irreducible, so either d or t is a unit.

Case 1. Suppose d is a unit, then $I = D$. This implies, $1 = ax + by \Rightarrow c = acx + bcy$. Here a divides acx and bcy, so a divides $acx + bcy$, which implies $a \mid c$.

Case 2. Suppose t is a unit, then $a = d \cdot t \Rightarrow \langle a \rangle = \langle d \rangle = I$. Since $b \in I$ (when $x = 0$, $y = 1$), so $b \in \langle a \rangle$. This implies, $b = a \cdot r \Rightarrow a \mid b$. Therefore, a is a prime element. $\qquad \square$

Consider the prime ideal $\langle x \rangle$ in $\mathbb{R}[x]$. Since $\mathbb{R}[x]/\langle x \rangle \cong \mathbb{R}$, which is a field, so $\langle x \rangle$ is maximal.

Recall from Chapter 4 that a prime ideal need not be a maximal ideal in an integral domain, D. So, in $D[x]$ every prime ideal need not be a maximal ideal. However, if D is a field then every prime ideal in $D[x]$ is a maximal ideal. For instance, the prime ideal $\langle x \rangle$ in $\mathbb{R}[x]$ is maximal as $\dfrac{\mathbb{R}[x]}{\langle x \rangle} \cong \mathbb{R}$, which is a field. This gives insights for the next proposition that in $F[x]$, prime ideals are indeed maximal, where F is a field.

Proposition 9.12 *Let F be a field. Then, a prime ideal in the ring $F[x]$ is a maximal ideal in $F[x]$.*

Proof Let I be a prime ideal in $F[x]$ and let M be a proper ideal that contains I. Then by Theorem 9.5, there exist elements $f(x), g(x)$ in $F[x]$ such that $M = \langle f(x) \rangle$ and $I = \langle g(x) \rangle$. By definition, $g(x)$ is prime and by Theorem 9.11, is irreducible over F. Now, since M contains I, there is an $h(x)$ in $F[x]$ such that $g(x) = f(x) \cdot h(x)$. Also, $g(x)$ is irreducible over F, therefore, $h(x)$ is a unit. Thus, $M = I$. $\qquad \square$

We know that in the integral domains $\mathbb{Z}$ and $\mathbb{Z}[x]$, every integer greater than 1 and every nonzero, non-unit polynomial can be uniquely factored into irreducibles (primes for integers and irreducible polynomials for $\mathbb{Z}[x]$). It's obvious to wonder if this unique factorization property holds for all integral domains. A unique factorization domain (UFD) is a type of integral domain where every element can be factored uniquely into irreducible elements, similar to the prime factorization of integers. However, in a ring like $\mathbb{Z}[-5]$, unique factorization does not hold since 6 can be factored as both $2 \cdot 3$ and $(1 + \sqrt{-5})(1 - \sqrt{-5})$. This concept generalizes the familiar unique factorization property from integers to more abstract algebraic structures.

Definition 9.13 (Unique Factorization Domain) An integral domain D is a *unique factorization domain* if

    (i) every nonzero element of D, which is not a unit can be written as a product of irreducibles of D, and

    (ii) the factorization into irreducibles is unique up to associates and the order in which the factors appear.

Example 9.14 A field $\langle F, +, \cdot \rangle$ is always a unique factorization domain as it contains no nonzero, non-unit element.

Example 9.15 The ring of formal power series over the complex numbers $\mathbb{C}$ is a unique factorization domain, but the subring of those that converge everywhere, in other words the ring of entire functions in a single complex variable, is not a unique factorization domain, since there exist entire functions with an infinity of zeros, and thus an infinity of irreducible factors.

Next theorem is based on the connection between principal ideal domain and unique factorization domain. In certain integral domains, such as the polynomial ring in infinitely many variables, $K[x_1, x_2, x_3, ...]$ over a field K, one can construct an infinite strictly increasing chain of ideals, $\langle x_1 \rangle \subset \langle x_1, x_2 \rangle \subset \langle x_1, x_2, x_3 \rangle \subset ...$. This example shows that ideal chains could be infinitely long, complicating factorization. In order to address this problem, a lemma has been established, allowing us to proceed with the proof of subsequent theorem.

Lemma 9.16 (Ascending Chain Condition for a PID) In a principal ideal domain, any strictly increasing chain of ideals $I_1 \subset I_2 \subset ...$ must be finite in length.

Proof Let D be the principal domain. Let $I_1 \subset I_2 \subset I_3 \subset ...$ be a strictly increasing chain of ideals. Let $I = \cup_{i \in I} I_i$.
First, we show that I is an ideal. Let, $a, b \in I$. This implies that $a \in I_i$ for some i and $b \in I_j$ for some j. Let $k = \max\{i, j\} \Rightarrow a, b \in I_k$
and I_k is an ideal. It follows that $a - b \in I_k \subset I$ which implies $a - b \in I$. $\hspace{2em}$ (9.5)
Let $r \in R$ and since $I_i \subset I$ is an ideal. So, $ra \in I_i$ and $ar \in I_i$. Also, we have $I = \cup_{i \in I} I_i$.
This implies that $ra, ar \in I$. $\hspace{2em}$ (9.6)
By (9.5) and (9.6), I is an ideal.
Now, we have to show that the chain terminates. Since D is a PID, there exists $a \in D$ such that $I = \langle a \rangle$. Now, $a \in \langle a \rangle$ which implies $a \in I = \cup_{i \in I} I_i$. So, $a \in I_j$ for some j. Therefore, $I \subseteq I_j$. $\hspace{2em}$ (9.7)
Now, $I = \cup_{i \in I} I_i$. This implies $I_i \subseteq I$ for every i. In particular, $I_j \subseteq I$. $\hspace{2em}$ (9.8)
From (9.7) and (9.8), we get $I = I_j \Rightarrow \cup_{i \in I} I_i = I_j$.
So, I_j must be last element of the chain and therefore the chain terminates.
Hence, in a principal ideal domain, any strictly increasing chain of ideals $I_1 \subset I_2 \subset ...$ must be finite in length. $\hspace{2em}$ $\square$

Consider the rings $\mathbb{Z}$, $\mathbb{Z}[x]$ (polynomials with integer coefficients) and $F_p[x]$ (polynomials over a finite field). These all rings are UFD. However, they also satisfy the ascending chain condition on ideals. In studying unique factorization domains, we recognize that these rings allow factorization into irreducibles in a unique way, up to units. This property of rings along with having ascending chain condition leads us to the concept of a Noetherian domain that has been discussed below.

Definition 9.17 (Noetherian Domain) An integral domain in which the increasing chain of ideals is finite is called a *Noetherian domain*.

Example 9.18 A finite ring will be Noetherian and so would be any field since a field F has only two ideal $\langle 0 \rangle$ and F.

Observe that in the ring $\mathbb{Z}$ of integers, an ideal $I = \langle 6 \rangle$ is contained in the maximal ideal $\langle 2 \rangle$ or $\langle 3 \rangle$. In the ring $\mathbb{C}[x]$, the ideal $\langle x^2 + 1 \rangle$ is contained in the maximal ideal $\langle x - i \rangle$. These examples show that the Noetherian property of a ring R is crucial in ensuring that every ideal is contained within a maximal ideal that has been proved in the following proposition.

Proposition 9.19 *Let R be a Noetherian ring. Then, any ideal $I \in R$ is contained in a maximal ideal of R.*

Proof If I itself is maximal, then, we have nothing to prove.

Suppose, I is not maximal, then there exists an ideal I_1 such that $I \subseteq I_1$. Then, if I_1 is maximal, we are done. If not, then there exists an another ideal I_2 such that $I \subseteq I_1 \subseteq I_2$ and continuing in same manner, we get an ascending chain of ideals which must become stationary after a finite number of steps. It follows that $I \subseteq I_1 \subseteq I_2 \subseteq ... \subseteq I_n = I_{n+1} = I_{n+2} ....$ Thus, I_n will be maximal. $\qquad\square$

Notice that the ring $\mathbb{Z}$ is a PID because every ideal is generated by some integer, and it is also a UFD because every integer can be uniquely factored into prime numbers. The polynomial ring $F[x]$ over the field F is a PID since every ideal is generated by some polynomial, and it is a UFD because each polynomial can be uniquely factored into irreducible polynomials. Similarly, the ring of Gaussian integers, $\mathbb{Z}[i]$ is a PID as every ideal can be generated by some Gaussian integer, and it is a UFD since every Gaussian integer has a unique factorization into prime elements. These examples show that in a PID, the property of being able to express any element as a product of irreducible elements (with this factorization being unique up to units) holds, thus confirming it as a UFD. This insight has been driven further to state the next theorem.

Theorem 9.20 *Every principal ideal domain is a unique factorization domain.*

Proof Let D be a principal ideal domain and let a_0 be any nonzero non-unit element in D. We show that a_0 is a product of irreducibles. In order to prove this, firstly, we show that a_0 has at least one irreducible factor. If a_0 is irreducible, then we are done.

Assume that $a_0 = b_1 \cdot a_1$, where neither b_1 nor a_1 is a unit and a_1 is nonzero. If a_1 is not irreducible, then a_1 can be written as $a_1 = b_2 \cdot a_2$, where neither b_2 nor a_2 is a unit and a_2 is nonzero.

Continuing in this manner, we obtain a sequence $b_1, b_2, ...$ of elements that are not units in D and a sequence $a_0, a_1, a_2, ...$ of nonzero elements of D with $a_n = b_{n+1} \cdot a_{n+1}$ for each n. Hence, $\langle a_0 \rangle \subset \langle a_1 \rangle \subset ...$ is a strictly increasing chain of ideals, which, by the preceding lemma, must be finite, say, $\langle a_0 \rangle \subset \langle a_1 \rangle \subset ... \langle a_r \rangle$. In particular, a_r is an irreducible factor of a_0. This implies that every nonzero non-unit in D has at least one irreducible factor.

Now, let, $a_0 = p_1 \cdot c_1$, where p_1 is irreducible and c_1 is not a unit. If c_1 is not irreducible, then it can be written as $c_1 = p_2 \cdot c_2$, where p_2 is irreducible and c_2 is not a unit.

Continuing in this manner, we obtain, a strictly increasing sequence $\langle a_0 \rangle \subset \langle c_1 \rangle \subset \langle c_2 \rangle \subset ...$ which must be finite in length. Let us say that the sequence ends with $\langle c_s \rangle$. Then, c_s is irreducible and $a_0 = p_1 \cdot p_2 ... p_s \cdot c_s$, where each p_i is also irreducible.

Therefore, every nonzero non-unit element of a principal ideal domain is a product of irreducibles.

Next, we show that the factorization is unique up to associates and the order in which the factors appear.

Let, some element a of D can be written as $a = p_1 p_2 ... p_r = q_1 q_2 ... q_s$, where p's and q's are irreducible and repetition is allowed. We show this by using induction on r.

If $r = 1$, then a is irreducible and, clearly, $s = 1$ and $p_1 = q_1$.

Assume that any element that can be expressed as a product of fewer than r irreducible factors can be so expressed in only one way (up to order and associates).

Since p_1 divides $q_1 q_2 ... q_s$, it must divide some q_i, say, $p_1 \mid q_1$. Then, $q_1 = u\, p_1$, where u is a unit of D.

Since $u\, p_1\, p_2 \ldots p_r = u\, q_1\, q_2 \ldots q_s = q_1\, (u\, q_2) \ldots q$ and $u\, p_1 = q_1$, then we have $p_2 \ldots p_r = (u\, q_2) \ldots q_s$.

Then, by induction hypothesis, these two factorizations are identical up to associates and the order in which the factors appear and the same is true about the two factorizations of a. Thus, the factorization is unique up to associates and the order in which the factors appear. Hence, every principal ideal domain is a unique factorization domain. $\qquad\square$

Example 9.21 Since $\mathbb{Z}$ is a principal ideal domain, hence $\mathbb{Z}$ is a unique factorization domain. In $\mathbb{Z}$, we have $12 = 2 \cdot 2 \cdot 3 = (-2) \cdot (-2) \cdot 3 = 2 \cdot (-2) \cdot (-3)$, where 2 and -2 are associates, 3 and -3 are associates. So, except for order and associates, the irreducible factors of 12 are same.

Remark Theorem 9.20 shows that every principal ideal domain is a unique factorization domain. However, the converse of this theorem is not true, i.e., a unique factorization domain need not be a principal ideal domain as the ring $\mathbb{Z}[x]$ is UFD but it is not a PID.

The following corollary is an immediate consequence of the Theorem 9.20.

Corollary 9.22 Let F be a field. Then, $F[x]$ is a unique factorization domain.

Proof Given that F is a field. Then, by Theorem 7.29, $F[x]$ is a principal ideal domain. By Theorem 9.20, every principal ideal domain is unique factorization domain, hence, $F[x]$ is a unique factorization domain. $\qquad\square$

In a sequence of applications of preceding corollary, Richard Singer gave an elegant proof of Eisenstein's criterion that has been discussed in the next theorem.

Theorem 9.23 (Another Proof of Eisenstein Criterion)   *Let* $f(x) = a_n x^n + a_{n-1} x^{n-1} + \ldots + a_0 \in \mathbb{Z}[x]$, *and suppose that p is prime such that $p \nmid a_n$, $p \mid a_{n-1}, \ldots, p \mid a_0$ and $p^2 \nmid a_0$. Then, $f(x)$ is irreducible over $\mathbb{Q}$.*

Proof Assume $f(x)$ is reducible over $\mathbb{Q}$. It follows that $f(x)$ is reducible over $\mathbb{Z}$. Then, there exist elements $g(x)$ and $h(x)$ in $\mathbb{Z}[x]$ such that $f(x) = g(x) \cdot h(x)$ and $1 \leq \deg g(x)$, $\deg h(x) < \deg f(x) = n$.

Let $\overline{f(x)}$, $\overline{g(x)}$ and $\overline{h(x)}$ be the polynomials in $\mathbb{Z}_p[x]$ obtained from $f(x)$, $g(x)$ and $h(x)$ by reducing all coefficients modulo p. Then, since p divides all the coefficients of $f(x)$ except a_n, we have $\overline{a_n} x^n = \overline{f(x)} = \overline{g(x)} \cdot \overline{h(x)}$.

Since $\mathbb{Z}_p$ is a field, $\mathbb{Z}_p[x]$ is a unique factorization domain. Thus, $x \mid \overline{g(x)}$ and $x \mid \overline{h(x)}$. So, $\overline{g(0)} = \overline{h(0)} = 0$ and, therefore, $p \mid g(0)$ and $p \mid h(0)$.

But, then, $p^2 \mid g(0) \cdot h(0) = f(0) = a_0$, which is a contradiction. So, our assumption was wrong and thus $f(x)$ is irreducible over $\mathbb{Q}$. $\qquad\square$

Till now, we have discussed principal ideal domain and unique factorization domain. Another important kind of integral domain is a Euclidean domain.

Definition 9.24 (Euclidean Domain) An integral domain D is called a *Euclidean domain* if there is a function d (called the *measure*) from the nonzero elements of D to the nonnegative integers such that

(i) $d(a) \le d(ab)$ for all nonzero a, b in D; and

(ii) if $a, b \in D$, $b \ne 0$, then there exist elements q and r in D such that $a = bq + r$, where $r = 0$ or $d(r) < d(b)$.

Example 9.25 The ring of integers, $\mathbb{Z}$ is a Euclidean domain with $d(a) = |a|$ (the absolute value of a) for $a \in \mathbb{Z}$. For all a, $b \ne 0 \in \mathbb{Z}$, we have $d(a) = |a|$ and $d(ab) = |ab|$. Now, $|ab| = |a| \cdot |b|$. Since, $|b| \ge 1$, so $|a| \le |a| \cdot |b| = |ab| \Rightarrow d(a) \le d(ab)$. Further, by division algorithm for integers a, $b\ (\ne 0) \in \mathbb{Z}$, there exist unique integers q and r such that $a = bq + r$, where $0 \le r < |b|$. Thus, $r < |b|$. Therefore, $\mathbb{Z}$ is a Euclidean domain.

Example 9.26 Let F be a field. Then $F[x]$ is a Euclidean domain with $d(f(x)) = \deg\, f(x)$. Let $a(x)$ and $b(x)$ be nonzero polynomials in $F[x]$. The degree of product of polynomials is the sum of the degrees of the factors, i.e., $\deg\,(a(x)b(x)) = \deg\,(a(x)) + \deg\,(b(x))$. Since $\deg\,(a(x)) \le \deg\,(a(x)) + \deg\,(b(x))$, this implies that $d(a(x)) \le d(a(x)b(x))$.

Further, by division algorithm for polynomials over a field, for $a(x)$, $b(x)\ (\ne 0) \in F[x]$, there exist unique polynomials, $q(x)$ and $r(x)$ in $F[x]$ such that $a(x) = b(x)q(x) + r(x)$, where either $r(x) = 0$ or $\deg\,(r(x)) < \deg\,(b(x))$. Thus, $F[x]$ is a Euclidean domain.

Example 9.27 Let D be a Euclidean domain with measure d. If a and b are associates in D, then $d(a) = d(b)$. Since a and b are associates, then they can be written as $a = u \cdot b$ where u is a unit.
$$\tag{9.9}$$
Since D is a Euclidean domain, so by definition of Euclidean domain $d(b) \le d(u \cdot b) = d(a)$. So, $d(b) \le d(a)$.
$$\tag{9.10}$$
Also, by Equation (9.9), we have, $b = u^{-1} \cdot a$ where u^{-1} is a unit. Given that D is a Euclidean domain, so by definition of Euclidean domain $d(a) \le d(u^{-1} \cdot a) = d(b)$. So, $d(a) \le d(b)$.
$$\tag{9.11}$$
From Equations (9.10) and (9.11), $d(a) = d(b)$.

Example 9.28 The ring of Gaussian integers, $\mathbb{Z}[i]$, is a Euclidean domain with $d(a + ib) = a^2 + b^2$. Define a function $d : \mathbb{Z}[i] \to \mathbb{Z}^+$ such that $d(a + ib) = a^2 + b^2$. We first show that $d(x) \le d(x \cdot y)\ \forall\ x, y \in \mathbb{Z}[i]$. Let, $x = (a + ib)$, $y = (c + id) \in \mathbb{Z}[i]$. Consider $x \cdot y = (a + bi) \cdot (c + di) = (ac - bd) + i(bc + ad)$.
Then, $d(x \cdot y) = (ac - bd)^2 + (bc + ad)^2 = a^2c^2 + b^2d^2 - 2abcd + b^2c^2 + a^2d^2 + 2abcd = (a^2 + b^2)c^2 + d^2(a^2 + b^2) = (a^2 + b^2) \cdot (c^2 + d^2) = d(x) \cdot d(y)$.
Now, since $d(x) \in \mathbb{Z}^+$, we have $d(x) < d(x) \cdot d(y) = d(x \cdot y)$ which implies, $d(x) < d(x \cdot y)$. Now, when $x = 0$, then $d(x) = d(x \cdot y)$. Therefore, $d(x) \le d(x \cdot y)$.
$$\tag{9.12}$$
Again, let $x, y \in \mathbb{Z}[i]$ such that $y \ne 0$, then $x \cdot y^{-1} \in \mathbb{Q}[i]$, the field of quotients. This implies, $x \cdot y^{-1} = s + ti$, where $s, t \in \mathbb{Q}$. Let, m and n be the integers nearest to s and t respectively. It follows that $|m - s| \le \dfrac{1}{2}$ and $|n - t| \le \dfrac{1}{2}$.
Now, $x \cdot y^{-1} = s + ti = (m - m + s) + (n - n + t)i = (m + ni) + [(s - m) + (n - t)i]$
$\Rightarrow x = (m + ni)y + [(s - m) + (n - t)i]y$
$$\tag{9.13}$$
Here, $q = (m + ni) \in \mathbb{Z}[i]$ and $r = [(s - m) + (n - t)i]y$. From Equation (9.13), we have $x = qy + r$ implies $r = x - qy$. So, $r \in \mathbb{Z}[i]$ since $q, y \in \mathbb{Z}[i]$.
Now, we show that $d(r) < d(y)$. Consider, $d(r) = d([(s - m) + (n - t)i]) \cdot d(y) =$
$[(s - m)^2 + (n - t)^2] \cdot d(y) < \left[\left(\dfrac{1}{2} \right)^2 + \left(\dfrac{1}{2} \right)^2 \right] \cdot d(y) = \left(\dfrac{1}{4} + \dfrac{1}{4} \right) \cdot d(y) = \dfrac{1}{2}d(y) < d(y) \Rightarrow$
$d(r) < d(y)$.

Thus, for x, $y \in D$ there exist q, $r \in D$ which satisfies $x = qy + r$ with $d(r) < d(y)$ (9.14)
Therefore, from (9.12) and (9.14), we get, $\mathbb{Z}[i]$ is a Euclidean domain with $d(a + bi) = a^2 + b^2$.

Notice that in the ring $\mathbb{Z}$, unit, $u = \pm 1$ and $d(u) = 1 = d(1)$ as $d(x) = |x|$. In the ring $\mathbb{Q}[x]$ (the polynomials over the rationals), the units are the nonzero constant polynomials. Then, as $d(f(x)) = \deg f(x)$, therefore $d(u) = d(1)$. In $\mathbb{C}[x]$ (the polynomials over the complex numbers), unit, $u = c(\neq 0)$, a constant polynomial and $d(u) = d(1)$. The subsequent proposition extends this concept in general for all Euclidean domains.

Proposition 9.29 *Let D be a Euclidean domain with measure d. Then, u is a unit in D if and only if $d(u) = d(1)$.*

Proof Let, $v \in D$. Here, $v = 1 \cdot v$ and since D is a Euclidean domain, so by definition of Euclidean domain, it follows that $d(1) \leq d(1 \cdot v) = d(v)$. So, $d(1) \leq d(v)$ for every $v \in D$. Let, u be a unit. So, $u \cdot u' = 1$ for some $u' \in D$. Then by definition of Euclidean domain D, $d(u) \leq d(u \cdot u') = d(1)$. So, $d(u) = d(1)$.
Conversely, if $d(u) = d(1)$, then we show that u is a unit. On applying division algorithm on u and 1, there exist q, $r \in D$ such that $1 = uq + r$ for some q, $r \in D$ where $r = 0$ or $d(r) < d(u)$. But $d(u) = d(1)$, which is minimum. So, $d(r) < d(u)$ is not possible. Therefore, $r = 0$. This implies that $1 = u \cdot q$. Hence, u is a unit. $\qquad \square$

Earlier, we have studied various rings like $\mathbb{Z}$, $F[x]$ (polynomials over a field F), $\mathbb{Z}[i]$ (Gaussian integers) that are Euclidean domains as well as PIDs. The next theorem draws a relation between a Euclidean domain and a principal ideal domain.

Theorem 9.30 *Every Euclidean domain is a principal ideal domain.*

Proof Let D be a Euclidean domain and let, I a nonzero ideal of D. Let $a \in I$, such that $d(a)$ is a minimum. Now, we show that $I = \langle a \rangle$.
Clearly, $\langle a \rangle \subseteq I$. Let $b \in I$, then there exist elements q and r such that $b = aq + r$, where $r = 0$ or $d(r) < d(a)$. This implies, $r = b - aq \in I$ and $d(a)$ is minimum. So, $d(r)$ cannot be less than $d(a)$.
Thus, $r = 0$ and $b \in \langle a \rangle$. Since, b is an arbitrary element of I, it follows that $I \subseteq \langle a \rangle$. Therefore, $I = \langle a \rangle$ and I being any arbitrary ideal of D, all the ideals of D are principal ideals. Hence, D is a principal ideal domain. $\qquad \square$

Remark The converse of the Theorem 9.30 is not hold true.

Consider an integral domain $\mathbb{Z}[w]$, where $w = \frac{1 + \sqrt{19}i}{2}$, which is a principal ideal domain but not a Euclidean domain.

Proposition 9.31 *For any non-trivial ideal, I of $\mathbb{Z}[i]$, the quotient ring, $\mathbb{Z}[i]/I$ is finite.*

Proof Consider the quotient ring, $\dfrac{\mathbb{Z}[i]}{I} = \{(a + ib) + I \mid a, b \in \mathbb{Z}\}$.

Let, $x = (a + ib) + I$ be any element of $\dfrac{\mathbb{Z}[i]}{I}$, where a, $b \in \mathbb{Z}$.

Since, $\mathbb{Z}[i]$ is a Euclidean domain and we know that every Euclidean domain is a principal ideal domain. Therefore, $I = \langle c + id \rangle$ for some c, $d \in \mathbb{Z}$.

Now, $(a + ib)$, $(c + id) \in \mathbb{Z}[i]$ and $\mathbb{Z}[i]$ is a Euclidean domain, then by definition of Euclidean domain, there exist some q, $r \in \mathbb{Z}[i]$ such that $(a + ib) = (c + id)q + r$ where either $r = 0$ or $d(r) < d(c + id)$.

Then, $(a + bi) + I = [(c + id)q + r] + I = r + I$ $\hfill (\because I = \langle c + id \rangle)$

Now, let $r = r_1 + r_2 i$ for some r_1, $r_2 \in \mathbb{Z}$, then $d(r) = r_1^2 + r_2^2 \geq 0$. Since $d(r) < d(c + id)$, therefore $0 \leq r_1^2 + r_2^2 < c^2 + d^2$.

Thus, any element of $\dfrac{\mathbb{Z}[i]}{I}$ is of the form $(r_1 + r_2 i) + I$, where $0 \leq r_1^2 + r_2^2 < c^2 + d^2$. It follows that r_1 and r_2 have finite choices. Hence, $\dfrac{\mathbb{Z}[i]}{I}$ is finite. $\hfill \square$

Next, we discuss an important result which is also a consequence of previous theorems, 9.20 and 9.30.

Corollary 9.32 Every Euclidean domain is a unique factorization domain.

Proof Every Euclidean domain is a principal ideal domain and every principal ideal domain is a unique factorization domain. Therefore, every Euclidean domain is a unique factorization domain. $\hfill \square$

Remark Now, let us summarize Theorems 9.20 and 9.30 and their remarks as follows: ED$\Rightarrow$PID$\Rightarrow$UFD. However, UFD$\nRightarrow$PID$\nRightarrow$ED.

Recall that in Chapter 7, we showed the property of unique factorization in $\mathbb{Z}[x]$. Also, observe that $\mathbb{Z}$ is UFD. Combining both these concepts and extending this result for general unique factorization domains, the next theorem is stated as follows.

Theorem 9.33 *If D is a unique factorization domain, then $D[x]$ is a unique factorization domain.*

Proof Since D is a UFD, every nonzero, non-unit element in D has a factorization into irreducibles, and this factorization is unique up to units and order of factors. We aim to show that $D[x]$ also has these properties.

First, we note that $D[x]$ is an integral domain because D is an integral domain, and the ring of polynomials over an integral domain is also an integral domain.

Next, we show that every nonzero, non-unit element of $D[x]$ can be factored into irreducible elements. This follows from the fact that $D[x]$ satisfies the ascending chain condition on principal ideals, meaning that any increasing sequence of principal ideals in $D[x]$ stabilizes, which implies that any nonzero element of $D[x]$ can be written as a product of irreducibles. Finally, the uniqueness of the factorization in $D[x]$ follows from the fact that the units in $D[x]$ are the same as the units in D, and the irreducibles in $D[x]$ are polynomials that cannot be factored further into products of non-units. Any two factorizations of a polynomial in $D[x]$ can be compared by reducing them to factorizations in D, where uniqueness is already guaranteed.

Since $D[x]$ satisfies both the existence and uniqueness of factorization, $D[x]$ is a unique factorization domain. $\hfill \square$

Illustrative Problems

Problem 1 In an integral domain, show that the product of an irreducible and a unit is an irreducible.

Solution Let a be an irreducible and u be a unit. Now we need to show that $a \cdot u$ is irreducible. Let $a \cdot u = x \cdot y$. Assume that neither x nor y is a unit. Since u is a unit, u^{-1} exists. It follows that $a = x \cdot y \cdot u^{-1} = x \cdot y'$ where $y' = y \cdot u$ and since y is a non-unit, y' is a non-unit. This implies, $a = x \cdot y'$ where neither x nor y' are units, which is a contradiction to given fact that a is irreducible. So, our assumption is wrong and hence $a \cdot u$ is irreducible.

Problem 2 Suppose that a and b belong to an integral domain where $b \neq 0$ and a is not a unit. Show that $\langle ab \rangle$ is a proper subset of $\langle b \rangle$.

Solution Clearly, $ab = a \cdot b \in \langle b \rangle \Rightarrow \langle ab \rangle \subseteq \langle b \rangle$. $\hspace{2cm}$ (9.15)

Now, we need to show that $\langle ab \rangle \neq \langle b \rangle$. Let, if possible, $\langle ab \rangle = \langle b \rangle$, it follows that $b = h \cdot ab \Rightarrow 1 = h \cdot a$ (using right cancellation). This implies, $h \cdot a = 1$. Therefore, a is a unit which is a contradiction to the given fact that a is a non-unit.

So, our assumption is wrong and therefore, $\langle ab \rangle \neq \langle b \rangle$ $\hspace{2cm}$ (9.16)

Hence, by Equations (9.15) and (9.16), $\langle ab \rangle \subset \langle b \rangle$.

Problem 3 Let D be an integral domain. Define $a \sim b$ if a and b are associates. Show that this defines an equivalence relation on D.

Solution A relation is equivalence if it is reflexive, symmetric and transitive. Given a relation $a \sim b$ if a and b are associates. We have to prove that '$\sim$' is an equivalence relation. Since a can be written as $a = 1 \cdot a$ where 1 is a unit, it implies $a \sim a$. So, $\sim$ is reflexive. Next, if $a \sim b$, then $a = u \cdot b$ where u is a unit. It follows that $b = u^{-1} \cdot a$ where u^{-1} is a unit since $(u^{-1})^{-1} = u$ exists. So, $b \sim a$. Therefore, $\sim$ is symmetric. Now, for transitivity, if $a \sim b$ and $b \sim c$, then $a = u \cdot b$ and $b = u_1 \cdot c$, where u and u_1 are units. Consider, $a = u \cdot b = u \cdot u_1 \cdot c = u_2 \cdot c$ and since u and u_1 are units, this implies, $u \cdot u_1 = u_2$ is unit. So, $a = u_2 \cdot c$ where u_2 is a unit. So, $a \sim c$. Therefore, $\sim$ is transitive. Hence, $\sim$ is an equivalence relation.

Problem 4 Let R be an integral domain with unity. If gcd $(a, b) = d$ for $a, b \in R$, then cd and gcd (ca, cb) are associates.

Solution Given that $\gcd(a, b) = d$ which implies $d \mid a$, so, $a = dk$. $\hspace{1cm}$ (9.17)

And, $d \mid b \Rightarrow b = dm$. $\hspace{1cm}$ (9.18)

Let, $\gcd(ca, cd) = d'$. Now, consider $ac = dk \cdot c = cdk$ (using (9.17)) which implies $cd \mid ca$. $\hspace{1cm}$ (9.19)

Similarly, consider $cb = c \cdot dm = cdm$ (using (9.18)), which implies $cd \mid cb$. $\hspace{1cm}$ (9.20)

Using Equations (9.19) and (9.20), we get, $cd \mid d'$ $\hspace{1cm}$ ($\because \gcd(ca, cb) = d'$.)

$\Rightarrow d' = cdt$ for some t $\hspace{1cm}$ (9.21)

Again, since $\gcd(ca, cb) = d'$, so, $d' \mid ca$ and $d' \mid cb$, which implies, $ca = d' \cdot s$ and $cb = d' \cdot n$ for some s and n respectively.

First, consider $ca = d' \cdot s \Rightarrow ca = cdts$ $\hspace{1cm}$ (by (9.21))

$\Rightarrow a = dts = (dt) \cdot s$ (using left cancellation law)

$\Rightarrow dt \mid a.$

Similarly, $dt \mid b$. It follows that $dt \mid d$ which implies $d = dtp$ for some p. This implies
$d(1 - tp) = 0 \Rightarrow 1 - tp = 0$ ($\because d \neq 0$.)

$\Rightarrow 1 = tp \Rightarrow t \cdot p = 1.$

So, t is a unit. Therefore, we have $\gcd(ca, cb) = d' = cd \cdot t$ where t is a unit. Hence, cd and d' are associates.

Problem 5 In $\mathbb{Z}[\sqrt{-5}]$, show that 21 does not factor uniquely as a product of irreducibles.

Solution We have, $21 = 3 \cdot 7$ (9.22)

Also, $21 = (1 + 2\sqrt{-5}) \cdot (1 - 2\sqrt{-5}).$ (9.23)

All factors $3, 7, (1 + 2\sqrt{-5}), (1 - 2\sqrt{-5})$ are irreducible over $\mathbb{Z}[\sqrt{-5}]$.

If 3 is reducible then, $3 = x \cdot y$ where neither x nor y is a unit. Thus, $3 = x \cdot y \Rightarrow N(3) = N(x \cdot y) \Rightarrow 9 = N(x) \cdot N(y) \Rightarrow 3 \cdot 3 = N(x) \cdot N(y) \Rightarrow N(x) = 3$ and $N(y) = 3$. It follows that there exist $a, b \in \mathbb{Z}$ such that $N(a + b\sqrt{-5}) = 3 \Rightarrow a^2 + 5b^2 = 3$ which does not hold since $a, b \in \mathbb{Z}$. So, our assumption is wrong. Either x or y is a unit. Therefore, 3 is irreducible.

Similarly, the above argument holds for 7. Therefore, 7 is irreducible.

Now, consider $(1 + 2\sqrt{-5})$. Let's assume $(1 + 2\sqrt{-5})$ is reducible. So, $(1 + 2\sqrt{-5}) = x \cdot y$ where neither x nor y is a unit. It follows that
$N(1 + 2\sqrt{-5}) = N(x \cdot y) \Rightarrow 1 + 5 \cdot 4 = N(x) \cdot N(y) \Rightarrow 21 = N(x) \cdot N(y) \Rightarrow 3 \cdot 7 = N(x) \cdot N(y)$. This implies $N(x) = 3, N(y) = 7$ or $N(x) = 7, N(y) = 3$.

Without loss of generality, assume $N(x) = 3$ and $N(y) = 7$. It follows that there exist $a, b, c, d \in \mathbb{Z}$ such that $N(a + b\sqrt{-5}) = 3 \Rightarrow a^2 + 5b^2 = 3$ and $N(c + d\sqrt{-5}) = 3 \Rightarrow c^2 + 5d^2 = 7$. Both the conditions do not hold since $a, b, c, d \in \mathbb{Z}$. So, our assumption is wrong. Either x or y is a unit. Therefore, $(1 + 2\sqrt{-5})$ is irreducible.

The above argument holds for $(1 - 2\sqrt{-5})$. Therefore, $(1 - 2\sqrt{-5})$ is irreducible.

Hence, 21 does not factor uniquely as a product of irreducibles in $\mathbb{Z}[\sqrt{-5}]$.

Problem 6 Show that $1 - i$ is an irreducible in $\mathbb{Z}[i]$.

Solution Define $N : \mathbb{Z}[i] \to \mathbb{N}$ such that $N(a + ib) = a^2 + b^2$.

Let's assume $(1 - i)$ is reducible. So, $(1 - i)$ can be written as $(1 - i) = x \cdot y$ where neither x nor y is a unit. It follows that $N(1 - i) = N(x \cdot y)$. Using properties of norm function, we get $2 = N(x) \cdot N(y) \Rightarrow 2 \cdot 1 = 2 = N(x) \cdot N(y)$. This implies that either $N(x) = 2, N(y) = 1$ or $N(x) = 1, N(y) = 2$.

Without loss of generality, assume $N(x) = 1$ and $N(y) = 2$. Then, x is a unit in $\mathbb{Z}[i]$.
Hence, $(1 - i)$ is irreducible in $\mathbb{Z}[i]$.

Problem 7 In $\mathbb{Z}[i]$, show that 3 is irreducible but 2 and 5 are not.

Solution Suppose 3 is reducible. So, $3 = x \cdot y$ where $x, y \in \mathbb{Z}[i]$ such that neither x nor y is a unit. It follows that $N(3) = N(x \cdot y) \Rightarrow 9 = N(x) \cdot N(y) \Rightarrow 3 \cdot 3 = N(x) \cdot N(y) \Rightarrow N(x) = 3, N(y) = 3$. It follows that there exist $a, b \in \mathbb{Z}$ such that $a^2 + b^2 = 3$ which does not hold. So, our assumption is wrong and either x or y is a unit. Therefore, 3 is irreducible.

Now, for next part, consider the factorizations $2 = (1+i) \cdot (1-i)$ and $5 = (2+i) \cdot (2-i)$. In order to prove that 2 and 5 are irreducibles, we will show that the factors $(1+i)$, $(1-i)$, $(2+i)$ and $(2-i)$ are not units in $\mathbb{Z}[i]$.

Let, if possible, $(1+i)$ is a unit. Then there exist $y = (a+bi) \in \mathbb{Z}[i]$ such that $(1+i) \cdot (a+bi) = 1 \Rightarrow a + ai + bi - b = 1 \Rightarrow (a-b) + (a+b) \cdot i = 1 + 0 \cdot i$. It follows that $a - b = 1$ and $a + b = 0$. On solving the equations, we get $a = \dfrac{1}{2} \notin \mathbb{Z}$, $b = \dfrac{-1}{2} \notin \mathbb{Z}$. So, these values of a and b are not possible. Therefore, $(1+i)$ is not a unit.

A similar, argument holds for $(1-i)$. Therefore, $(1-i)$ is not a unit.

Now, consider $(2+i)$. Let, if possible, $(2+i)$ is a unit. Then, there exist $y' = (c+di) \in \mathbb{Z}[i]$ such that $(2+i) \cdot (c+di) = 1 \Rightarrow 2c + 2di + ci - d = 1 \Rightarrow (2c-d) + (2d+c) \cdot i = 1 + 0 \ i$. It follows that $2c - d = 1$ and $2d + c = 0$. On solving the equations, we get $c = \dfrac{2}{5} \notin \mathbb{Z}$, $d = \dfrac{-1}{5} \notin \mathbb{Z}$. So, these values of c and d are not possible. Therefore, $(2+i)$ is not a unit.

Similarly, $(2-i)$ is not a unit.

Hence, 2 and 5 are not irreducibles.

Problem 8 Find out all units of $Z[i]$.

<u>Solution</u> Let $(a+ib)$ be a unit of $\mathbb{Z}[i]$ and by definition of norm function, we know that x is a unit if and only if $N(x) = 1$. It follows that $N(a+ib) = 1 \Rightarrow a^2 + b^2 = 1$ which is possible when $a = \pm 1, b = 0$ and $a = 0, b = \pm 1$. Therefore, the units of $\mathbb{Z}[i]$ are $1, -1, i$ and $-i$.

Problem 9 Show that $\mathbb{Z}[x]$ is not a principal ideal domain.

<u>Solution</u> Let us take $I = \{f(x) \in \mathbb{Z}[x] \mid f(0) \text{ is even}\}$. Now, we will show that I is an ideal in $\mathbb{Z}[x]$. Since the identity function $f(x) = x$ satisfies the condition $f(0) = 0$, therefore, $I \neq \emptyset$.

Consider, $f(x), g(x) \in I$ such that $f(0)$ and $g(0)$ are even. Then, $f(x) - g(x) \in I$ since the difference between two evens, $f(0) - g(0)$ is even. Also, if $r(x) \in \mathbb{Z}[x]$ then $r(x)f(x) \in I$ since the product of two evens (when r is even) and the product of an odd and an even (when r is odd) are even. Similarly, $f(x) \cdot r \in I$. Therefore, I is an ideal in $\mathbb{Z}[x]$.

Now, we have to show that I is not generated by any function $h(x) \in \mathbb{Z}[x]$. Let, if possible, $I = \langle h(x) \rangle$. Now, the functions $f(x) = x \in I$ since $f(0) = 0$, which is even and $g(x) = 2 \in I$ since $g(0) = 2$, which is even. It follows that the functions x and 2 belong to I and $I = \langle h(x) \rangle$. So, $x \in \langle h(x) \rangle$ and $2 \in \langle h(x) \rangle \Rightarrow x = f(x) \cdot h(x)$ \hfill (9.24)

and $2 = g(x) \cdot h(x)$ for some $f(x), g(x) \in \mathbb{Z}[x]$. \hfill (9.25)

Using degree rule in Equations (9.24) and (9.25), we get, $\deg x = \deg f(x) + \deg h(x) = 1$ and $\deg 2 = \deg g(x) + \deg h(x) = 0$. It follows that $\deg g(x) = \deg h(x) = 0$ since degree is always greater than zero and $\deg f(x) = 1$. Therefore, $g(x) = c_1$ and $h(x) = c_2$ where c_1 and c_2 are constant and without loss of generality, assume $f(x) = x$.

On putting the values of $f(x)$, $g(x)$ and $h(x)$ in Equation (9.25), we get $c_1 \cdot c_2 = 2$ which implies either $c_1 = \pm 1, c_2 = \pm 2$ or $c_1 = \pm 2, c_2 = \pm 1$. If $c_2 = \pm 1$, then $h(x) = \pm 1$ for which $h(0) = \pm 1$, which is not even. This is a contradiction. So, $c_2 = \pm 2$ implies $h(x) = \pm 2$. (9.26)

Now, on putting the value of $h(x)$ and $f(x)$ in Equation (9.24), we get $x = \pm 2x \Rightarrow 1 = \pm 2$. This is a contradiction. So, our assumption is wrong.

Therefore, $I \neq \langle h(x) \rangle$. Hence, $\mathbb{Z}[x]$ is not a principal ideal domain.

Problem 10 In $\mathbb{Z}[\sqrt{-5}]$, prove that $1 + 3\sqrt{-5}$ is irreducible but not prime.

<u>**Solution**</u> Define a function $N : \mathbb{Z}[\sqrt{-5}] \to \mathbb{N}$ such that $N(a + b\sqrt{-5}) = a^2 + 5b^2$. Now, assume $1 + 3\sqrt{-5}$ is reducible. Then, $(1 + 3\sqrt{-5}) = x \cdot y$ where neither x nor y is a unit. It follows that $N(1 + 3\sqrt{-5}) = N(x \cdot y) \Rightarrow 1 + 5 \cdot 9 = N(x) \cdot N(y)$ $(\because N(x \cdot y) = N(x) \cdot N(y).)$

$\Rightarrow 46 = N(x) \cdot N(y) \Rightarrow 2 \cdot 23 = N(x) \cdot N(y)$.

Without loss of generality, assume that $N(x) = 2$ and $N(y) = 23$. Then by definition of norm function, there exist a, b, c, $d \in \mathbb{Z}$ such that $a^2 + 5b^2 = 2$ and $c^2 + 5d^2 = 23$ which is not possible. So, our assumption is wrong. Either x or y is a unit. Therefore, $(1 + 3\sqrt{-5})$ is irreducible.

Now, we have, $(1 + 3\sqrt{-5}) \cdot (1 - 3\sqrt{-5}) = 46 = 2 \cdot 23$. This follows that $(1 + 3\sqrt{-5}) \mid 2 \cdot 23$. Then if it is prime, then either $(1 + 3\sqrt{-5}) \mid 2$ or $(1 + 3\sqrt{-5}) \mid 23$.

Case 1. If $(1 + 3\sqrt{-5}) \mid 2$, then there exists $(a + b\sqrt{-5}) \in \mathbb{Z}[\sqrt{-5}]$ such that $2 = (1 + 3\sqrt{-5}) \cdot (a + b\sqrt{-5}) = a + b\sqrt{-5} + 3a\sqrt{-5} - 15b = (a - 15b) + (3a + b)\sqrt{-5}$.
On comparing the coefficients, we get $a - 15b = 2$ and $3a + b = 0$. On solving these equations, we get $a = \dfrac{1}{23} \notin \mathbb{Z}$ and $b = \dfrac{-3}{23} \notin \mathbb{Z}$. So, this is not possible.
Therefore, $(1 + 3\sqrt{-5})$ does not divide 2.

Case 2. If $(1 + 3\sqrt{-5}) \mid 23$, then there exists $(c + d\sqrt{-5}) \in \mathbb{Z}[\sqrt{-5}]$ such that $23 = (1 + 3\sqrt{-5}) \cdot (c + d\sqrt{-5}) = c + d\sqrt{-5} + 3c\sqrt{-5} - 15d = (c - 15d) + (3c + d)\sqrt{-5}$.
On comparing the coefficients, we get $c - 15d = 23$ and $3c + d = 0$. On solving these equations, we get $c = \dfrac{1}{2} \notin \mathbb{Z}$ and $b = \dfrac{-3}{2} \notin \mathbb{Z}$. So, this is not possible.
Therefore, $(1 + 3\sqrt{-5})$ does not divide 23.
Thus, we have $(1 + 3\sqrt{-5}) \mid 2 \cdot 23$ but, neither $(1 + 3\sqrt{-5}) \mid 2$ nor $(1 + 3\sqrt{-5}) \mid 23$. Hence, $(1 + 3\sqrt{-5})$ is not prime.

Problem 11 In $\mathbb{Z}[\sqrt{5}]$, prove that $1 + \sqrt{5}$ is irreducible but not prime.

<u>**Solution**</u> Let, if possible, $(1 + \sqrt{5})$ is reducible in $\mathbb{Z}[\sqrt{5}]$. Then, $(1 + \sqrt{5}) = x \cdot y$ where neither x nor y is a unit. It follows that $N(1 + \sqrt{5}) = N(x \cdot y) \Rightarrow 6 = N(x) \cdot N(y) \Rightarrow 2 \cdot 3 = N(x) \cdot N(y)$.

So, without loss of generality, assume $N(x) = 2$ and $N(y) = 3$. Since $x, y \in \mathbb{Z}[\sqrt{5}]$, let $x = (a + b\sqrt{5})$, $y = (c + d\sqrt{5})$ where a, b, c, $d \in \mathbb{Z}$ such that $N(a + b\sqrt{5}) = 2$ and $N(c + d\sqrt{5}) = 3$ which implies $a^2 + 5b^2 = 2$ and $c^2 + 5d^2 = 3$ respectively. The equations do not hold since they contradict the fact that a, b, c, $d \in \mathbb{Z}$. So, our assumption is wrong and hence $(1 + \sqrt{5})$ is irreducible.

We know that, $(1 + \sqrt{5}) \cdot (1 - \sqrt{5}) = 6 = 2 \cdot 3$. So, we have $(1 + \sqrt{5}) \mid 2 \cdot 3$. Now, if $(1 + \sqrt{5})$ is prime, then either $(1 + \sqrt{5}) \mid 2$ or $(1 + \sqrt{5}) \mid 3$.

Case 1. If $(1 + \sqrt{5}) \mid 2$, then there exists $(a + b\sqrt{5}) \in \mathbb{Z}[\sqrt{5}]$ such that $2 = (1 + \sqrt{5}) \cdot (a + b\sqrt{5}) = a + b\sqrt{5} + a\sqrt{5} + 5b = (a + 5b) + (a + b)\sqrt{5}$. On comparing the coefficients, we get $a + 5b = 2$ and $a + b = 0$. On solving these, we have $a = \dfrac{-1}{2} \notin \mathbb{Z}$ and $b = \dfrac{1}{2} \notin \mathbb{Z}$. This is a contradiction. So, no such $a + b\sqrt{5}$ exists.
Therefore, $(1 + \sqrt{5}) \nmid 2$.

Case 2. If $(1 + \sqrt{5}) \mid 3$, then there exists $(c + d\sqrt{5}) \in \mathbb{Z}[\sqrt{5}]$ such that $3 = (1 + \sqrt{5}) \cdot (c + d\sqrt{5}) = c + d\sqrt{5} + c\sqrt{5} + 5d = (c + 5d) + (c + d)\sqrt{5}$. On comparing the coefficients, we get $c + 5d = 3$ and $c + d = 0$. On solving these, we have $c = \dfrac{-3}{4} \notin \mathbb{Z}$ and $d = \dfrac{3}{4} \notin \mathbb{Z}$. This is a contradiction. So, no such $c + d\sqrt{5}$ exists. Therefore, $(1 + \sqrt{5}) \nmid 3$.

Thus, we have $(1 + \sqrt{5}) \mid 2 \cdot 3$ but neither $(1 + \sqrt{5}) \nmid 2$ nor $(1 + \sqrt{5}) \nmid 3$. Hence, $(1 + \sqrt{5})$ is not prime.

Problem 12 Find all the units of $\mathbb{Z}[\sqrt{-5}]$.

<u>Solution</u> Suppose $a + b\sqrt{-5}$ is a unit in $\mathbb{Z}[\sqrt{5}]$. Then, $(a + b\sqrt{-5}) \cdot (c + d\sqrt{-5}) = 1 + 0 \cdot \sqrt{-5}$ for some $c, d \in \mathbb{Z}$. On taking conjugates both sides, we get $(a - b\sqrt{-5}) \cdot (c - d\sqrt{-5}) = \bar{1} = 1$ which implies $(a^2 + 5b^2) \cdot (c^2 + 5b^2) = 1$ in $\mathbb{Z}$. It follows that $a^2 + 5b^2 = 1$, so, $a = \pm 1$, $b = 0$. Thus, $a + b\sqrt{-5} = \pm 1$ are the units in $\mathbb{Z}[\sqrt{-5}]$.

Problem 13 For a commutative ring with unity we may define associates, irreducibles, and primes exactly as we did for integral domains. With these definitions, show that 2 is prime element in Z_6 but 2 is not irreducible.

<u>Solution</u> Consider the ring $\mathbb{Z}_6 = \{0, 1, 2, 3, 4, 5\}$ mod 6. First, we will show that 2 is prime element in $\mathbb{Z}_6$. Clearly, $2 \neq 0$ and is a non-unit. Let, $2 \mid a \otimes_6 b$. Then, in order to show that 2 is prime, we have to show that either $2 \mid a$ or $2 \mid b$. In $\mathbb{Z}_6$, we can write ab as $ab = 6q + a \otimes b$ for some q. Now, since $2 \mid 6q$, $2 \mid a \otimes_6 b$, therefore, $2 \mid ab \Rightarrow 2 \mid a$ or $2 \mid b \Rightarrow 2 \mid a$ or $2 \mid b$ in $\mathbb{Z}_6$. Hence, 2 is a prime element in $\mathbb{Z}_6$. Again, as we know, $2 \otimes_6 4 = 2$, where neither 2 or 4 is a unit. So, 2 is not irreducible in $\mathbb{Z}_6$.

Problem 14 Let R be a Euclidean domain and let A be an ideal of R, then show that there exists $a_0 \in A$ such that $A = \{a_0 \cdot x \mid x \in R\}$.

<u>Solution</u> If $A = \{0\}$, then $a_0 = 0$. Suppose $A \neq \{0\}$, then there exists at least one $0 \neq a \in A$. Let, $a_0 \in A$ be such that $d(a_0)$ is minimal. Note that its existence is ensured by the well ordering principle which states that every non-empty subset of nonnegative integers has a least element.

Now, we will show that $A = \langle a_0 \rangle$. Let, $a \in A$, $a \neq 0$, then by definition, there exist t, $r \in R$, such that, $a = a_0 t + r$, where either $r = 0$ or $d(r) < d(a_0)$.

Suppose $r \neq 0$. Then, $a_0 \in A$, $t \in R$, implies, $ta_0 \in A$. Thus, $a \in A$, $ta_0 \in A \Rightarrow a - ta_0 \in A \Rightarrow r \in A$.

But $d(a_0)$ is the smallest d-value in A and $d(r) < d(a_0)$, which leads to a contradiction. Hence, $r = 0$ implies $a = ta_0$. Thus, any $a \in A$ can be written in the form ta_0. This implies, $A \subseteq \{a_0 x \mid x \in R\}$. But $\{a_0 x \mid x \in R\} \subseteq A$ as $a_0 \in A$, $xa_0 \in A$ for all $x \in R$ since A is an ideal. Hence, $A = \{a_0 x \mid x \in R\}$.

Problem 15 Show by an example that it is possible to find two elements a, b in a Euclidean domain such that $d(a) = d(b)$, but a, b are not associates.

<u>Solution</u> Consider $D = \{a + ib \mid a, b \in \mathbb{Z}\} = \mathbb{Z}[i]$, the ring of Gaussian integers, where $d(a + ib) = a^2 + b^2$. Then, using proposition 9.31, D is a Euclidean domain. Here, $d(2 + 3i) = 13 = d(2 - 3i)$, but $2 + 3i$ and $2 - 3i$, are not associates.

Since, by Problem 8, Chapter 9, the units of D are $\pm 1, \pm i$ and thus an associate of $2 + 3i$ can be $(2 + 3i) \cdot 1, (2 + 3i) \cdot (-1), (2 + 3i)i, (2 + 3i) \cdot (-i)$. Therefore, the associates of $2 + 3i$ are $2 + 3i, -2 - 3i, -3 + 2i, 3 - 2i$, which are all different from $2 - 3i$.

Problem 16 Show that in a principal ideal domain, every nonzero prime ideal is maximal.

Solution Let $P = \langle p \rangle$, where $p \neq 0$, be a nonzero prime ideal in a principal ideal domain R. Suppose $P \subseteq Q = \langle q \rangle \subseteq R$. Then, $p \in P \subset Q = \langle q \rangle$ implies $p = q \cdot r$. So, $q \cdot r \in P$. It follows that either $q \in P$ or $r \in P$.

If $q \in P$, then $x \cdot q \in P \ \forall \ x \in R \Rightarrow Q \subseteq P$ and thus $Q = P$.

If $r \in P$, then $r = p \cdot t$ implies $r = qrt \Rightarrow r \cdot (1 - qt) = 0 \Rightarrow 1 = q \cdot t (r \neq 0)$.

But $q \in Q$, $t \in R$ implies $q \cdot t \in Q$, so, $1 \in Q$ and hence $Q = R$. Note $r = 0$ would mean $p = q \cdot 0$ implies $p = 0$, which is not true. Hence, in a principal ideal domain, every nonzero prime ideal is maximal.

Problem 17 Let R be a principal ideal domain which is not a field, then an ideal $A = \langle a_0 \rangle$ is a maximal ideal if and only if a_0 is an irreducible element.

Solution Let $A = \langle a_0 \rangle$ be a maximal ideal. We will first prove that $a_0 \neq 0$. Suppose $a_0 = 0$. Since R is not a field, there exists $0 \neq b \in R$ such that b^{-1} does not exist. Let $B = \langle b \rangle$ and as $a_0 = 0$, $A = \langle 0 \rangle$ and $\langle 0 \rangle \subseteq B \subseteq R$ implies $A \subseteq B \subseteq R$.

Now, $B \neq A$ as $b \in B$, $b \neq 0$, and $A = \langle 0 \rangle$. Also, $B \neq R$ as $1 \in R$, but $1 \notin B$. Since if $1 \in B = \langle b \rangle$ then some x such that $1 = bx$ shows that b is invertible which is not true. Hence, $a_0 \neq 0$.

Next, we will show that a_0 is not a unit. Suppose a_0 is a unit, then $a_0 \cdot a_0^{-1} = 1$. Since $a_0 \in A$, $a_0^{-1} \in R$, $a_0 \cdot a_0^{-1} \in A \Rightarrow 1 \in A \Rightarrow A = R$, which is not possible as A is maximal. Thus, a_0 is not a unit.

Let now, $a_0 = bc$ for some $b, c \in R$. We show either b or c is a unit. Let, $B = \langle b \rangle$. Since $a_0 = bc$, $a_0 \in B$, therefore $x \cdot a_0 \in B \ \forall \ x \in R$. This implies that $A \subseteq B$. But A is maximal thus either $B = R$ or $B = A$.

Case 1. If $B = R$, then $1 \in B = \langle b \rangle$ as $1 \in R$. It follows that $1 = xb$ for some x. Therefore, b is a unit.

Case 2. If $B = A$, then $b \in A = \langle a_0 \rangle$. It follows that $b = ya_0$ for some y. This implies, $a_0 = bc = ya_0 c \Rightarrow a_0 - ya_0 c = 0 \Rightarrow a_0 \cdot (1 - yc) = 0 \Rightarrow 1 - yc = 0.$ $(\because a_0 \neq 0.)$ Therefore, c is a unit. Hence a_0 is irreducible.

Conversely, let a_0 be an irreducible element. We show $A = \langle a_0 \rangle$ is maximal. Let I be any ideal such that $A \subset I \subseteq R$. Since R is a principal ideal domain, $I = \langle x \rangle$ for some $x \in R$. Now, $x \notin A$ as if $x \in A$, then $\langle x \rangle \subseteq A$ which implies $I \subseteq A$, but $A \subseteq I$ means $A = I$ which is not true. Thus $x \notin A$.

Again, $A = \langle a_0 \rangle \subset I$ implies $a_0 = xy$ for some y. Since, a_0 is irreducible, so either x or y is a unit. If y is a unit, then $y \cdot y^{-1} = 1$ and $a_0 = xy$. Then, $a_0 \cdot y^{-1} = x$. But $a_0 \in A$, $y^{-1} \in R$ implies $a_0 \cdot y^{-1} \in A$ So, $x \in A$, which is not true. Thus, y is not a unit. Therefore, x is a unit and $x \cdot x^{-1} = 1$.

Now, $x \in I$, $x^{-1} \in R$ and I is an ideal. So, $x \cdot x^{-1} \in I$ implies $1 \in I$, therefore $I = R$. Hence, A is maximal ideal of R.

Problem 18 Show by an example that subring of a Noetherian ring may not be Noetherian.

Solution Let $\mathbb{Q}$ be the field of rational numbers, then $\mathbb{Q}$ is a Noetherian ring and thus $\mathbb{Q}[x]$ is Noetherian. Let $S = \{f(x) \in \mathbb{Q}[x] \mid f(x) = a_0 + a_1 x + a_2 x^2 + \ldots + a_n x^n, a_0 \in \mathbb{Z},$

$a_i \in \mathbb{Q} \ \forall \ i \geq 1\}$. Let $f(x) = a_0 + a_1 x + a_2 x^2 + ... + a_n x^n$ and $g(x) = b_0 + b_1 x + b_2 x^2 + ... + b_n x^n \in S$.

Consider, $f(x) - g(x) = (a_0 - b_0) + (a_1 - b_1)x + (a_2 - b_2)x^2 + ... + (a_n - b_n)x^n$. Now, since a_0, $b_0 \in \mathbb{Z}$, it implies $(a_0 - b_0) \in \mathbb{Z}$. Also since a_i, $b_i \in \mathbb{Q} \ \forall \ i \geq 1$, it follows that $(a_i - b_i) \in \mathbb{Q} \ \forall \ i \geq 1$. So, $f(x) - g(x) \in S$.

Next, consider $f(x) \cdot g(x) = (a_0 + a_1 x + a_2 x^2 + ... + a_n x^n) \cdot (b_0 + b_1 x + b_2 x^2 + ... + b_n x^n)$
$= (a_0 \cdot b_0) + (a_1 \cdot b_0 + a_0 \cdot b_1)x + (a_2 \cdot b_0 + a_1 \cdot b_1 + a_0 \cdot b_2)x^2 + ... + (a_n \cdot a_0 + a_{n-1} \cdot b_1 + ... + a_0 \cdot b_n)x^n$, where $(a_0 \cdot b_0) \in \mathbb{Z}$ $(\because a_0, b_0 \in \mathbb{Z})$ and $a_i \cdot b_j \in \mathbb{Q} \ \forall \ i \geq 1$, or $j \geq 1$ $(\because c_i, b_j \in \mathbb{Q} \ \forall \ i, j \geq 1)$.

So, $f(x) \cdot g(x) \in S$. Then, S is a subring of $\mathbb{Q}[x]$.

Here, the chain $\langle x \rangle \subset \langle x/2 \rangle \subset \langle x/4 \rangle \subset ...$ is an ascending chain of ideals in S which does not terminate after finite number of steps.

Suppose for instance, equality holds at $\langle x \rangle = \langle x/2 \rangle$, then $x/2 \in \langle x \rangle \Rightarrow x/2 = h(x)x$ for some $h(x) = c_0 + c_1 x + c_2 x^2 + ... + c_m x^m$, where $c_0 \in \mathbb{Z}$

$$\Rightarrow \frac{x}{2} = c_0 x + c_1 x^2 + c_2 x^3 + ... + c_m x^{m+1}$$

$$\Rightarrow 0 + \frac{1}{2}x + 0 \cdot x^2 + ... = 0 + c_0 x + c_1 x^2 + c_2 x^3 + ... + c_m x^{m+1}$$

$\Rightarrow c_0 = \dfrac{1}{2} \notin \mathbb{Z}$. Therefore, $\langle x \rangle \subset \langle x/2 \rangle$. Continuing in this fashion, we get that the equality does not hold at any step i.e., $\langle x \rangle \subset \langle x/2 \rangle \subset \langle x/4 \rangle \subset ....$ So, the chain never terminates and hence there exists a subring S of a Noetherian ring which is not a Noetherian ring.

Problem 19 Prove that 7 is irreducible in $\mathbb{Z}[\sqrt{6}]$, even though $N(7)$ is not prime.

Solution Clearly, $7 \neq 0$ and is a non-unit. Let, if possible, 7 is reducible. Then, $7 = x \cdot y$, where neither x nor y is a unit. So, using the definition of norm function, $N(7) = 49$
$\Rightarrow N(x \cdot y) = 49 \Rightarrow N(x) \cdot N(y) = 7 \cdot 7$ $\qquad (\because N(x \cdot y) = N(x) \cdot N(y).)$ $\qquad \Rightarrow N(x) = 7$, $N(y) = 7$ $(\because x$ and y are non-units, so, $N(x) \neq 1$, $N(y) \neq 1.)$
Let, $x = a + b\sqrt{6} \in \mathbb{Z}[\sqrt{6}]$ and we have, $N(x) = 7$ so that $N(a + b\sqrt{6}) = 7$ which implies $a^2 - 6b^2 = 7$. The equation does not hold since $a, b \in \mathbb{Z}$. Therefore, 7 is irreducible in $\mathbb{Z}[\sqrt{6}]$.

Now, we know $N(7) = 49 \mid 49 = 7 \cdot 7$. It follows that $N(7) \mid 7 \cdot 7$, but $N(7) = 49 \nmid 7$. Hence, $N(7)$ is not prime.

Problem 20 Prove that $\mathbb{Z}[\sqrt{5}]$ is not a unique factorization domain.

Solution Consider the following factorizations, $4 = ((-1) + \sqrt{5}) \cdot (1 + \sqrt{5})$ and $4 = 2 \cdot 2$. Now, in order to prove that $\mathbb{Z}[\sqrt{5}]$ is not a unique factorization domain, we prove that all the factors are irreducibles so that it cannot be uniquely represented as product of irreducibles. Here, by Problem 11, Chapter 9, $(1 + \sqrt{5})$ is irreducible in $\mathbb{Z}[\sqrt{5}]$.

Next, consider 2. Let, if possible, 2 is reducible. Then, $2 = x \cdot y$ where neither x nor y is a unit. This implies, $N(2) = N(x \cdot y) = N(x) \cdot N(y) \Rightarrow 2 \cdot 2 = N(x) \cdot N(y) \Rightarrow N(x) = 2$, $N(y) = 2$. It follows that there exist $a, b \in \mathbb{Z}$ such that $N(a + b\sqrt{5}) = 2$ which implies $a^2 - 5b^2 = 2$. The equation does not hold in $\mathbb{Z}$. So, our assumption is wrong. Therefore, 2 is irreducible.

Now, consider $-1 + \sqrt{5}$. Let, if possible, $-1 + \sqrt{5}$ is reducible. Then, $-1 + \sqrt{5} = x \cdot y$ where neither x nor y is a unit. This implies, $N(-1 + \sqrt{5}) = N(x \cdot y) = N(x) \cdot N(y) \Rightarrow 4 = N(x) \cdot$

$N(y) \Rightarrow 2 \cdot 2 = N(x) \cdot N(y) \Rightarrow N(x) = 2, N(y) = 2$. It follows that there exist $a, b \in \mathbb{Z}$ such that $N(a + b\sqrt{5}) = 2$ which implies $a^2 - 5b^2 = 2$. The equation does not hold in $\mathbb{Z}$. So, our assumption is wrong. Therefore, $-1 + \sqrt{5}$ is irreducible.

Thus, we have $4 = ((-1) + \sqrt{5}) \cdot (1 + \sqrt{5})$ and $4 = 2 \cdot 2$ where all the factors $(-1 + \sqrt{5})$, $1 + \sqrt{5}$, 2 are irreducibles. Hence, $\mathbb{Z}[\sqrt{5}]$ is not a unique factorization domain.

Problem 21 Let R be a unique factorization domain. Then show that every prime element in R generates a prime ideal.

Solution Let p be a prime element of a unique factorization domain R. Let $S = \langle p \rangle$ be the ideal of R generated by p. Then, we show that S is a prime ideal. Suppose ab is an element of S where $a, b \in R$. So, we have $ab \in S$ implies $ab = kp$ for some $k \in R$. This implies,
$$p \mid ab \Rightarrow p \mid a \text{ or } p \mid b \qquad (\because p \text{ is a prime element of } R.)$$
This implies $a = sp$ or $b = tp$ for some $s, t \in R$. It follows that $a \in \langle p \rangle$ or $b \in \langle p \rangle$. Hence, $\langle p \rangle$ is a prime ideal of R.

Problem 22 Let d be an integer less than -1 that is not divisible by the square of a prime. Prove that the only units of $\mathbb{Z}[\sqrt{d}]$ are 1 and -1.

Solution Suppose that $a + b\sqrt{d} \in \mathbb{Z}[\sqrt{d}]$ is a unit in $\mathbb{Z}[\sqrt{d}]$. Then, it follows that $N(x) = 1 \Rightarrow N(a + b\sqrt{d}) = 1 \Rightarrow a^2 - db^2 = 1$

Now, given that $d < -1$, so $-d > 1$. Therefore, the above condition holds only when $b = 0$ and $a = \pm 1$.

Hence, the only units of $\mathbb{Z}[\sqrt{d}]$ are 1 and -1.

Problem 23 In $\mathbb{Z}[\sqrt{2}] = \{a + b\sqrt{2} \mid a, b \in \mathbb{Z}\}$, show that every element of the form $(3 + 2\sqrt{2})^n$ is a unit, where n is a positive integer.

Solution In $\mathbb{Z}[\sqrt{2}]$, consider $(3 + 2\sqrt{2}) \cdot (3 - 2\sqrt{2}) = 9 - 2 \cdot 4 = 1 \Rightarrow [(3 + 2\sqrt{2}) \cdot (3 - 2\sqrt{2})]^n = 1^n = 1 \Rightarrow (3 + 2\sqrt{2})^n \cdot (3 - 2\sqrt{2})^n = 1$. Thus, $(3 + 2\sqrt{2})^n$ is a unit.

Problem 24 If a and b belong to $\mathbb{Z}[\sqrt{d}]$, where d is not divisible by the square of a prime and ab is a unit, prove that a and b are units.

Solution Define a norm function, $N : \mathbb{Z}[\sqrt{d}] \to \mathbb{Z}^+$ such that $N(a + b\sqrt{d}) = a^2 - db^2$. Given that ab is a unit. By proposition 9.8, we know that x is a unit if and only if
$$N(x) = 1. \qquad (9.26)$$
It follows that $N(ab) = 1 \Rightarrow N(a) \cdot N(b) = 1 \qquad (\because N(a \cdot b) = N(a) \cdot N(b))$
$\Rightarrow N(a) \cdot N(b) = 1 \cdot 1 \Rightarrow N(a) = 1, N(b) = 1$. So, using result (9.26), a and b are units.

Problem 25 In $\mathbb{Z}[\sqrt{-7}]$, show that $N(6 + 2\sqrt{-7}) = N(1 + 3\sqrt{-7})$ but $6 + 2\sqrt{-7}$ and $1 + 3\sqrt{-7}$ are not associates.

Solution Consider a function N, called the norm, from $\mathbb{Z}[\sqrt{-7}]$ into the nonnegative integers defined as $N(a + b\sqrt{-7}) = a^2 + 7b^2$. Then, it follows that $N(6 + 2\sqrt{-7}) = 36 + 7 \cdot 4 = 36 + 28 = 64$ and $N(1 + 3\sqrt{-7}) = 1 + 7 \cdot 9 = 1 + 63 = 64$. Therefore, $N(6 + 2\sqrt{-7}) = N(1 + 3\sqrt{-7}) = 64$.

Suppose, if possible, $6 + 2\sqrt{-7}$ and $1 + 3\sqrt{-7}$ are associates. Then, $6 + 2\sqrt{-7} = u \cdot (1 + 3\sqrt{-7})$ where u is a unit.
$$\qquad (9.27)$$

By Problem 22, if d is an integer less than -1 that is not divisible by the square of a prime, then the only units of $\mathbb{Z}[\sqrt{d}]$ are 1 and -1. Therefore, we have 1 and -1 are only units in $\mathbb{Z}[\sqrt{-7}]$. Therefore, u can be 1 or -1. On putting it in Equation (9.27), it follows that $6 + 2\sqrt{-7} = 1 \cdot (1 + 3\sqrt{-7}) = (1 + 3\sqrt{-7})$ or $6 + 2\sqrt{-7} = (-1) \cdot (1 + 3\sqrt{-7}) = (-1 - 3\sqrt{-7})$. This is a contradiction since both the equations are not true. So, our assumption is wrong and hence, $6 + 2\sqrt{-7}$ and $1 + 3\sqrt{-7}$ are not associates.

Problem 26 Show that $3x^2 + 4x + 3 \in \mathbb{Z}_5[x]$ factors as $(3x + 2)(x + 4)$ and $(4x + 1)(2x + 3)$. Explain why this does not contradict the corollary of Theorem 9.20.

Solution In $\mathbb{Z}[\sqrt{5}]$, consider, $(3x + 2) \cdot (x + 4) = 3x^2 + 12x + 2x + 8 = 3x^2 + 14x + 8$ mod $5 = 3x^2 + 4x + 3$. Also, $(4x + 1) \cdot (2x + 3) = 8x^2 + 12x + 2x + 3 = 8x^2 + 14x + 3$ mod $5 = 3x^2 + 4x + 3$. Therefore, $3x^2 + 4x + 3 \in \mathbb{Z}_5[x]$ factors as $(3x + 2)(x + 4)$ and $(4x + 1)(2x + 3)$.
Now, in $\mathbb{Z}_5[x]$, consider $2 \cdot (x + 4) = 2x + 8$ mod $5 = (2x + 3) \Rightarrow 2(x + 4) = 2x + 3$ and we know that 2 is a unit in $\mathbb{Z}_5$.
Again, consider $3 \cdot (3x + 2) = 9x + 6$ mod $5 = 4x + 1 \Rightarrow 3(3x + 2) = 4x + 1$ and we know that 3 is a unit in $\mathbb{Z}_5$.
So, $(x + 4)$ and $(2x + 3)$ are associates. Also, $(3x + 2)$ and $(4x + 1)$ are associates. Therefore, it does not contradict the uniqueness of the factorization in the uniqueness of unique factorization domain, $\mathbb{Z}_5[x]$.

Problem 27 Show that an integral domain with the property that every strictly decreasing chain of ideals $I_1 \supset I_2 \supset \dots$ must be finite in length is a field.

Solution Let, R be an integral domain with the property that every strictly decreasing chain of ideals $I_1 \supset I_2 \supset \dots$ must be finite in length is a field.
For any $r(\neq 0) \in R$, consider the chain of ideals $\langle r \rangle \supset \langle r^2 \rangle \supset \dots$. According to the given condition, this chain of ideals is finite length, we have, $\langle r \rangle \supset \langle r^2 \rangle \supset \dots \langle r^n \rangle = \langle r^{n+1} \rangle = \dots$ for some $n \in \mathbb{N}^+$. It follows that $\langle r^n \rangle \subseteq \langle r^{n+1} \rangle \Rightarrow r^n \in \langle r^{n+1} \rangle \Rightarrow r^n = r^{n+1} \cdot s$ for some $s \in R$. This implies, $r^n \cdot (rs - 1) = 0 \Rightarrow rs - 1 = 0$ $(\because r \neq 0.)$
This implies $rs = 1$ and hence, r is a unit. Since, r is an arbitrary nonzero element of R, therefore, R is a field.

Problem 28 An integral domain R is said to satisfy the ascending chain condition if every strictly increasing chain of ideals $I_1 \subset I_2 \subset \dots$ must be finite in length. Then show that an integral domain R satisfies the ascending chain condition if and only if every ideal of R is finitely generated.

Solution Let, if possible, I is an ideal in R which is not finitely generated. Let, $a_1 \in I$. Then, $\langle a_1 \rangle \neq I$ because I is not finitely generated. Now, pick $a_2 \in I$, but $a_2 \notin \langle a_1 \rangle$. Then, we have $\langle a_1 \rangle \subset \langle a_1, a_2 \rangle$, but $\langle a_1 \rangle \neq \langle a_1, a_2 \rangle \neq I$. Continuing this process, we get a infinite ascending chain of ideals $\langle a_1 \rangle \subset \langle a_1, a_2 \rangle \subset \langle a_1, a_2, a_3 \rangle \subset \dots$, which is a contradiction. So, our assumption is wrong. Therefore, every ideal of R is finitely generated.

Conversely, for any ascending chain of ideals $I_1 \subseteq I_2 \subseteq \dots$, consider the ideal $\left\langle \bigcup_{j=1}^{\infty} I_j \right\rangle$. Since

$\langle \bigcup_{j=1}^{\infty} I_j \rangle$ is finitely generated, suppose $\langle \bigcup_{j=1}^{\infty} I_j \rangle = \langle a_1, a_2, \ldots, a_n \rangle$. Suppose that $a_1 \in I_{k_1}, a_2 \in I_{k_2}, a_3 \in I_{k_3}, \ldots, a_n \in I_{k_n}$.

Take $k = \max\{k_1, k_2, \ldots, k_n\}$. Then, $I_s = I_{s+1}$ for all $s \geq k$ and hence the chain of ideal $I_1 \subseteq I_2 \subseteq \ldots$ is finite length. Thus, $I_1 \subseteq I_2 \subseteq \ldots \subseteq I_k = I_{k+1} = \ldots$.

Exercise

(1) Let $A = \langle a \rangle$, $B = \langle b \rangle$ be two ideals in an integral domain with unity. Show that $A = B$ iff a and b are associates. Further, show that if $\langle a \rangle + \langle b \rangle = \langle d \rangle$, then $d = \gcd(a, b)$.

(2) Let D be a principal ideal domain. Show that every proper ideal of D is contained in a maximal ideal of D.

(3) Let R be a PID. Show that
    (i) two nonzero elements a, b are co-prime iff $x, y \in R$ s.t. $ax + by = 1$.
    (ii) if $a \mid bc$ and a, b are co-prime then $a \mid c$.

(4) In a Euclidean domain R with measure d, show that
    (i) $d(a) = d(-a)$ for all $0 \neq a \in R$.
    (ii) if $a \mid b$ and $d(a) = d(b)$ then a, b are associates.
    (iii) if for $0 \neq a \in R$, $d(a) = 0$ then a is a unit.
    (iv) if $a \mid b$ and a is not an associate of b then $d(a) < d(b)$. $(a, b \neq 0)$.

(5) In a commutative ring R with unity, prove that associate of a prime (irreducible) element is prime (irreducible).

(6) If p, q are prime elements in an integral domain with unity such that $p \mid q$, then show that p, q are associates.

(7) Prove that the ring of polynomials over a field F is a Euclidean domain. Hence, or otherwise deduce that any nonzero polynomials $f(x), g(x) \in F[x]$ have a greatest common divisor.

(8) Prove that if p is a prime in $\mathbb{Z}$ that can be written in the form $a^2 + b^2$, then $a + bi$ is irreducible in $\mathbb{Z}[i]$. Find three primes that have this property and the corresponding irreducibles.

(9) In $\mathbb{Z}[\sqrt{5}]$, prove that 2 is irreducible but not prime.

(10) Give an example of a unique factorization domain with a subdomain that does not have a unique factorization.

(11) Show that quotient ring of a principal ideal domain is a principal ideal domain and the same is true for its homomorphic image.

(12) Prove that in $\mathbb{Z}/\langle 8 \rangle$, 2 is a prime element but not irreducible.

(13) Prove or disprove that a subdomain of a Euclidean domain is a Euclidean domain.

(14) Let p be a prime divisor of a positive integer n. Prove that p is irreducible in $\mathbb{Z}_n$ if and only if p^2 divides n.

(15) Let K be the field of quotients of a unique factorization domain R then show that primitive polynomials $f, g \in R[x]$ are associates in $R[x]$ iff f, g are associates in $K[x]$.

(16) In $\mathbb{Z}[\sqrt{-7}]$, show that $N(6 + 2\sqrt{-7}) = N(1 + 3\sqrt{-7})$ but $6 + 2\sqrt{-7}$ and $1 + 3\sqrt{-7}$ are not associates.

(17) Find the inverse of $1 + \sqrt{2}$ in $\mathbb{Z}[\sqrt{2}]$. What is the multiplicative order of $1 + \sqrt{2}$?

(18) Express both 13 and $5 + i$ as products of irreducibles from $\mathbb{Z}[i]$.

(19) Prove that for every field F, there are infinitely many irreducible elements in $F[x]$.

(20) Let $R = \mathbb{Z} \oplus \mathbb{Z} \oplus \dots$ (the collection of all sequences of integers under component-wise addition and multiplication). Show that R has ideals $I_1, I_2, I_3, \dots$ with the property that $I_1 \subset I_2 \subset I_3 \subset \dots$. (Thus R does not have the ascending chain condition.)

(21) Show that $\mathbb{Z}[\sqrt{-6}]$ is not a unique factorization domain.

(22) Let R be an integral domain with field of fractions F and $p(x) \in R[x]$ monic. Assume $p(x) = a(x)b(x)$ with $a(x), b(x) \in F[x]$ monic with smaller degree than $p(x)$. If $a(x) \notin R[x]$ then R is not a unique factorization domain.

(23) Prove that $R = \{p(x) \in F[x] \mid \text{coefficient of } x \text{ is } 0\}$ is a subring of $F[x]$ but is not a unique factorization domain.

(24) Show that $R = \mathbb{Z} + x\mathbb{Q}[x]$ is not a unique factorization domain.

(25) Show that any ring which is a quotient ring of a polynomial ring over $\mathbb{Z}$ (or a field F) is Noetherian.

INDEX